COGNITIVE ECOLOGY II

Cognitive Ecology II

EDITED BY REUVEN DUKAS
& JOHN M. RATCLIFFE

THE UNIVERSITY OF CHICAGO PRESS CHICAGO AND LONDON

REUVEN DUKAS is associate professor of psychology, neuroscience, and behavior and a member of the Animal Behavior Group at McMaster University.
JOHN RATCLIFFE is a postdoctoral fellow at the Center for Sound Communication at the Institute of Biology of the University of Southern Denmark.

The University of Chicago Press, Chicago 60637
The University of Chicago Press, Ltd., London

Printed in the United States of America
18 17 16 15 14 13 12 11 10 09   1 2 3 4 5

ISBN-13: 978-0-226-16935-4 (cloth)
ISBN-13: 978-0-226-16936-1 (paper)
ISBN-10: 0-226-16935-9 (cloth)
ISBN-10: 0-226-16936-7 (paper)

Library of Congress Cataloging-in-Publication Data
Cognitive ecology II / edited by Reuven Dukas and John M. Ratcliffe
p. cm.
Includes bibliographical references and index.
ISBN-13: 978-0-226-16935-4 (cloth : alk. paper)
ISBN-13: 978-0-226-16936-1 (pbk : alk. paper)
ISBN-10: 0-226-16935-9 (cloth : alk. paper)
ISBN-10: 0-226-16936-7 (pbk : alk. paper)
1. Cognition in animals. 2. Animal ecology. 3. Animal behavior—Evolution. I. Dukas, Reuven. II. Ratcliffe, John M.
QL785.C512 2009
591.5'13—dc22

2009016599

♾ The paper used in this publication meets the minimum requirements of the American National Standard for Information Sciences—Permanence of Paper for Printed Library Materials, ANSI Z39.48-1992.

CONTENTS

ACKNOWLEDGMENTS

We thank the Animal Behaviour Society for sponsoring the 2007 symposium "The Evolutionary Ecology of Learning, Memory and Information Use," which brought together most of the authors; our editor, Christie Henry, for her support and encouragement; and the authors, whose scholarship and dedication made our task interesting and enjoyable.

1 Introduction

REUVEN DUKAS & JOHN M. RATCLIFFE

> I have endeavoured in this chapter briefly to show that the mental qualities of our domestic animals vary, and that the variations are inherited.
>
> CHARLES DARWIN, *The Origin of Species* (1859)

Cognitive ecology focuses on the ecology and evolution of "cognition," defined as the neuronal processes concerned with the acquisition, retention, and use of information. That is, animals rely on a wide variety of information to interact with physical and biotic elements in their surrounding environment, and these interactions determine individual "fitness," defined as lifetime reproductive success. Individuals within a species exhibit large variation in fitness, which is determined in part by heritable variation in their cognitive traits. Consequently, cognitive traits are subjected to evolution by natural selection. Hence, we ought to rely on ecological and evolutionary knowledge for studying cognition. Likewise, because the ways in which animals use available information influence their fitness, we must consider cognitive traits when we examine animal ecology and evolution (Dukas 1998d, 2004a).

An evolutionary approach to understanding cognition is as old as evolutionary biology itself. It received considerable attention by Darwin, first in *The Origin of Species* (1859), especially in chapter 8 (see above quotation), and later in his analyses of emotion and sexual selection (1871, 1872). Darwin's followers also integrated evolution into their writings on psychology (Romanes 1883; Baldwin 1893; James 1890; L. Morgan 1890). For much of the twentieth century, however, evolutionary ecology and psychology were quite detached. The emergence of ethology (Tinbergen 1963; Lorenz 1970) and, later, behavioral ecology (Krebs and Davies 1978) helped bring together biologists and experimental psychologists (Kamil and Sargent 1981) and gradually led to greater integration of cognition, ecology, and evolution. For illuminating historical perspectives, see R. Richards 1987, Houck and Drickamer 1996, Kruuk 2003, and Burkhardt and Richard 2005. Evidence for the growing interest in integrative research on the ecology and evolution of cognition was explicitly summarized in three books at the end of the twentieth century (Balda et al. 1998; Dukas 1998a; Shettleworth 1998; see also Shettleworth 2009). Much new research building on these foundations has been conducted in the past decade, and some of this exciting work is presented here.

Cognition can be divided into several interrelated and perhaps inseparable components. "Perception" is defined as the translation of environmental signals into neuronal representations. "Learning" is the acquisition and maintenance of neuronal representations of new information. "Working memory" comprises a small set of neuronal representations active over some short duration. "Attention" refers to the neuronal representations activated at any given time. "Long-term memory" consists of passive, stored representations of previously learned information. Finally, "decision making" involves the determination of action given the available information about relevant environmental features and experience (Platt 2002; Dukas 2004a).

At the mechanistic level, cognitive traits are determined and modulated by genes, neurons, and hormones. At the functional level, cognitive abilities are evolved features that are shaped by natural selection based on their relative contribution to animal fitness and that are subject to a variety of mechanistic and phylogenetic constraints (Dukas 1998b). Since research pertinent to cognitive ecology is conducted in several scientific disciplines, we have to integrate a vast amount of knowledge from all these fields in order to understand the ecology and evolution of animal cognition (Ratcliffe et al. 2006). The following chapters represent a variety of attempts at achieving this challenging goal.

Our definition of "cognition" at the start of this chapter focuses on the acquisition, retention, and use of information. Organisms can perform rather well only by perceiving and responding to external information. Most or all animals, however, also acquire and retain information, and in chapter 2, Dukas discusses the ecological and evolutionary significance of this ability. The mechanistic research on animal learning and memory is typically conducted under the necessary controlled laboratory conditions using a few animal species whose ecology and behavior in the wild are not well known (e.g., the domestic chicken, *Gallus gallus domesticus,* and the Norway rat, *Rattus norvegicus*). A notable exception is the honeybee, *Apis mellifera,* which has been studied extensively in the field (Seeley 1996; von Frisch 1967). The honeybee is an ideal model system for integrating mechanistic knowledge on genes, neurons, and hormones with whole-animal information on behavior and ecology, as Fahrbach and Dobrin elegantly illustrate in chapter 3.

Part II presents new work on established and emergent research programs relating cognition to avian ecology. Song learning in birds is one of the few exceptional cases where a central behavior has been thoroughly examined from the key angles of neurobiology and ecology. In chapter 4, Beecher and Burt present an update on a long-term research program examining the rules that young song sparrows (*Melospiza melodia*) use to decide which songs to learn, retain, and later use and the role that the social ecology of song sparrows has played in shaping these rules. Searcy and Nowicki (chapter 5) take a fresh look at bird song learning by examining the recent nutritional stress hypothesis. They suggest that conditions during early development have long-term effects on features of the male song, which females can readily perceive, and on other aspects of male quality, which females cannot easily assess. Hence, females rely on male song as a reliable indicator of overall male quality. Searcy and Nowicki review the recent data, which are mostly in agreement with the nutritional stress hypothesis.

Pravosudov (chapter 6) also addresses effects of nutritional stress during development, but his focus is on the hippocampus and spatial memory. Work on spatial memory in birds has provided a clear link between the ecological need to store food, a relatively enhanced spatial memory used to retrieve the cached food, and the relative volume of the hippocampus, the brain region processing spatial memory. Given the importance of spatial memory for the survival of certain bird species, Pravosudov and colleagues predicted that hippocampal development and hence spatial memory would remain intact even under nutritional stress. Their data reviewed in chapter 6, however, refute that prediction and suggest that constraints during development preclude the insulation of certain brain regions from nutritional stress. The chapters by Searcy and Nowicki and Pravosudov add the important dimension of development to understanding animal cognitive ecology.

In chapter 7, Sol relies on recent data to address the old question of why some animals have large brains relative to body size even though such brains incur substantial costs in terms of delayed maturation and high maintenance. Sol then reviews recent studies providing support for the cognitive-buffer hypothesis, which states that a relatively large brain is associated with an enhanced ability to handle novel situations and hence with increased probability of survival in novel or altered environments.

With few exceptions, all animals have to make decisions within the four general categories of feeding, predator avoidance, interactions with competitors, and sexual behavior. Part III focuses on cognitive aspects of decisions made within two of these behavioral categories: reproduction and antipredator behavior. Female choice of mates has been studied for a long time, and as Ryan, Akre, and Kirkpatrick note in chapter 8, it is well established that females choose mates based on their perceived quality. It is also clear that such female choice has generated sexual selection, which is responsible for the evolution of many features of male mating signals. Less apparent, however, are the ways in which an animal's cognitive biology contribute to the patterns of mate choice in the wild. Ryan et al. analyze how cognitive mechanisms could affect female mate choice and identify exciting directions in this relatively unexplored avenue of research. In chapter 9, Phelps and Ophir focus on male sexual behavior within a unique mammalian model that relies on natural variation in the mating systems of voles of the genus *Microtus*. Using their thorough knowledge of voles' natural history and behavior, Phelps and Ophir discuss the neuroanatomy, neurochemistry, and genetics underlying males' pair-bonding and their corresponding use of space.

Little attention had been devoted to cognition at the embryonic stage, but recent experiments reviewed by Warkentin and Caldwell (chapter 10) clearly indicate that embryos possess sophisticated abilities to assess and respond to cues of predation. As Warkentin and Caldwell note, embryonic cognition is an essential yet relatively neglected aspect of cognitive research even though it is relevant for many animals. In chapter 11, Ratcliffe takes a fresh look at the historically well-studied system of bats and moths. This classic model system focuses on the auditory domain used by bats to detect insects and other potential prey and by many moths to perceive and attempt to avoid impending predation. Because bats produce echolocation calls to perceive their environment and because some moths generate sounds to deter attacking bats, researchers eavesdropping on these signals have a unique opportunity to examine how the available auditory information translates into decisions made by both predator and prey. The last part of the volume is devoted to social information. First, Manser (chapter 12) links antipredatory and social behavior by analyzing the alarm calls of meerkats (*Suricata suricatta*). These cooperatively breeding animals employ referential alarm calls, which indicate the approach of specific predators and cause receivers to show an appropriate escape response to these predators. Kendal, Coolen, and Laland (chapter 13) review the current knowledge on social learning. From humans' biased perspective, social learning is a basic way of life. Most animals, however, do not rely on social learning, whereas some species use it only conditionally. Kendal et al. organize their discussion around the two key questions of when individuals should rely on social learning and whom they should learn from. Finally, Federspiel, Clayton, and Emery (chapter 14) illustrate how integrating knowledge about animals' natural behavior, ecology, and evolutionary history with the powerful empirical techniques of experimental psychology has helped us understand the use of social information by different bird species. The volume concludes with a short evaluation of what we have achieved, current shortcomings, and promising avenues of future research.

PART I

Learning

ULTIMATE AND PROXIMATE MECHANISMS

2 Learning: Mechanisms, Ecology, and Evolution

REUVEN DUKAS

2.1. Introduction

Learning in humans and other animals was closely examined in a variety of scientific disciplines during the twentieth century. Recently, with the growing awareness that there are remarkable similarities among species ranging from fruit flies to humans, there has been increased integration of ideas and data across traditional disciplines. I attempt to present such an integrative approach to the study of learning in this chapter. I focus on five major questions. First, what is learning? I provide a definition, discuss the difficulties associated with quantifying learning, and briefly outline key genetic and neurobiological mechanisms underlying learning. Second, why learn? To answer, I commence with analyzing life without learning and then detail fitness benefits of learning. Third, who learns? While I suggest that the answer is probably all animals with a nervous system, there is still no empirically based answer to this question. Therefore, I instead focus on examining what an animal has to possess in order to learn and what costs are involved. Fourth, what do animals learn? To illustrate the broad reliance on learning among animals, my answer focuses on a single species, the fruit fly *Drosophila melanogaster*, which learns much more than we had thought. Finally, is learning important? I discuss effects of learning on ecological interactions, the dominant role of learning in the life history of numerous species, and effects of learning on evolutionary change. Although there is a large body of literature on a multitude of features of learning, my interdisciplinary analysis identifies topics requiring further investigation. I thus conclude with a list of such promising future directions.

2.2. What is learning?

2.2.1. DEFINITION AND MEASUREMENTS

Learning may be defined as the acquisition of neuronal representations of new information. That information is then retained for at least a short period (short-term memory) and often for long durations (long-term memory) and typically influences relevant decisions and behaviors. Examples of learning include neuronal representations of new (*i*) spatial environmental configurations, (*ii*) sensory information including visual, auditory, olfactory, taste, and tactile features, (*iii*) associations between stimuli and environmental states, and (*iv*) motor patterns, for example, the sequence of body movements involved in manipulating a novel food (Dukas 2008a). My above definition refers to explicit neuronal representation of information and hence excludes habituation and sensitization, which are typically considered simple forms of learning.

Although learning involves changes in neuronal activity and configurations, it can currently be quantified only through its effects on behavior. That is, owing to the large number of neurons involved and the distributed nature of neuronal activity, one cannot quantify learning directly through measuring changes in neuronal activity. One can, however, examine the neuronal mechanisms underlying learning and memory (section 2.2.2). Unfortunately, even the quantification of learning through behavior is not a straightforward task. On the one hand, numerous claims for learning in a variety of species have been based on inadequate experiments. Major weaknesses include protocols that could be biased due to the employment of observers not blind to treatments (e.g., R. Rosenthal and Fode 1963) and lack of proper control treatments used to verify that either the presentation of stimuli alone or environmental states alone do not generate behavioral biases that may be misinterpreted as learning (e.g., Alloway 1972). On the other hand, claims for lack of learning in certain species are also problematic because they could merely reflect subjects' failure to engage in behaviors that indicate learning owing to inadequate experimental settings rather than a true inability to learn.

2.2.2. GENETICS AND NEUROBIOLOGY OF LEARNING

The biochemical and genetic architecture underlying learning has been examined primarily in several model systems, including the soil nematode *Caenorhabditis elegans*, the aquatic snail *Aplysia californica*, the fruit fly *Drosophila melanogaster*, the honeybee *Apis mellifera*, and the mouse *Mus musculus*. Obviously, reviewing the immense knowledge acquired over the past few decades is beyond the scope of this chapter. Rather, I will briefly outline key concepts. At the mechanistic level, learning can be perceived as a basic cellular process involving changes in the synaptic properties of neurons. Immediate changes are mediated by neurotransmitters, whereas long-term changes, which involve both biochemical and physical changes in synaptic properties, involve gene expression. The biochemical and structural changes associated with learning and memory can be studied at the level of individual neurons. Indeed, a few neurons may be sufficient for exhibiting learning in tiny animals such as *C. elegans* (section 2.4.1). Typically, however, learning involves synchronous modulation of numerous neurons, each with specific sensitivities to a variety of environmental features (Kandel et al. 1995; Dubnau et al. 2003).

The dynamics of learning has been elegantly dissected using genetic mutants in fruit flies, but remarkably similar mechanisms underlie learning in all animals, including humans. There are two mechanistically distinct forms of consolidated memory. Anesthesia-resistant memory (ARM) can be formed even after a single training session. It does not require protein synthesis and is relatively short lasting (~24 hours). The path leading to long-term memory (LTM) begins with learning that requires spaced training, meaning that flies require a series of training sessions separated by breaks. The flies then form short-term memory (STM) and medium-term memory (MTM) lasting about 1 and 5 hours respectively. That memory consolidates via a process involving protein synthesis to LTM lasting several days (Dubnau et al. 2003).

In short, many of the genes, biochemical pathways, and structural changes underlying learning and memory have been elucidated. At the level of individual neurons, learning is controlled by several hundred genes and a similar number of biochemicals. Typically, neuronal representations of the environment require the synchronous activity of numerous neurons. This aspect of learning is challenging to quantify owing to the large numbers of cells and connections involved.

2.3. Why learn?

A variety of misleading assertions have linked learning to variation and unpredictability. To clarify the confusion on this topic and answer the question of when animals should learn, I will start by examining the null model of life without learning.

2.3.1. LIFE WITHOUT LEARNING

Most or all organisms experience variation and unpredictability in their external and internal environment. The most fundamental and universal mechanism for handling variation and unpredictability is gene regulation. For example, *Escherichia coli* bacteria can alter gene expression to generate energy from the locally available sugar. If glucose becomes unavailable, the bacteria can activate alternate sets of genes that allow them to exploit several other sugars as their energy source. Furthermore, when their environment lacks a necessary amino acid, *E. coli* can produce the required enzymes for synthesizing that amino acid (Pierce 2002). In addition to staying stationary and adjusting to the changing environment, *E. coli* and other bacteria can also respond to environmental changes through chemotaxis. This involves a sophisticated system of information processing and behavioral machinery that enables bacteria to move toward energy sources and away from noxious chemicals (Koshland 1980; Eisenbach and Lengeler 2004).

Bacteria can obviously benefit from responding to a change in the environment by seeking better conditions. They would also benefit, however, from modulating their response if they subsequently fail to locate better settings. Indeed, the genetic networks underlying chemotaxis are sensitive to change in the environment rather than to the absolute condition. For example, transferring bacteria from a dish with high glucose concentration to a dish with low glucose concentration causes a change in their movement pattern, which, after a short period, rebounds to the baseline level (Koshland 1980). That is, the bacteria adjust to the new conditions.

Larger and more mobile single-cell organisms such as *Paramecium* can sense and respond to a broader range of environmental variables and exhibit chemotaxis, thermotaxis, geotaxis, and thigmotaxis (movement in response to touch) (Jennings 1906; Saimi and Kung 1987). For example, a *Paramecium* accelerates its forward movement if touched from behind, and it stops and alters its swimming direction if its forward movement is obstructed. The perception of environmental variables and the control of movements are mediated by electrical signaling generated by ion movement across the cell membrane. Similar electrical signaling is also employed by the nervous systems of all multicellular organisms (Eckert 1972; Shelton 1982; P. Anderson 1989; Greenspan 2006; Meech and Mackie 2007). Behavior mediated via electric signaling can also be modulated. Neuronal modulation involves short- and long-term changes in synaptic properties mediated by neurotransmitters and gene expression. Simple forms of neuronal modulation allow animals to either habituate or sensitize to some environmental change. Such modulation can be seen as an ancestral type of learning.

In sum, genetic regulation and behavioral modification underlie organismal response to environmental variation. Whereas bacteria, owing to their small cell size, can rely on chemical diffusion for behavioral coordination and decisions, large single-cell organisms also employ electrical communication to allow fast responses to environmental changes. Multicellular organisms rely heavily on intercellular electrical communication provided by their nervous system to coordinate responses to environmental variation. Both the genetic and electrically mediated behavioral responses can be modulated. That is, even organisms that do not learn are highly adept at responding to and modulating their responses to variable and unpredictable environments.

2.3.2. ADAPTIVE SIGNIFICANCE OF LEARNING

The fundamental biological ability to modulate genetic and electrical activity (section 2.3.1) is the precursor for learning, which can be perceived as a coordinated neuronal modulation resulting in neuronal representation of information (section 2.2.1). Learning enables individuals to exploit environmental features that are unique to a certain time and place. Animals' ability to learn about such features expands the type and amount of information they can respond to and, consequently, their behavioral repertoire. For example, a bee can acquire a neuronal representation of her nest location, record the spatial location, odor, and color of the best flowers to forage on, and learn a new motor pattern for handling these flowers. And individuals in many species gain from learning to identify their parents, neighbors, potential mates, offspring, and competitors. One can readily imagine how every organism can benefit from learning. Indeed, learning is probably a universal property of all or most animals with a nervous system.

Although one would expect learning to be adaptive, little research has been devoted to quantifying the fitness consequences of learning. In an experiment examining this issue, grasshoppers (*Schistocerca americana*) were assigned to two groups, each receiving two synthetic foods. One food contained the optimal proportion of all nutrients essential to grasshopper growth, while the other was carbohydrate deficient. Subjects in the learning group could associate each of the two foods with distinct tastes, colors, and spatial locations. Subjects in the random group, however, had the food-cue associations assigned randomly twice a day so they could not learn to associate the cues with specific food. The learning grasshoppers rapidly learned to restrict their visits to the nutritionally balanced food, whereas the random grasshoppers kept visiting each food dish at equal frequencies. The random grasshoppers, however, gradually increased the proportion of time spent feeding on the balanced food, suggesting that they relied on a nonlearning mechanism such as a change in taste-receptor sensitivity (Abisgold and Simpson 1988; Simpson and Raubenheimer 2000). Nevertheless, the overall feeding duration on the balanced diet was over 99% for the learning grasshoppers but only 87% for the random grasshoppers (fig. 2.1a). Moreover, the random grasshoppers did not feed on the balanced food as regularly as the learning grasshoppers did, because they approached equally often the dish with the deficient diet. Such an approach typically resulted in brief feeding followed by resting. The behavioral differences between the treatments translated into a 20% higher growth

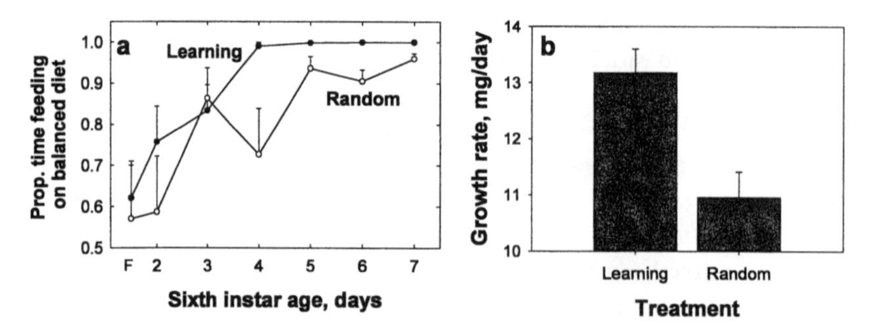

FIGURE 2.1. a. The proportion of time (mean ± 1 SE) spent at the dish containing nutritionally balanced food by sixth-instar grasshoppers belonging to the random and learning treatments ("F" on the x-axis refers to the first recorded meal). b. The average growth rate of grasshoppers belonging to the random and learning treatments. Data from Dukas and Bernays 2000.

rate in the learning grasshoppers (fig. 2.1b). It is likely that the fitness benefit from learning would be significant also in natural settings, where learning could also translate into less travel and hence lower mortality due to predation (Dukas and Bernays 2000).

2.4. Who learns?

My definition of learning (section 2.2.1), which includes the term "neuronal representations," conveniently restricts it to animals with nervous systems. Among such animals, learning is perhaps a universal property. It would be difficult to conclude that some animals with nervous systems do not possess learning abilities because of the high odds of obtaining negative results in experiments to detect learning, especially in species we are not very familiar with (section 2.2.1). Because we cannot state which animals with nervous systems do not learn, I will focus instead on examining two features that can determine the prevalence of learning among animals with nervous systems. These features are the biological requirements for learning and the costs of learning.

2.4.1. THE HARDWARE REQUIREMENTS FOR LEARNING

The most essential prerequisites for learning are the abilities to sense some features of the environment and to modulate cellular responses to these features. Because all or most organisms possess these two characteristics (section 2.3.1), all animals with nervous systems may have the potential to learn. This includes even organisms with a small number of neurons such as the soil nematode *C. elegans,* which was chosen over 30 years ago as a simple model system for examining the structure of a whole nervous system and the way it generates behavior (Brenner 1974). Each hermaphrodite *C. elegans* worm comprises 959 cells, of which 302 are neurons and 56 are glial and other types of support cells. The morphology of each neuron and its chemical synapses and gap junctions has been mapped, and most neurons have been assigned a presumed function. The neurons contain many of the neurotransmitters known in other animals, including serotonin, dopamine, glutamate, acetylcholine, and gamma-aminobutyric acid (GABA) (de Bono and Maricq 2005).

Researchers began to document learning in *C. elegans* in the 1990s. For example, Wen et al. (1997) exposed *C. elegans* worms to either sodium ions or chloride ions with food (a suspension of *E. coli* bacteria) and the alternate ions with no food. In the test following training, worms exhibited a strong preference for the ions previously associated with food, whereas worms in the control groups showed no preference (fig. 2.2). In further experiments, Wen et al. (1997) documented aversion learning and, in two mutants, loss of learning. Using an ecologically relevant assay, Y. Zhang et al. (2005) documented aversive learning to pathogenic bacteria in *C. elegans* and examined the neuronal and biochemical pathways that underlie this learning. Their analyses suggest that changes in a few neurons may be sufficient for the expression of learning (Quinn 2005).

In sum, *C. elegans* illustrates that a nervous system consisting of only 302 neurons can generate learning. The exact set of all biochemical requirements for learning is unknown. However, the neurotransmitters and enzymes that have been determined to mediate learning are involved in other behaviors and cellular processes and are widespread among animals (Bargmann 1998; de Bono and Maricq 2005). Hence, it seems that all animals with nervous systems have the potential to learn. To convincingly falsify the prediction that

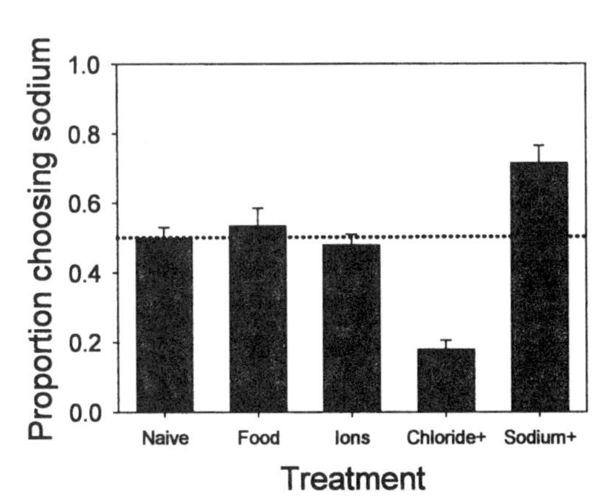

FIGURE 2.2. Learning in *C. elegans* worms. Naive worms, worms exposed to food alone, and worms exposed to sodium and chloride ions alone exhibited no significant preference for either chemical in the following test, whereas worms exposed to one ion with food and the other with no food significantly preferred the ion associated with food in the test ($P < 0.05$). Data from Wen et al. 1997.

all animals with nervous systems learn, we would require strong converging behavioral and neurogenetic evidence, with the latter identifying the precise features that allow an animal to generate complex behavior but no learning.

2.4.2. COSTS OF LEARNING

The acquisition, retention, and use of information by animals cannot be cost free. Because the cost of biological information has been primarily studied in the context of gene regulation, I will briefly discuss errors and their control in the genetic system. There are key similarities between the genetic and nervous systems. Most notably, both genetic regulation and neuronal control involve coordinated networks of individual units with no central management, and genes and neurons may belong to more than one network, each regulating a distinct function. Furthermore, because genes determine neuronal structure and activity, genetic errors can directly translate into errors in neuronal networks (Dukas 1999a).

The maintenance of genetic information is prone to error because DNA is subjected to high rates of damage, which could interfere with DNA replication and transcription (Kirkwood et al. 1986; Bernstein and Bernstein 1991). Both gene transcription and translation into proteins are also prone to error, owing mostly to the stochastic nature of biochemical reactions that depend on infrequent events involving a small number of molecules. A few mechanisms that help reduce either the errors or damage caused by errors include extensive redundancy, active enzymatic correction of DNA damage, a variety of feedback loops, and optimal rates of transcription and translation that minimize error (Raser and O'Shea 2005). All the mechanisms just mentioned come at a cost. Redundancy implies that cells produce and maintain more DNA, genes, RNA, and enzymes than the minimum required. Similarly, possessing regulatory circuits increases the number of genes, RNA, and enzymes. Enzymatic correction of DNA damage is energetically expensive, and because frequent transcription followed by inefficient translation results in lower noise than infrequent transcription and efficient translation, cells incur an extra energetic cost associated with excess production of mRNA (Rao et al. 2002; Raser and O'Shea 2005).

Perhaps the most vivid illustration of the trade-off between accuracy and cost comes from research on variants of DNA polymerase in phage T4. First, an increase in DNA polymerase accuracy is associated with an exponential increase in energy expenditure and a decrease in the rate of DNA synthesis. Second, wild-type polymerase is not as accurate as a few available mutants, suggesting that the wild type possesses the optimal balance between accuracy and cost (Bessman et al. 1974; Galas and Branscomb 1978; Galas et al. 1986).

In sum, the genetic system is prone to errors and contains a variety of costly error correction mechanisms, which reduce but do not eliminate errors.

Learning results in at least STM and often LTM. STM involves chemical changes in synaptic properties, and LTM requires gene transcription and translation into proteins (section 2.2.2). As just discussed, these processes are affected by genetic errors. It is possible that gene regulation in some parts of the brain is subjected to higher standards of accuracy coupled with higher metabolic costs. An alternative means of reducing the effect of errors is through redundancy in neuronal networks, which is also associated with higher costs of developing and maintaining additional, metabolically expensive brain tissue (Dukas 1999a). Also, synaptic transmission is unreliable, meaning that only fewer than half of the presynaptic impulses arriving in a synapse produce a postsynaptic response. The limiting factor is the stochastic nature of neurotransmitter release processes, which involve either a small number of release sites or a low probability of neurotransmitter release per site (Allen and Stevens 1994). The simplest way of compensating for synaptic unreliability is through extensive redundancy. For example, in pyramidal neurons of the rat hippocampus, the synchronous activity of several dozen synapses is necessary to produce a spike train (Allen and Stevens 1994).

The only direct evidence for the costs of learning and memory comes from a series of experiments with fruit flies (*D. melanogaster*). In one study, Mery and Kawecki (2003) found that artificial selection on learning ability in adult flies, which increased learning scores in the selected lines, was associated with reduced larval competitive ability when food was limited. In another study, Mery and Kawecki (2004) exposed food-limited adult flies to alternating substrate conditions, which required use of learning for substrate choice every other day. Under these conditions, flies from lines selected for improved learning ability exhibited lower egg-laying rates than flies from unselected lines. Finally, Mery and Kawecki (2005) documented in flies selected for enhanced learning ability that, compared with a few control treatments, flies that were subjected to a training regime that produced LTM showed earlier death under starvation and dehydration (fig. 2.3). This set of experiments suggests that, at least in flies with artificially selected enhanced learning ability, both learning and memory have significant physiological costs.

In addition to the structural and physiological costs just discussed, learning has ecological costs associated with inexperience. That is, in many species, inexperienced, typically young individuals incur very high mortality rates owing to a combination of deficient feeding techniques and antipredatory behavior (reviewed in Dukas 1998c). For example, recently independent juvenile yellow-eyed juncos (*Junco phaeonotus*) were about one-third as proficient

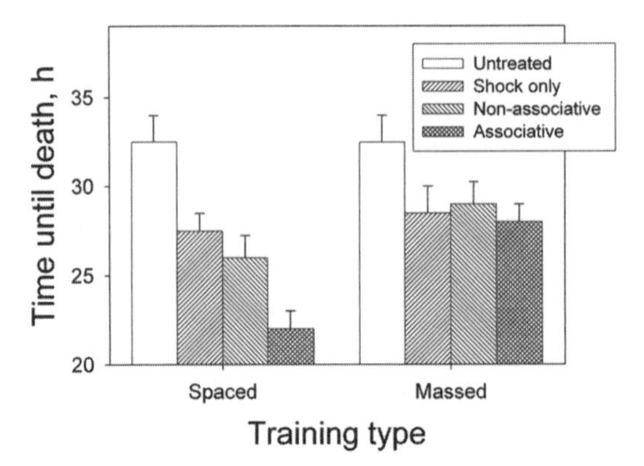

FIGURE 2.3. Costs of long-term memory. Female fruit flies subjected to training that produced long-term memory (associative conditioning with spaced training) had higher mortality rates when kept with no food and water than all other treatments that produced either anesthesia-resistant memory (with massed training) or no memory (untreated, shock only, and nonassociative conditioning with spaced training). Data from Mery and Kawecki 2005.

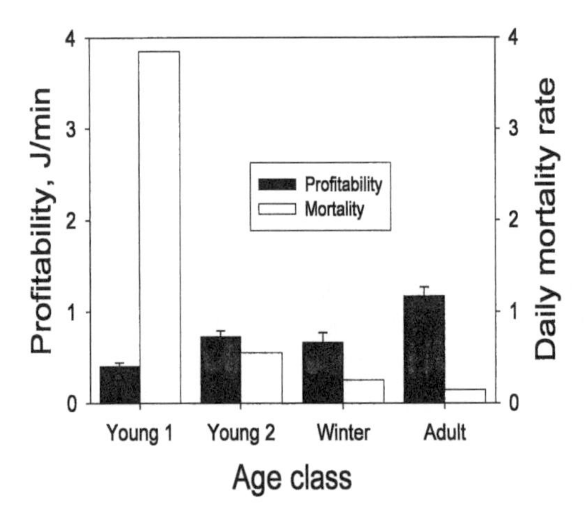

FIGURE 2.4. Effects of experience on feeding proficiency (the profitability of large mealworms) and mortality rate in yellow-eyed juncos. Age classes: recently independent juveniles (Young 1), young experienced juveniles (Young 2), old experienced juveniles during the fall and winter (Winter), and adults over a year old (Adults). Data from K. Sullivan 1988a, 1988b, 1989.

at handling large mealworms as adults. The juveniles spent over 90% of the day foraging and had mortality rates that were over 20 times higher than those of adults, who fed for only 30% of the day (K. Sullivan 1988a, 1988b, 1989). The gradual improvement in feeding proficiency and, probably, antipredatory behavior lasted several months (fig. 2.4). Similarly, independent juvenile European shags (*Phalacrocorax aristotelis*) compensated for low proficiency in capturing fish by spending up to twice as much time as adults foraging until constrained by short day length in late fall. Insufficient feeding during the short winter days was probably the major cause for the high juvenile winter mortality rate, which was five times that of adults (Daunt et al. 2007). The ecological cost of inexperience, however, has to be viewed in perspective: learning allows animals to exploit environmental features, behavioral repertoire, and niches that cannot be used otherwise (section 2.3.2). So the initial period of inexperience can be seen as a necessary component of a life history that relies on and ultimately benefits from learning (section 2.6.2).

2.5. What do animals learn?

Early studies of animal behavior emphasized the importance of instinct in most nonhuman animals and insects in particular (Fabre et al. 1918; Tinbergen 1951). Such views still dominate in many disciplines of ecology and evolution. Two related events that have been instrumental in changing our current understanding of learning are the highly successful establishment of fruit flies as a model system for research on the neurogenetics of learning and the realization that the genetic and cellular mechanisms controlling learning and memory are remarkably similar across diverse taxa. Intriguingly, although evolutionary biologists have studied fruit flies for about a century (Kohler 1994), it was neurogeneticists who critically documented robust learning in *Drosophila melanogaster* (Quinn et al. 1974; Davis 2005). The neurogenetic work on learning in fruit flies has required the development of behavioral protocols for quantifying learning. Consequently, a variety of ingenious procedures have been developed in the past few decades, which indicate that fruit flies rely on learning for all major life activities.

I will focus on fruit flies in this section because they possess a few characteristics that raise the question as to whether they should learn at all. First, fruit flies *D. melanogaster* are tiny, short-lived animals. The adult female is only about 2.5 millimeters long, and the male is slightly smaller. Larval development takes about 4 days (Ashburner et al. 1976), and adults in the wild probably have a life expectancy of only several days. Second, the larval and adult brains contain approximately 20,000 and 200,000 neurons respectively. These are remarkably small numbers compared with vertebrates. Finally, fruit fly life history appears straightforward and does not readily indicate reliance on learning. In fact, we still do not know to what extent fruit flies depend on learning in natural settings because all research relevant to learning has thus far been conducted in the laboratory.

Female fruit flies lay their eggs on decaying fruit containing yeast. Hence, upon emergence, the larvae just have to commence feeding, which is their sole task. Even though the larval task seems easy and they possess limited sensory and information-processing abilities, larvae are able to learn. In the first documentation of learning in fruit fly larvae, Aceves-Pina and Quinn (1979) allowed groups of third-instar larvae to experience three 30-second pulses of one odor together with the application of an electric shock. The larvae also experienced three 30-second pulses of another odor not associated with shock. The two treatments were alternated and separated by 90-second breaks. In a subsequent choice test, the larvae exhibited significant avoidance of the odor associated with shock. Control treatments employing the application of only odorants or only shocks indicated no effects of these nonassociative treatments on odorant choice. Furthermore, larvae from learning-deficient mutant lines failed to show associative learning (Aceves-Pina and Quinn 1979).

The documentation of learning in fruit fly larvae raised the question of whether such learning ability may contribute to fitness in natural settings. To explore this issue, I tested for associative learning of ecologically relevant tasks in fly larvae. Groups of larvae learned to prefer odors associated with high-quality food and to avoid odors associated with disturbance caused by simulated predation. The larvae, however, did not show significant learning of odors associated with optimal temperature (Dukas 1999b). In further experiments, Gerber et al. (2004) associated two illumination conditions (light and dark) with sugar and one of two negative reinforcers: quinine and table salt. Experienced larvae, which were tested individually, preferred the illumination associated with sugar. In sum, fruit fly larvae can learn to associate either odors or light conditions with the two types of environmental states most relevant to the larval stage, which are food quality and danger. One can readily imagine that there is substantial variation in food quality and danger within a fruit and between adjacent fruits in nature, that such variation is associated with odor or lighting, and that larvae can gain from learning about and seeking stimuli associated with higher growth rate and survival.

Like the larvae, adult fruit flies can learn to avoid odors associated with electric shock (Quinn et al. 1974) and to prefer odors associated with food (Tempel et al. 1983). The adults can also learn to avoid light sources of distinct frequencies associated with aversive states (shock or violent shaking) (Quinn et al. 1974; Folkers 1982) and to avoid flying toward visual patterns associated with high temperature (Wolf and Heisenberg 1991; G. Liu et al. 2006). In short, adult fruit flies can learn about odors, colors, and visual patterns associated with either positive or negative outcomes.

Both male and female fruit flies also learn in the context of sexual behavior. The original protocol for learning in the context of courtship involved males that courted recently mated, unreceptive females for one hour. Compared to inexperienced males, the experienced males exhibited reduced courtship of immobilized virgin females (Siegel and Hall 1979). Further experiments indicated that the males learn to associate the failure to mate with specific female pheromones (Ejima et al. 2005). Experiments involving more naturalistic settings and no immobilized test females indicated that male learning is adaptive: experience with courting unreceptive, recently mated females caused males to selectively reduce subsequent courtship of mated females but not virgin females. The experienced males were also faster to approach virgin females during the test and slower to respond to mated females than inexperienced males

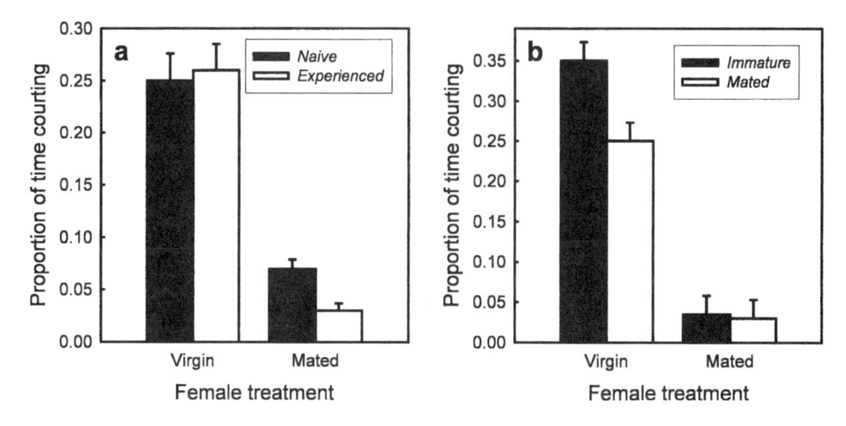

FIGURE 2.5. The proportion of time (mean ± 1 SE) male fruit flies with distinct experiences spent courting either virgin or recently mated females. In panel a, the males were either naive (black bars) or experienced at courting mated females (white bars). In panel b, the males experienced either immature females (black bars) or mated females (white bars). Data from Dukas 2005a.

were. In another experiment, experience with courting unreceptive, immature females caused males to selectively increase subsequent courtship of virgin females but not mated females (fig. 2.5). In this case, the information about the imminent availability of virgin females caused the experienced males to approach either female type faster than inexperienced males (Dukas 2005a).

Much of the work on learning in the context of courtship in fruit flies has been restricted to males. Females, however, have ample opportunities for learning about potential mates and rely on such learning to improve mate choice. Most notably, although it takes females one to two days to reach sexual maturity, during that time immature females are courted vigorously by many males (Dukas 2006). When immature female fruit flies experienced courtship only by small males, which are less desirable mates than large males, the females subsequently were more likely to mate with small males than females that had experienced courtship by large males (Dukas 2005b). That is, when the females learned that only small males were available, they were less likely to reject small males as mates.

Male fruit flies also seem to learn in the context of aggression. In the field, males defend territories containing decaying fruit and females. Larger males are more likely to hold territories and have a higher mating success (Markow 1988). In laboratory trials, males rapidly establish dominance hierarchy based on fighting, with the winner remaining at a food cup containing a female while the loser retreats. When a loser in one match was allowed to rest for 30 minutes and then either rematched with the familiar winner from the first match or matched with an unfamiliar winner from another match, losers lunged significantly more often toward unfamiliar than familiar winners (Yurkovic et al. 2006). Individual recognition could be beneficial for both males and females also in the context of courtship and mate choice. For example, a male could benefit from learning to avoid a female that has recently rejected him. Published evidence to date, however, indicates only that male fruit flies can learn to distinguish among categories of females of distinct reproductive state (Ejima et al. 2005). Nevertheless, individual recognition is known in other insects (Dukas 2008a) and invertebrates (e.g., Gherardi and Tiedemann 2004; Detto et al. 2006) and is ubiquitous in vertebrates (see Beecher and Burt, chapter 4 in this volume).

The extensive work on fruit fly learning is highly illuminating because they employ learning in all four central behavioral categories of feeding, predator avoidance, aggression, and sexual behavior. It is almost certain that fruit flies do not possess exceptional learning abilities relative to other animals. Rather, they have been studied closely. Hence, it would be sensible to assume that most other animals also rely on learning in all central domains of life. Similarly, it is likely that vertebrates such as birds, which have larger brains and life expectancies of a few years, possess elaborate learning and memory abilities. Overall, it appears that the magnitudes of learning and memory abilities in nonhuman animals have been underestimated despite their potentially broad influences on animal behavior, ecology, and evolution.

2.6. Is learning important?

Learning affects all major ecological and evolutionary processes in animals, but this has been underappreciated owing to the difficulty of quantifying learning and its influence.

2.6.1. ECOLOGICAL SIGNIFICANCE OF LEARNING

As illustrated in section 2.5, most ecological interactions in most animals involve learning. Learning affects patterns of competition, predation, and antipredatory behavior, and it can determine levels of immigration and emigration. Perhaps most fundamentally, the ability to learn has opened up numerous niches unavailable otherwise. For example, unlike many solitary wasps and bees, which specialize on one or a few food types, social hymenoptera have several generations per year. This means that, owing to seasonal variation, individuals must learn about the best food available at their place and time. That is, the ability to learn was probably a precondition for the evolution of sociality in ants, wasps, and bees. Extensive learning is also required in all species consuming prey that is challenging to capture or handle. This category includes many species of birds and mammals and numerous other taxa. Examples include fish-eating birds (e.g., Daunt et al. 2007), oystercatchers (Norton-Griffiths 1969), and cheetahs (Caro 1994). In sum, learning is a precondition allowing the occupation of many niches, and it determines key ecological interactions such as competition and predation.

2.6.2. LIFE HISTORY OF LEARNING

Life history research has focused on the effects of three physical factors—growth, effort, and senescence—on reproductive success (Stearns 1992; Roff 2002). The effects of learning have been noted, especially in a few long-term avian studies (Nol and Smith 1987; Wooler et al. 1990; Black and Owen 1995; Rattiste 2004), but the relative contribution of learning to performance throughout the life span has not been examined closely. In animals that reach final growth before sexual maturity, the three major contributors to reproductive success are effort, physiology, and learning. Reproductive effort is defined as investment in current reproduction that decreases future reproduction or survival. Although it is commonly asserted that effort should increase with age, theoretical analyses emphasize that effort could also decrease or plateau with age (Fagen 1972; Charlesworth and Leon 1976; Taylor 1991; Roff 2002). The empirical data are mixed, with some studies suggesting increased effort with age (Pugesek 1981; Clutton-Brock 1984; Candolin 1998; Poizat et al. 1999) and others documenting no change with age (W. Reid 1988).

Major physiological attributes such as muscle power and endurance increase early in the life of some species. From sexual maturity onward, physiology is subjected to senescence, defined as a decrease in body condition, associated with decreased fertility and survival rates, with increased age (Rose 1991; Kirkwood and Austad 2000). As with reproductive effort, however, theory and data indicate that patterns of senescence can diverge from the classical pattern of exponential increase in mortality rates with age (Abrams 1993; Reznick et al. 2004; Williams et al. 2006). Overall, we know relatively little about lifetime patterns of physiological performance in nonhuman animals in the wild.

Learning is somewhat similar to physical growth. Hence, the investment in learning may be highest before animals reach sexual maturity. Unlike physical growth, however, some tasks can be learned only by performing them, a feature referred to as "learning by doing" in the economic literature (Arrow 1962). Consequently, learning may contribute to a gradual increase in performance throughout an individual's life as long as it is not impaired by senescence (Dukas 2008d).

The lifetime pattern of performance in forager honeybees is remarkably similar to that of many other species, including humans: performance is quite low initially, gradually increases to a peak at about midlife, followed by a steady decrease into old age (reviewed in Dukas 1998c, 2008d; Helton 2008). To estimate the contribution of learning to the observed increase in performance, I quantified the foraging success of bees collecting nectar from an artificial feeder placed 400 meters from the hive and bees foraging on wild flowers in a natural forest. Unlike the natural settings, which require learning a wide range of tasks, nectar collection from the artificial feeder requires little learning. Indeed, feeder bees exhibited no significant change in the net rate of nectar collection from the feeder, whereas natural foragers showed a fourfold increase in the net rate of food delivery to the colony over their first few days as foragers (fig. 2.6). The major contributors to that increase were decreases in departure weights and increases in arrival weights of foragers with experience (Dukas 2008c, 2008d).

Whereas learning seems to be the major contributor to the observed increase in performance with bee experience, physiological analyses revealed that most of the flight muscle enzymes were at their peak before bees started foraging. Proteomic analyses, however, suggested that structural changes in bees' flight muscles could translate into some increase in performance with flight experience, but this possibility has not been tested (Schippers et al. 2006). Similarly, in the feeder study just mentioned, there may have been a small, significant increase in effort with forager experience (Dukas 2008d). Overall though, the data indicate a dominant contribution of learning to forager performance with experience and minor, though perhaps significant, roles

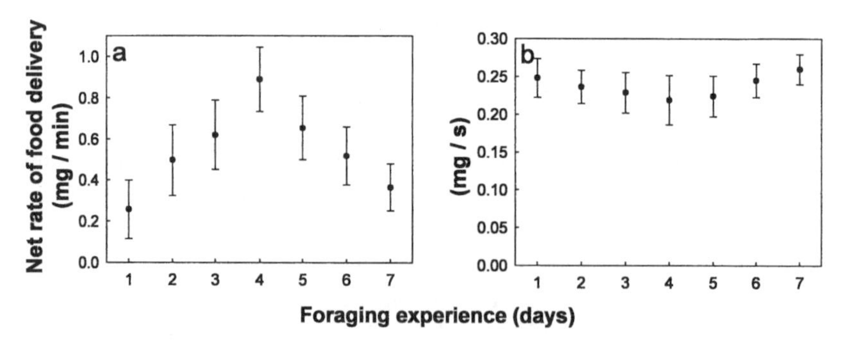

FIGURE 2.6. The net (±SE) rate of food delivery as a function of experience in honeybees foraging (a) in a natural forest (in mg/min) and (b) at a feeder providing 2.5 M sugar water (in mg/s). Bees in the challenging natural settings exhibited gradual improvement in performance, most likely owing to learning, but not at the trivial feeder. Data from Dukas 2008c, 2008d.

of physiological improvements and effort. This conclusion may be relevant for numerous species in which continuous learning over a large proportion of individuals' lives can translate into substantial increases in performance. In the human literature, such a cumulative effect of learning is referred to as expertise (reviewed in Ericsson et al. 2006).

2.6.3. EVOLUTIONARY SIGNIFICANCE OF LEARNING

Learning has contributed to evolutionary change in at least two major ways. First, learning can help animals cope with environmental change. That is, populations of animals that learn to survive and reproduce despite a dramatic ecological change can have the opportunity to adapt to the new environment over generations. Learning, however, can also decrease the rate of evolutionary change if individuals can maximize fitness by adjusting behaviorally to new environments such that no genetic change follows (reviewed in B. Robinson and Dukas 1999; Huey et al. 2003; Price et al. 2003). At least in birds, an aspect of learning ability, feeding innovation (Sol, chapter 7 in this volume), seems to have influenced evolutionary change. Feeding innovation in birds is positively correlated with (*i*) the number of species per taxon (Nicolakakis et al. 2003), (*ii*) the number of subspecies per species (Sol et al. 2005c), (*iii*) invasion success (Sol et al. 2002), and (*iv*) survival in novel environments (Sol et al. 2005a). The limited data for mammals, however, show no evidence that enhanced cognitive abilities increased the rate of morphological evolution in either great apes or hominoids (M. Lynch and Arnold 1988).

Second, the other major way in which learning has influenced evolutionary change is through its effect on assortative mating that contributes to reproductive isolation and speciation. In species in which mate choice is based on innate rules rather than learning, selection would act against divergence in secondary traits used for mate choice because novel traits would typically confer lower fitness. Such interference would not occur if divergence in secondary traits, such as color pattern, song, and odor, is accompanied by young learning to prefer the novel traits (Price 2008). This effect of learning is best documented in birds, in which young indeed learn from their parents and perhaps neighboring conspecifics about some of the desired characteristics of future mates (Lorenz 1970; ten Cate and Vos 1999). For example, males of the two allopatric subspecies of the zebra finch, *Taeniopygia guttata guttata* and *Taeniopygia guttata castanotis*, which reside in Indonesia and Australia respectively, differ in their plumage and song. When nestlings of both sexes were cross-fostered to the other subspecies, they all preferred to breed with the foster parents' subspecies, forming 100% hybrid pairs (reviewed in Clayton 1990).

The positive effects of learning on assortative mating may be largest in species with biparental care, in which young of either sex can learn from their parents about the desired characteristics of future mates. Learning, however, can promote assortative mating also in species with either uniparental care or no care at all. In mallards (*Anas platyrhynchos*), only the mother cares for her chicks. The young males learn the female-specific characteristics and, after sexual maturity, seek similar females. The females appear to mate with the males that court them most (Kruijt et al. 1982). It is likely, however, that the females also rely on innate mate choice mechanisms. In parasitic brown-headed cowbirds (*Molothrus ater*), young birds join conspecific birds probably based on innate cues. Both sexes learn population-specific courtship behavior from adults, and the males also perfect their courtship behavior based on feedback from the females (West and King 1988; Freeberg 1998; Freeberg et al. 2002).

Both male and female fruit flies (*D. melanogaster*) learn in the context of mate choice (section 2.5), and such learning could promote assortative mating (Dukas 2004b). To examine the effects of learning on assortative mating, I studied the closely related species pair *D. persimilis* and *D. pseudoobscura*, which have been widely used in research on speciation. These two species, which are sympatric along the Pacific West Coast, are visually indistinguishable but differ in their pheromonal composition and parameters of the male courtship song. Males of the two species indiscriminately court inter- and intraspecific females, but the females exhibit partial preference to intraspecific males. Hybridization is rare in the field but frequent in the laboratory, especially between male *D. persimilis* and female *D. pseudoobscura*. The hybrid daughters are fertile but hybrid sons are infertile (e.g., Mayr 1946; Noor 1995; Machado et al. 2002; Ortiz-Barrientos et al. 2004). Interspecific courting and mating are costly for both females and males. Females that mate interspecifically produce only half as many fertile offspring, and males waste time and energy courting interspecific females, which often reject them. Hence, learning in the context of sexual behavior could be adaptive for both females and males if it can lead to reduced interspecific matings and courting respectively.

In a series of experiments, I found that female *D. pseudoobscura* that had long-term experience with conspecific males were significantly less likely to mate with male *D. persimilis* than inexperienced females were. Males of both species that engaged in courtship and subsequent rejection by heterospecific females exhibited significantly lower levels of interspecific courtship than inexperienced males did. Furthermore, the reduced courtship was also associated with fewer interspecific matings by experienced male *D. persimilis* than by naive *D. persimilis* (fig. 2.7). Overall, it seems that the innate partial female preference for intraspecific males is increased with female experience. Males'

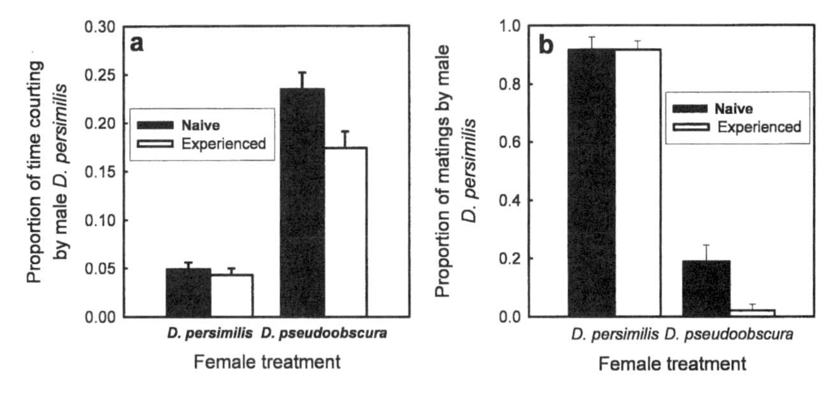

FIGURE 2.7. a. The proportion of time either naive male *D. persimilis* (black bars) or *D. persimilis* males experienced at courting female *D. pseudoobscura* (white bars) spent courting either two female *D. persimilis* or two female *D. pseudoobscura*. Each bar depicts the mean ± 1 SE for 48 males, with a total of 192 males. b. The proportion of matings in vials containing either naive male *D. persimilis* (black bars) or *D. persimilis* males experienced at courting female *D. pseudoobscura* (white bars). Each bar represents the mean ± 1 SE proportion of matings in each of 8 replicates of 24 vials each, with a total of 192 males. Data from Dukas 2008b.

responses to rejection by interspecific females amplify the female selectivity and further increase the levels of assortative mating. Hence, learning in the context of mate choice and courtship could contribute to increased levels of assortative mating that leads to speciation (Dukas 2008b, 2009).

Effects of learning on assortative mating have also been documented in fish (Magurran and Ramnarine 2004; Verzijden and ten Cate 2007) and spiders (Hebets 2003). Given the broad taxonomic distribution of the above studies and the theory indicating learning's potential importance in incipient speciation (Lachlan and Servedio 2004; Beltman et al. 2004; Beltman and Metz 2005; Price 2008), further empirical studies on this topic are needed.

2.7. Prospects

We understand better than ever some basic properties of learning and memory but learning is still not well integrated within ecology and evolution. Perhaps the most fundamental unanswered question concerns the evolution of learning. Do all animals with nervous systems learn (section 2.3.2)? Can we identify specific cellular mechanisms associated with the evolution of learning? Have physiological costs of learning and memory played a crucial role in shaping the evolution of learning (section 2.4.2)? There have been thus far only rudimentary explorations of such questions (e.g., W. Wright et al. 1996). Increased understanding of the neurogenetic mechanisms underlying learning and memory has led to the realization that there is great similarity in these mechanisms across all animals. This means that neurogenetic tools developed for one model species may be employed for addressing evolutionary ecological questions regarding learning in other species as well (Fitzpatrick et al. 2005; Smid et al. 2007).

Learning is a key factor in the life history of most animals (section 2.6.2), yet it has not been well integrated into the life history literature, which has focused on physical traits such as growth, effort, and senescence (e.g., Stearns 1992). There are a few well-studied subdisciplines, including spatial memory (chapter 6 in this volume), song learning (chapters 4 and 5 in this volume), and social learning (chapter 13 in this volume) and a variety of other studies relating behavior or ecology to brain-region size (chapter 7 in this volume; Clutton-Brock and Harvey 1980; Healy and Guilford 1990; Barton et al. 1995). But we still do not possess a coherent view of the life history trade-offs determining relative investments in learning and memory among animals.

Finally, recent work on mechanisms of speciation makes it clear that learning can play an important role in population divergence (Price 2008). Despite the traditional focus on birds in work on learning and speciation, learning may be as important in other taxa, including the most commonly used model system for speciation, fruit flies *Drosophila* spp. (section 2.6.3). We especially need a large body of empirical work that examines the role of learning in speciation in particular and evolutionary change in general. As usual, evaluating what we already know helps us choose what to learn next.

ACKNOWLEDGMENTS

I thank Lauren Dukas and John Ratcliffe for comments on the manuscript. My work has been supported by the Natural Sciences and Engineering Research Council of Canada.

3 The How and Why of Structural Plasticity in the Adult Honeybee Brain

SUSAN E. FAHRBACH & SCOTT DOBRIN

3.1. Introduction

What do London taxi drivers and foraging honeybees have in common? Both must navigate complex environments to reach various destinations and then return at the end of the workday to a fixed base of operations (garage or hive). Both improve their navigational skills with practice. Both also exhibit changes in the volume of specific brain regions that are correlated with navigational experience: the longer the time on the job, the greater the extent of the changes.

Brain data from London taxi drivers have been collected using noninvasive brain imaging (Maguire et al. 2000, 2003, 2006). There is a significant positive correlation between years on the job as a taxi driver and the volume of the posterior hippocampus as estimated using noninvasive brain-imaging techniques. Because many experimental studies have linked the hippocampus to the ability to solve tasks requiring spatial reasoning, a plausible interpretation of these data is that an enlarged posterior hippocampus is associated with superior navigational performance within the familiar terrain (Goel et al. 2004). Indirect evidence for this assertion is found in a study that compared the performance of London taxi drivers with the performance of London bus drivers. Bus drivers drive as many miles as do taxi drivers but, unlike taxi drivers, navigate predictable, fixed routes rather than the whole of London. On tests of their ability to recognize London landmarks and to judge proximity of London landmarks, the taxi drivers, who on average had larger posterior hippocampi than the bus drivers, scored significantly higher (Maguire et al. 2006).

Studies of the much smaller honeybee brain have been based primarily on the generation of volume estimates from histological sections (Withers et al. 1993; Durst et al. 1994; S. Farris et al. 2001). Relative to less experienced foragers, highly experienced foragers have a larger volume of neuropil associated with a brain structure called the mushroom bodies (Withers et al. 1993; Ismail et al. 2006). It is often assumed that the enlarged mushroom bodies of the highly experienced forager yield superior foraging performance, but at present there is no direct evidence for this assumption.

The brains of taxi drivers and honeybees are linked at present only as preeminent examples of experience-dependent structural plasticity in adult brains. Until the cellular and molecular bases of selective changes in regional brain volume are defined, it will not be sensible to address topics such as conservation or parallel evolution of mechanisms for neural plasticity during the evolution of lineages leading to modern-day chordates and arthropods. The goals of this chapter are more modest: to summarize the current status of studies of structural plasticity in the mushroom bodies of adult worker honeybees; to survey related taxa for evidence of comparable brain plasticity; and to use information from studies of honeybee behavior to suggest how future studies can tackle the provocative question of the meaning of changes in size of a region of the brain within the lifetime of the individual. Two challenges in particular engaged us. The first was the desire to incorporate existing literature on worker honeybee foraging performance into a neuroscience framework. The economic importance of the honeybee as a generalist pollinator means that such studies are scattered across diverse sources, including apicultural and agricultural journals rarely read by neuroscientists. The second challenge was to consider specialization as an alternative to the simple "bigger is better" paradigm that dominates analysis of the functional significance of regional brain growth in adults.

3.2. The honeybee as a model for the study of neural plasticity

Humans have long exploited the European honeybee, *Apis mellifera,* as a globally effective pollinator, as a source of honey and wax, and as a model for studying the evolution and regulation of complex societies. There is also a substantial research tradition of comparative animal psychology that uses honeybees to investigate learning, memory, and specialized aspects of cognition such as navigation and group decision making (von Frisch 1967; Menzel 1985; Seeley 1996; Giurfa 2007). Studies in this tradition have described the behavioral plasticity of the honeybee. More recently, the Honeybee Genome Project and the associated projects that paved the way for the genome effort, including a major honeybee brain-based EST (expressed sequence tag) project and several quantitative trait loci (QTL) analyses, have positioned the honeybee for analysis of how the environment orchestrates gene expression in the brain and how gene expression is in turn coupled to behavior (Whitfield et al. 2002, 2003; Rüppell et al. 2004; Cash et al. 2005; Honey Bee Genome Sequencing Consortium 2006; Hunt et al. 2007). These studies are building our knowledge of the plasticity of gene expression in the honeybee brain.

During the years immediately prior to the postgenomic era of honeybee research, however, the intersection of honeybee behavioral ecology with comparative studies of cognition created favorable conditions for the development of honeybee neurobiology. Numerous studies were published in the 1980s and 1990s describing in detail the neuroanatomy and neurochemistry of the honeybee brain (e.g., Mobbs 1982; Mercer et al. 1983; Rybak and Menzel 1993). Two papers published in 1993 and 1994 built directly on this knowledge base by describing the neuroanatomical plasticity of the honeybee brain (Withers et al. 1993; Durst et al. 1994). Both of these papers described an increase in the volume of the mushroom body neuropils correlated with foraging, a behavior typically performed by adult worker bees 3 weeks of age and older (Winston 1987). Subsequent studies established that it is the most experienced foragers (e.g., bees with 2 weeks' foraging experience or foragers with exceptional amounts of wing wear) that have the largest volume of neuropil associated with the mushroom bodies (S. Farris et al. 2001; Ismail et al. 2006). Other regions of the honeybee brain did not increase in volume in association with foraging experience.

The initial reports of enlarged mushroom bodies in foragers were contemporaneous with a growing appreciation of the extent of the plasticity retained by adult nervous systems of vertebrates (e.g., S. White and Fernald 1997; Buonomano and Merzenich 1998) and the dissemination of efficient tools for the estimation of regional brain volumes, including computerized image analysis and the Cavalieri method (Michel and Cruz-Orive 1988). These papers also built on two earlier publications on structural plasticity at the level of individual neurons in the adult honeybee brain, as well as efforts to identify factors that trigger the transition to foraging (Coss et al. 1980; Brandon and Coss 1982; G. Robinson et al. 1989; G. Robinson 1992).

The 1993 and 1994 papers inspired numerous follow-up studies. Another region of the honeybee brain, the antennal lobes, has been shown to display experience-dependent changes in structure in adults, but in a pattern related to olfactory experience rather than to foraging (Winnington et al. 1996; Sigg et al. 1997). Other studies sought, and found, evidence for changes in the volume of the mushroom body neuropil in other hymenopteran insects and in *Drosophila melanogaster* (Heisenberg et al. 1995; Gronenberg et al. 1996; O'Donnell et al. 2004, 2007; Withers et al. 2008). Studies of the honeybee using the Golgi technique (to reveal cytoarchitecture) and treatment with pilocarpine, a muscarinic cholinergic agonist, have revealed that the observed volume changes reflect growth of the dendritic arborizations of individual neurons, and that this growth is likely triggered in vivo by signaling via receptors for acetylcholine (S. Farris et al. 2001; Ismail et al. 2006). A comparison of the brains of different species of scarabaeid beetles revealed that a larger calycal compartment of the mushroom body neuropil is associated with a generalist feeding strategy, suggesting an evolutionary association of larger mushroom bodies with complexity of feeding ecology (S. Farris and Roberts 2005). No studies, however, have directly addressed the functional significance of growth of the mushroom body neuropil during the life of an individual honeybee.

3.3. Mushroom bodies: Neuroanatomy

In the honeybee and other insects, the mushroom bodies are paired structures located in the dorsal protocerebrum (Strausfeld et al. 1998). The mushroom bodies receive information from primary sensory neuropils, especially the antennal lobes and the optic lobes, and send information to pre-motor regions in the undifferentiated protocerebrum (Fahrbach 2006). Reciprocal connections couple the mushroom bodies on the right and left sides of the brain (Mobbs 1984).

The term "mushroom bodies" has its origin in the distinctive shape of its neuropils: two stalks, each topped with a cup-shaped rind of neuropil called the calyx (pl. "calyces"). Each calyx contains the intrinsic neurons of the mushroom bodies (fig. 3.1a). These neurons are referred to as Kenyon cells in honeybees and other hymenopterans (Kenyon 1896). Of the just under one million neurons that constitute the honeybee brain, approximately one-third are Kenyon cells (Witthöft 1967; Fahrbach 2006). The dendritic arborizations of the Kenyon cells form the calycal neuropils that surround the Kenyon cells; each Kenyon cell projects a bifurcating axon to form lobed output neuropils via the peduncle (fig. 3.1b). Other important modulatory inputs have been identified, but the major path of information flow through the honeybee mushroom bodies is to the Kenyon cells in the calyces and from the Kenyon cells in the lobes (Fahrbach 2006).

The calyces of the honeybee are divided into three major regions apparent in sectioned material as differences in the texture of the neuropil (fig. 3.1b). These three regions are also functionally defined by the sensory inputs they receive (Gronenberg 2001). The uppermost region (lip) is innervated by olfactory projection neurons found in the antennal lobes. The midsection of the calyx (collar) is innervated by interneurons located in the medulla and lobula regions of the optic lobes (Ehmer and Gronenberg 2002). The base of the calyx, the basal ring, is innervated by projections from both the antennal (olfactory and mechanosensory) and optic lobes. A small region of the calyces adjacent to the lip receives gustatory inputs originating in the subesophageal ganglion (Schröter and Menzel 2003). The hymenopteran mushroom bodies are similar overall to those of other insects but are notable for the complexity of their calyces and the extent to which the visual world is represented.

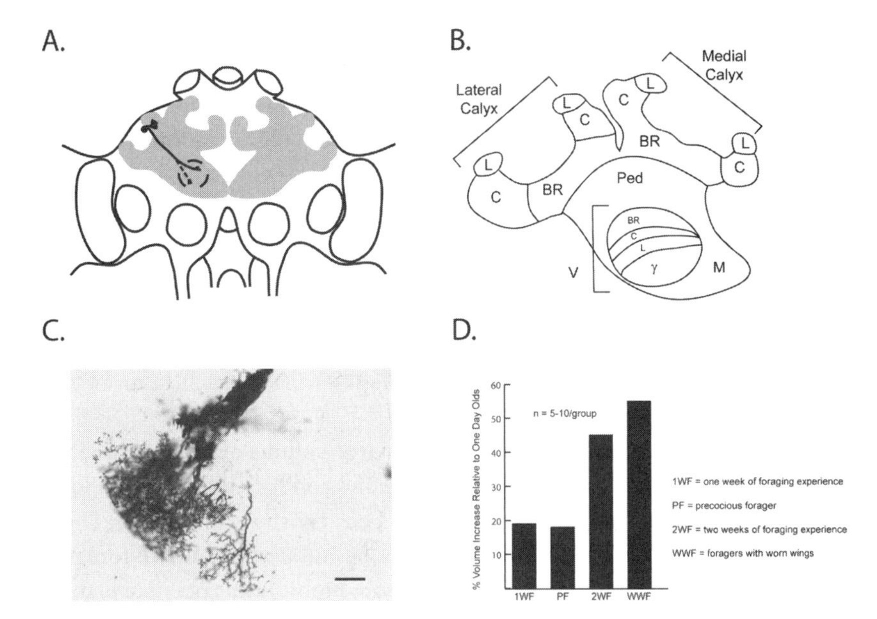

FIGURE 3.1. a. Drawing of a cross section through the adult honeybee brain. The neuropil of the mushroom bodies is shown in gray. A single Kenyon cell (not to scale) is depicted. The dendritic arborizations of the Kenyon cells form the calycal neuropils. The branched axons of the Kenyon cells form the lobe neuropils. For a full description of the heterogeneous population of Kenyon cells in the honeybee brain, see Fahrbach 2006. b. Diagram of the subdivisions of the neuropils that constitute the mushroom bodies of the honeybee. The lateral and medial calyces on each side of the brain unite to form a single peduncle, which branches to form two lobes. The segregation of sensory input in the calyces is maintained in the vertical lobe. For details, see Fahrbach 2006. Abbreviations: BR, basal ring; C, collar; L, lip; M, medial lobe; V, vertical lobe; γ, gamma lobe. c. Dendritic arborizations of Golgi-impregnated Kenyon cells in the mushroom body calyx of an adult worker honeybee. The rightmost profile depicts the arborization of a single Kenyon cell, which can be used to characterize the number, length, and branching patterns associated with different amounts of foraging experience. The left cluster of several adjacent impregnated Kenyon cells is a more realistic representation of the density of neuronal processes in the calyces of the mushroom bodies. Scale bar, 10 μm. d. Compiled results from several studies (redrawn using a common axis), showing the typical increase in volume of the mushroom body neuropils associated with extensive foraging experience in the honeybee. Wing wear is considered a reliable marker of foraging experience (Higginson and Barnard 2004). For original data, see Withers et al. 1993 and S. Farris et al. 2001.

In other insects, such as dipterans, the mushroom bodies are frequently described as olfactory structures, because antennal lobe projections provide the majority of sensory inputs to this region (Davis 2005; Fahrbach 2006).

3.4. How does foraging experience change the structure of the honeybee mushroom bodies?

In honeybees, foragers have a larger volume of mushroom body neuropil than same-aged nest-mates that perform hive tasks, and more experienced foragers have larger mushroom body neuropils than age-matched foragers with fewer days of foraging experience (fig. 3.1d; Withers et al. 1993; Durst et al. 1994; S. Farris et al. 2001; Ismail et al. 2006). The volume of specific brain regions is typically determined using an unbiased method of volume estimation based on systematic random sampling (the Cavalieri method) of histological sections of known thickness (Gundersen et al. 1988). Worker honeybees from a single colony have very low variation in body size, so that regional brain volumes are typically published without correction for body size (Waddington 1989). The observed changes in volume reflect growth of existing elements, as the neuroblasts (stem cells) that produce the Kenyon cells undergo programmed cell death during the pupal stage of development (Fahrbach et al. 1995b; Ganeshina et al. 2000).

The association between foraging and a large volume of mushroom body neuropil has been observed in numerous studies conducted in different locations using different populations of honeybees (Withers et al. 1993; Durst et al. 1994). Because worker honeybees can be induced to initiate foraging precociously by manipulations of colony demography, it was possible to demonstrate that behavioral status (being a forager or not being a forager) was a better predictor of mushroom body volume than age. Strong evidence of a primary role for foraging itself as a driver of growth of the mushroom bodies in the bee is found in studies in which workers with 2 weeks of foraging experience were found to have a significantly greater volume of mushroom body neuropil than age-matched foragers from the same colony that had only 1 week of foraging experience (Ismail et al. 2006). It has been argued that the life of the forager is more visually oriented, learning based, and memory taxing than that of a hive bee. A positive feedback loop has been envisioned, sometimes implicitly and sometimes explicitly, in which foraging experience promotes the growth of the mushroom body neuropil, which in turn supports the efficient performance of foraging tasks (Fahrbach and Robinson 1995).

Studies of individual neurons impregnated by the Golgi method demonstrated that at least some portion of foraging experience–dependent volume change reflects the growth of individual Kenyon cells that have dendritic arborizations in the collar region of the calyces (S. Farris et al. 2001). Administration of pilocarpine, a muscarinic cholinergic receptor agonist, to caged bees stimulated an increase in volume of the mushroom body neuropil equal to that produced by a week of foraging under natural conditions (Ismail et al. 2006). As will be described below, the absolute requirement of such studies that volume comparisons be restricted to age-matched groups of honeybees led to the development of a "field-to-lab" method of rearing that is particularly appropriate for studies of the functional significance of mushroom body growth. Studies in the ant *Pheidole dentata*, another social hymenopteran that displays foraging-related growth of the mushroom bodies, have provided evidence that there are more synapses in larger calyces (Seid et al. 2005). It is likely that the same is true of enlarged honeybee calyces, although this has not been proved. Given that Golgi studies revealed that spine density was not decreased in foragers relative to nurses (S. Farris et al. 2001), longer, more branched dendrites almost certainly reflect more synapses. Studies designed to uncover the molecular correlates of foraging-dependent structural plasticity in the honeybee mushroom bodies are already in progress. A preliminary DNA microarray analysis of gene expression in the mushroom bodies of experienced foragers and of bees treated with pilocarpine to mimic the stimulation of muscarinic receptors putatively induced by foraging revealed that only a small number of genes showed quantitative changes in expression (Lutz et al. 2007). This finding suggests that the coupling of experience to dendritic branching in the calyces might rely primarily upon activity-dependent, post-translational modification of constitutively expressed proteins. This conclusion is predicated on a cDNA microarray based on a honeybee brain EST project (Whitfield et al. 2002). It will be exceedingly interesting to validate these initial observations and to extend the analysis using different sampling schemes and a more complete microarray based on the full honeybee genome (Honey Bee Genome Sequencing Consortium 2006).

3.5. What is the function of the honeybee mushroom bodies?

Our knowledge of the function of the mushroom bodies in insects is based primarily on analyses of the effects of genetic perturbations of these structures in the brains of the fruit fly *Drosophila melanogaster*. Some findings from the fruit fly have yet to be validated in the honeybee. For example, there appears to be no equivalent in the honeybee brain of the amnesiac gene-expressing DPM neurons that innervate the lobes of the mushroom bodies (Waddell et al. 2000). Conversely, important findings from honeybee research, such as the role of the octopaminergic innervation of the calyces in olfactory association learning, are not prominent features of the fruit fly model (Heisenberg 2003). Here we review what is known directly from studies of honeybees, concluding by noting points of intersection with the fly literature. A major difference between honeybee and fruit fly studies on this topic is that the larger honeybee is typically studied as an individual, while fruit flies are typically trained and tested in groups (Giurfa 2007).

It is not possible at present to create transgenic lines of honeybees with gene knockouts, although knockdown of protein kinase A by injection of short antisense oligonucleotides into the brain and disruption of an octopamine receptor and vitellogenin (a yolk protein) using RNA interference have been achieved (Fiala et al. 1999; Farooqui et al. 2003; Guidugli et al. 2005; Amdam et al. 2006). This means that the primary tools available for study of mushroom body function in bees are the classic tools of behavioral neuroscience: lesions and pharmacological manipulations.

There are few studies of the effects of lesions or drug treatments on the naturally expressed behaviors of foraging bees or bees in the hive. A classic laboratory test of association learning using bees restrained in small harnesses is the proboscis extension reflex, or PER (Bitterman et al. 1983). Honeybees (and other insects) will reflexively extend their proboscises to drink when sucrose (the unconditioned stimulus, or US) is applied to the antennae. If an odor or other conditioned stimulus (CS), such as a textured plate that can be touched by an antenna, is paired with the application of sucrose to the antenna and the opportunity to drink a sucrose solution, an association is formed between the CS and the food reward. After training, the CS alone elicits the PER. In honeybees, associations with the CS develop quickly, in some cases resulting in formation of a stable memory after as few as three pairings. The basic appetitive-learning PER paradigm is simple in design but can be adapted for use in more complex studies of learning and memory (Giurfa 2007). More recently, a related aversive-learning task has been developed for use with honeybees. This task is referred to as the sting extension reflex, or SER (Vegoz et al. 2007). The SER exploits the fact that harnessed bees reflexively extend their stings in response to a small electric shock (US). If this unconditioned response is paired with an odor (CS), bees can learn to extend their stings in response to the CS alone.

Alternative appetitive-learning paradigms used to study learning in honeybees involve training flying honeybees (either in the field or in a flight room) to approach a target to receive a food reward, such as the opportunity to drink a sucrose solution (Lehrer 1997), or inducing bees to fly through a maze to reach a food reward (S. Zhang et al. 2000). To date, these methods have been used infrequently to assess the effects of physical or chemical manipulations of brain function on bee behavior (for an exception, see Si et al. 2005). Other assays based on naturally occurring bee behaviors, such as the rapid learning of cues imparted to bees by wax comb (Breed et al. 1995), have played a surprisingly small role in studies of the brain regulation of behavior in bees.

Our appreciation of the function or functions of the mushroom bodies in the bee is therefore based primarily on assessment of the impact of disruption of mushroom body function on tasks using PER. Confident interpretation of the results from such experiments relies on the use of control assays that demonstrate equivalent sucrose response thresholds in the experimental groups and the maintenance of the unconditioned reflex during the entire period of testing (Giurfa 2007).

If carefully done, local cooling of brain regions produces a reversible brain lesion (Horel 1991). Cooling was combined with PER conditioning in the 1970s and 1980s to establish a role for the honeybee mushroom bodies in the storage of olfactory memories that are resistant to retrograde amnesia induced by electric shocks, whole-body cooling, or exposure to carbon dioxide (Erber et al. 1980). Alternatively, permanent lesions of the mushroom bodies of the honeybee can be induced by feeding the DNA synthesis inhibitor hydroxyurea to first-instar larvae (Timson 1975). If applied at the appropriate stage of very early larval development, hydroxyurea treatment results in the death of a substantial subset of the mushroom body neuroblasts (Malun et al. 2002b). The absence of the neuroblasts and the Kenyon cell progeny they would have produced results in shrunken or missing mushroom bodies. In most cases, one or both of the median calyces is absent. Such lesions did not result in defects in acquisition or retention of either mechanosensory or olfactory associations assayed using PER in 5- to 7-day-old hive-reared honeybees (Malun et al. 2002a), although subtle deficits in performance were detected when multiple odorant stimuli were presented (Komischke et al. 2005). In contrast to the cooling studies, these findings suggested that the mushroom bodies may not be essential for the formation of associations that occurs during PER training, although other explanations of this important negative result, including incomplete ablation of redundant pathways, can also be given.

Cognitive functions that require information processing in the mushroom bodies are predicted to be disrupted or facilitated by the application of drugs that block or mimic the action of neurotransmitters or intracellular second-messenger systems within the mushroom bodies. Injection of picrotoxin, an inhibitor of the GABAA receptor, into the antennal lobes disrupts odor-induced, local field potential oscillations in the calyces of the mushroom bodies; bees with disrupted oscillatory activity displayed reduced ability to discriminate between structurally similar odorants in a PER paradigm when the picrotoxin was present, although less demanding odorant discriminations could still be made (Stopfer et al. 1997). This finding suggests, as in the case of the hydroxyurea lesions, that the mushroom bodies are not essential for the learning of associations during PER training but might instead permit comparison of sensory stimuli with stored representations of similar cues.

The mushroom bodies are innervated by cholinergic fibers from the antennal lobes and, possibly, the optic lobes; they also contain glutamate and glutamate transporters (Bicker 1999; Si et al. 2004). Treatment of the mushroom bodies of the bee with cholinergic blockers such as scopolamine and with glutamate receptor blockers such as MK-801 impairs retrieval of learned olfactory associations (Gauthier et al. 1994; Si et al. 2004), assayed by using PER training. These data, in contrast to the hydroxyurea and picrotoxin studies cited above, suggest that the mushroom bodies are important for the normal display of PER.

Efforts to identify the neural correlates of PER resulted in the identification of VUMmx1, a mushroom body afferent, and PE1, a mushroom body efferent, or "extrinsic neuron" (Hammer 1993; Hammer and Menzel 1995; Rybak and Menzel 1998). These neurons place the mushroom bodies within a circuit that mediates learning by bringing neural representations of the US (sucrose) together with neural representations of the CS (the odorant that predicts the appearance of the sucrose reward), ultimately directing this information to pre-motor brain regions.

The VUMmx1 neuron has an unusual morphology. The soma is in the subesophageal ganglion, but its processes arborize bilaterally in the antennal lobe, the lateral protocerebrum, and the mushroom body calyx (Hammer 1993). Pairing of an odor with suprathreshold electrical stimulation of VUMmx1 was found to be as effective in inducing PER training as when the odor was paired with a sucrose reward (Hammer and Menzel 1995). The VUMmx1 neuron is octopaminergic, and injections of octopamine into the calyx or antennal lobe were able to substitute for sucrose in appetitive odor conditioning.

The PE1 neuron also has an unusual morphology. PE1 has a single large dendrite that arborizes throughout the vertical lobe and projects to the putatively pre-motor lateral and medial protocerebrum. After PER training, PE1 responses to the CS are strengthened (Rybak and Menzel 1998). The identification of VUMmx1 and PE1 activity as significant for PER suggests that the mushroom bodies deserve their designation as a brain module that makes PER training possible. It is intuitively appealing to assume that PER training is a reflection of how bees transform a world of odors into a predictable set of signals that nectar or honey is available, but the link between modulators of PER training and honeybee performance in the field has yet to be demonstrated.

Numerous studies of odor-related learning in *Drosophila melanogaster* have led to the conclusion that some, but not all, forms of such behavior require the mushroom bodies (Heisenberg 2003; Davis 2005). Instead of PER, the typical task used in studies of mushroom body function is a form of olfactory discrimination learning in which flies learn to jump in the presence of an odor previously paired with an electric shock to the feet (Margulies et al. 2005). Flies trained to associate an odorant (CS) with an electric shock (US) learn to avoid an expected shock when presented with the odorant alone. The retrieval of memories of odors appears to require synaptic output from the Kenyon cells in the lobes (McGuire et al. 2001). Flies treated as larvae with hydroxyurea so that they have no mushroom bodies do not make the association of odorant and shock (deBelle and Heisenberg 1994). But these flies appear otherwise normal, at least within the confines of laboratory life, until their ability to recall odors is tested.

This striking result—near normal behavior in the absence of the mushroom bodies—seems inconsistent with the apparent dominance of the protocerebrum by the mushroom body neuropils. Heisenberg, inspired both by his own experiments with flies and by the classic studies of Huber in the 1950s and 1960s on orthopterans, suggested that the function of the mushroom bodies might be better sought by thinking in terms of cognitive functions of even higher order than multimodal sensory integration and association learning (F. Huber 1960). In a thoughtful review published in 2003, he noted that flies lacking functional mushroom bodies often appear to be unable to "make up their minds," particularly in situations that provide conflicting information or require switching behavioral responses (Heisenberg 2003). This insight has not yet generated new experiments using honeybees, possibly because current studies of bees, including our own, are either anchored within the framework of behavioral development ("becoming a forager" or "acquiring more foraging experience") or are dependent upon PER training.

3.6. Why are the mushroom bodies larger in experienced foragers?

Assuming that the increase in the volume of mushroom body neuropil is not a meaningless epiphenomenon in the life of a worker honeybee, we think that there are three possible explanations for foraging-dependent growth of the mushroom bodies. First, it is possible that this change in brain structure enables the bee to perform a new task or cognitive operation that was not performed earlier in life. Second, growth of the mushroom bodies may improve the performance of a previously performed task. Third, the change in brain structure could inhibit performance of behaviors characteristic of bees with smaller mushroom bodies, resulting in covert morphological and overt behavioral specialization of experienced foragers. We will consider each of these possibilities in turn after a brief review of the nature of foraging. They are not necessarily mutually exclusive (fig. 3.2).

Our aforementioned London taxi driver (section 3.1) could likely relate well to a forager honeybee. In both cases, the ability to earn a living is dependent upon the quality and possibly the number of cognitive spatial maps. In

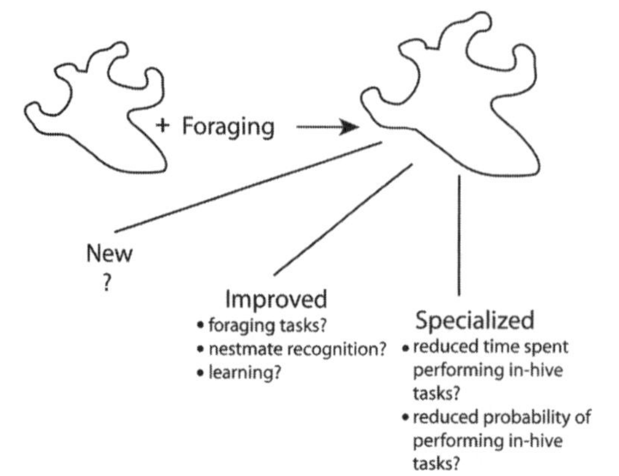

FIGURE 3.2. At present, the functional significance of foraging-induced expansion of the mushroom body neuropil in worker honeybees is unknown. This diagram depicts three possible changes in behavior that may result from brain plasticity and that are discussed in this chapter. These possibilities are not mutually exclusive.

both cases, no two workdays are ever identical. In addition to displaying navigational prowess, however, the foraging honeybee also collects nectar, pollen, water, and/or propolis (plant resins), carries these resources to the hive, and transfers them to the appropriate in-hive workers (Winston 1987). Foragers also interpret the messages of other foragers on the dance floor so that they can exploit new food sources.

We believe that it is unlikely that any entirely new behavior is a result of the foraging-induced growth of the mushroom bodies. This reflects the fact that we are studying a form of structural brain plasticity in foragers that occurs after foraging has been initiated. By definition, any brain growth that occurs prior to the initiation of foraging is not foraging dependent. This does not therefore appear to be a productive line of investigation. This conclusion would need to be revisited if, for example, other types of experience (e.g., training to perform a particular task, such as PER, in the laboratory) are demonstrated to result in comparable growth of the mushroom body neuropil.

The second possibility is that foraging experience makes a bee a better forager. This hypothesis assumes that the mushroom bodies support the cognitive processing required by foraging: to date, no evidence contradicts this assumption, although bees with chemically ablated mushroom bodies have not been tested for foraging ability, as the hydroxyurea-treated bees studied by Giurfa and colleagues were tested as preforagers at 5–7 days of age (Komischke et al. 2005). Improved individual foraging performance would have a positive effect on the fitness of a colony. A reduction in flower-handling time, a better choice of routes, the ability to learn associations between floral odors and nectar rewards, and the ability to retain memories of rewarding resource patches longer could all contribute to greater foraging efficiency. Better foragers might also achieve a superior balance between resource faithfulness and resource switching, which could optimize the contribution of the individual worker bee to exploitation of available resources (Laverty and Plowright 1988; Laverty 1994). A better forager might communicate more efficiently with other bees via the dance language or be more sensitive to pheromones that influence foraging, such as brood pheromone (Pankiw 2004). Alternatively, a better forager might simply have a lower threshold for leaving the hive and thereby put in a longer workday.

Although beekeepers anecdotally report differences in productivity among their colonies, there are few studies of honeybee foraging efficiency measured in the field that take the experience levels of individual foragers into account. In a study published in 1994, bees were tagged on the first day of adult life and then returned to their colony (Dukas and Visscher 1994). A modified entrance permitted each departing and returning bee to be temporarily detained for the purpose of recording each identified bee's weight. The time of each measurement was also recorded. These investigators found that bees with more foraging experience returned with larger foraging loads (and, other than for the oldest bees, there was actually a nonsignificant decline in trip duration), although performance eventually declined in the most experienced (oldest) foragers, resulting in an overall profile of lifetime foraging performance shaped like a shallow inverted U. Roughly similar results were reported in a subsequent study (Schippers et al. 2006). In both studies, observations were made at the hive entrance. It is therefore impossible to know if the more experienced foragers increased their loads by visiting more flowers or by extracting more resources from individual flowers. A follow-up study cleverly dissociated physiological parameters and learning by examining the food delivery rate and hourly visitation rate of bees trained to visit a sugar water feeder located 400 meters from the hive. In contrast to the results of the earlier studies in which bees were required to extract resources from flowers, extended foraging experience at a feeder was not associated with improved performance over the course of a forager's career (Dukas 2008d).

There is a need for focused studies on this topic using individually tagged bees of known age and known foraging experience. Despite the labor-intensive nature of these studies, all of the necessary methods are already present in the tool kit of the bee behavioral biologist. Some simple measures from the apicultural literature, such as observations of flower-handling time (Stubbs and Drummond 2001) or direct measures of flight activity (Ellis et al. 2003), could be profitably paired with neuroanatomical analyses. Two experimental strategies appear especially promising. The first is to use groups of known-age bees with predictably different volumes of mushroom body neuropil generated using the field-to-lab method in behavioral tests, both in the field and in the laboratory (Ismail et al. 2006). The second is to study the relationship between foraging experience and foraging performance in a sheltered environment (screen house or "bee dome") in which the complexity and effort required to extract resources are controlled by the experimenter.

The absolute requirement that comparisons of mushroom body neuropil volume be restricted to age-matched bees (to ensure that the results truly reflected quantity of foraging experience rather than chronological age) led our research group to develop a field-to-lab method of rearing worker honeybees that is particularly well suited for the study of the functional consequences of the growth of the mushroom body neuropil relatively late in adult life (Ismail et al. 2006). In this method, worker honeybees complete metamorphosis and emerge from their cells in an incubator. On the first day of adult life, each bee is marked with a paint dot and then returned to a field colony, where she undergoes normal adult behavior development under field conditions. Observations at the hive entrance allow each worker to be marked a second time when she initiates foraging. Cohorts of marked bees are allowed to forage for a week in the field. At the end of a week of foraging, they are divided into field control, cage control, and experimental groups. The field control group is allowed to remain in the field for a second week of foraging; the cage control group is caged in the laboratory in an incubator for a week and fed sugar water; experimental groups are also caged in the laboratory, where they are treated with neurotransmitter agonists and antagonists, as described in Ismail et al. (2006); other manipulations are also possible. An essential aspect of all such experiments will be to control both age and foraging experience in bees subjected to behavioral analysis. It is also highly desirable to combine behavioral and neuroanatomical analyses in the same study.

Given the small number of field studies of foraging performance, one might ask if there is any evidence that bees with greater amounts of foraging experience (which predicts that they will have enlarged mushroom bodies) excel in laboratory tests of learning. A recent study examined the performance of workers between 26 and 52 days of adult life on laboratory tests of light responsiveness, sucrose responsiveness, and associative learning using PER training. Because the bees were housed in a typical colony with access to the outdoors, it is likely that the older bees had accumulated more foraging experience. The oldest bees in this sample displayed a slight enhancement in positive phototaxis but did not differ from younger bees in their ability to associate an odor with a sucrose reward (Rüppell et al. 2007). A similar study that controlled not only age but also extent of foraging experience produced the unexpected result that the more experienced bees performed less well than less experienced bees on acquisition of a conditioned PER response but, once an olfactory association had been acquired, were less likely to erroneously generalize their response to a different odor (Behrends et al. 2007). This finding suggests that one possible interpretation of growth of the mushroom bodies is improved ability to inhibit inappropriate responses.

In mammals, changes in the complexity of dendritic arbors (of hippocampal or cortical) neurons are correlated with performance on cognitive tasks. For example, rats housed in a stimulating (enriched) environment have increased complexity of dendritic arbors in layer III pyramidal neurons of the parietal cortex and perform better than controls on laboratory tests of spatial memory (Leggio et al. 2005). Conversely, the complexity of the dendritic arbors of hippocampal pyramidal neurons can be reduced by exposure to hypoxic conditions (Titus et al. 2007). Rats displaying hypoxia-induced cytoarchitectural changes in the CA3 region of the hippocampus have impaired performance in a radial-arm maze test. This body of literature argues in support of the concept that growth of neuropil is associated with improved performance. If this is also the case in the honeybee, the challenge for researchers is to define the role of the mushroom bodies in the everyday life of the forager. A recent set of studies conducted in our laboratory suggests a promising target for future research (fig. 3.3). We used a well-established laboratory assay to assess possible cholinergic mediation of odor-based nest-mate recognition (Breed 1983; Breed et al. 1988). Treatment of young worker honeybees with the muscarinic agonist pilocarpine improved their ability to discriminate nest-mates from non-nest-mates (Ismail et al. 2008). Although the subjects in this study were younger than normal-age foragers, this simple assay should be adaptable for the study of context-specific olfactory discrimination in foragers.

The third possibility is that growth of the mushroom body neuropil leads, not to improved performance on foraging-related tasks, but rather to specialization. The concept of specialization is entrenched in behavioral ecology (e.g., Ferry-Graham et al. 2002; Irschick et al. 2005). For example, a published definition (Irschick et al. 2005, 405) of a "functional specialist" is "a species whose morphology (or physiology) constrains it to a subset of available resources." It is unconventional to think of the honeybee, a species renowned as a generalist pollinator, as a specialist, but the age polyethism (age-related behavioral castes) of adult worker honeybees can be fitted to a specialist framework if one thinks beyond visible differences in morphology. At a minimum, time spent performing a specialized task is not available for performance of other tasks. A neurobiological view of specialization draws attention to the inhibition of circuits controlling the behavior discarded or reduced in frequency of performance when specialization occurs.

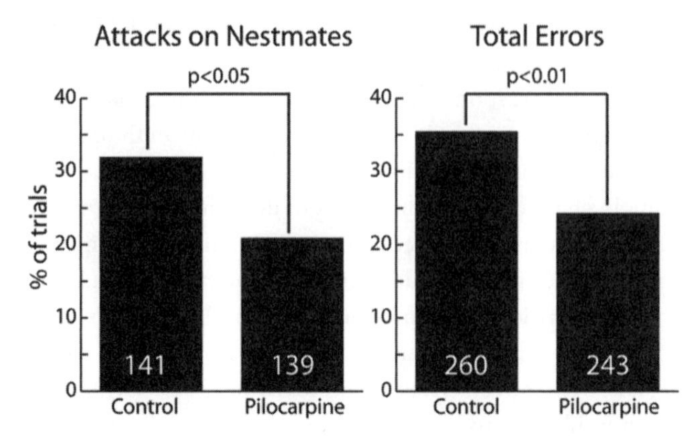

FIGURE 3.3. Treatment of young worker honeybees with pilocarpine, a muscarinic agonist, significantly reduced the number of errors made in a simple laboratory assay of nest-mate recognition. Bees were housed in groups of 10–12 nest-mates in plastic Petri dishes (90 mm diameter, 10 mm deep, with an 8 mm hole drilled in the lid for feeding-tube access) and treated orally with a 1:1 solution of sucrose and water (control) or pilocarpine dissolved in sucrose (10^{-3} M) for 5 days. Each dish contained a small piece of comb (25 mm^2) from the bees' natal colony. On the fifth day after the dishes were established, a single sucrose-fed nest-mate or non-nest-mate was added to each dish. The number of aggressive interactions was recorded over a 4-minute period. The figure shows the percentage of trials resulting in attacks directed toward an unfamiliar introduced stimulus bee who was a nest-mate (left panel) and a summary of responses to all introductions (right panel; nest-mates and non-nest-mates). The G^2 statistic was used to analyze contingency tables. Pilocarpine treatment reduced both the percentage of trials in which a nest-mate was attacked (one possible type of error) and the percentage of trials in which any error (rejection of nest-mate, acceptance of non-nest-mate) occurred. Numbers in the bars indicate the number of dishes tested. Adapted from data published in Ismail et al. 2008.

If one examines time budgets of forager and nonforager workers in a honeybee colony (as in Seeley 1982), it is notable that younger workers in a colony with normal age demography perform numerous tasks within the hive but do not forage. Conversely, foragers are significantly less likely to perform in-hive tasks than nonforagers. Ethograms developed by using glass-walled observation hives have indicated that the primary activities of foragers in an undisturbed colony are foraging and resting (Seeley 1982). These observations raise the rarely considered possibility that the growth of the mushroom body neuropil reflects the addition of inhibitory synapses to the calyces, and that the expansion of this neuropil results, not in enhanced foraging performance, but rather in a decreased probability of performing other tasks. Such a change could reflect either a reduced response to the cues that facilitate responses to other behaviors, such as nursing or attending the queen, or a decreased ability to switch from one task to another (Heisenberg 2003). Again, the use of groups of known-age bees with predictably different volumes of mushroom body neuropil in behavioral and neuroanatomical studies, generated using the field-to-lab method, could provide evidence regarding this hypothesis.

Another experimental approach to the study of specialization of the forager is the examination of the probability that an established forager will revert to nursing (tending larval brood) within the hive if the nurse population is artificially depleted. There is evidence that older foragers (assumed to have more foraging experience than younger foragers, and therefore to have larger mushroom bodies) are less likely to display behavioral reversion (R. E. Page et al. 1992). This preliminary observation is consistent with the hypothesis that foragers are specialized relative to hive bees.

A direct example of an association between dendritic growth and behavioral inhibition is found in the mammalian literature. Deer mice reared in enriched environments have higher dendritic spine densities in layer V pyramidal neurons in motor cortex and are better able to inhibit stereotyped motor behavior than controls reared under standard conditions (Turner et al. 2003).

3.7. Studies of experience-dependent plasticity in the mushroom bodies of other insects

Studies of the ant *Camponotus floridanus* and the orchard bee *Osmia lignaria* have shown that a larger volume of mushroom body neuropil is associated with foraging experience in these species (Gronenberg et al. 1996; Withers et al. 2008). Foraging, however, may belong to a larger set of stimuli that induce growth of the mushroom body neuropil. Studies of the eusocial paper wasps *Polybia aequatorialis* and *Mischocyttarus mastigophorus* support the idea that behavioral specialization may be a critical factor. Like the honeybee, paper wasps exhibit age-based division of labor. Young paper wasp workers perform in-nest jobs, progress to activities on the exterior of the nest, and then finally initiate foraging. The wasp mushroom body, which is similar to the honeybee mushroom body in terms of gross morphology and inputs (Ehmer and Hoy 2000), increases in volume as the wasp progresses from in-nest behaviors to exterior behaviors (O'Donnell et al. 2004). In nests with multiple females, social dominance is positively correlated with mushroom body size (O'Donnell et al. 2007). The dominant female wasp is highly aggressive and primarily remains on the nest to defend and care for her brood, while subordinate wasps spend more time foraging. If foraging experience alone induces mushroom body plasticity, then the mushroom bodies of the subordinate wasps would be predicted to have the largest volumes, but the opposite was found to be true.

It is possible that the plasticity seen in both the wasp and the honeybee results from behavioral specialization and the changed social environment of the specialized individual. An alternative argument is that both the dominant paper wasp and the forager honeybee engage in more cognitively challenging tasks (recognition of subordinate individuals in the paper wasp, navigation in the foraging honeybee), and that the growth of the mushroom bodies provides a general readout of the complexity of tasks performed. An evolutionary counterpoint to a simple complexity argument may be found in the mushroom bodies of solitary wasps. Species of social wasps in general have larger mushroom bodies than solitary species (Ehmer et al. 2001). A solitary wasp must care for the brood as well as forage but lacks the social interactions experienced in a multifemale nest, and thus, we may speculate that the maintenance of the dominance hierarchy through social interactions is a primary driver for mushroom body growth in adult female paper wasps (O'Donnell et al. 2007).

Honeybee foragers, of course, also participate in complex social interactions within the hive (Winston 1987). There is, however, strong evidence that our emphasis on the link between foraging experience and growth of the mushroom bodies in honeybees is not misplaced (Ismail et al. 2006). Foraging bees had a larger volume of mushroom body neuropil than did age-matched bees confined to the hive after a week of foraging experience: the confined bees had already matured as foragers and could therefore experience the social interactions characteristic of "being a forager" inside the hive but were unable to leave the hive to visit flowers.

Eusocial species offer the opportunity for comparative studies across castes; it is also possible to compare the plasticity of adult male and female brains within a species. The neuropils associated with the mushroom bodies of honeybee queens increase in volume during the first 2 weeks of life independent of whether they were naturally mated, instrumentally inseminated, or never left the queen bank (Fahrbach et al. 1995a). This indicates that flight, or even leaving the hive, is not necessary to trigger the changes seen.

A similar early phase of experience-independent growth of the mushroom body neuropils has also been observed in honeybees in the behavioral caste of nurse bees (Withers et al. 1993), in young workers confined to the hive (Withers et al. 1995), and in young workers reared under conditions of social isolation and sensory deprivation (Fahrbach et al. 1998). In drones, the mushroom body neuropils are larger in males aged 10 days and older than in younger males (Fahrbach et al. 1997). Although interpretation of the drone study is confounded because all males studied were permitted to take mating flights, taken together these results help us place the foraging-dependent growth of the mushroom bodies into a broader context of multiple life stage–appropriate variants of brain plasticity. At present, no examples other than that of the forager have been investigated at the cytoarchitectural or molecular levels of analysis.

3.8. Specific future directions

Progress in understanding the cellular and molecular consequences of foraging experience in the honeybee brain is all but inevitable. But what ways of investigating the functional significance of the changes in brain structure that are a normal part of the life of every worker in a honeybee colony are most likely to lead to progress on the why question?

One strategy is to use what we know about the details of the observed neuroanatomical changes to guide our research. For example, several studies have drawn attention to foraging-dependent changes in the collar region of the calyces. This is the calycal region that receives projections from the optic lobes. These observations suggest that a productive strategy will be to search for enhanced performance on tasks that depend specifically on responses to visual signals. Perhaps PER training using odor cues alone is unlikely to be a sensitive indicator of changes in brain structure. At a minimum, visual cues can be incorporated into PER testing, and the SER can be used in parallel with PER-based studies of learning (see section 3.5). Another example comes from the demonstration that the dendritic arborizations of the Kenyon cells of experienced foragers are characterized by increased branching at the distal tips (S. Farris et al. 2001). What inputs target this subset of synaptic sites? The details of the neuroanatomy can provide powerful guides to our behavioral analyses. One element that is currently completely missing from our analyses is a consideration of possible changes in the cytoarchitecture of the main mushroom body output regions, the lobes. More studies using the well-established Golgi method of stochastic impregnation of neurons are needed if we are to understand the synaptic meaning of mushroom body plasticity (Strausfeld 1980).

A second strategy is based on the use of age-matched bees with varying amounts of foraging experience (1 week vs. 2 weeks) in a broad range of tasks, including studies in the field at flowers as well as at the hive entrance. The field-to-lab method described in preceding sections makes this possible.

It is critical, however, to design experiments that can detect decrements as well as improvements in performance. The longer a London tax driver has been on the job, the better his performance on tests related to knowledge of London becomes. One reason for the improved performance is possibly the previously described experience-dependent increase in the volume of the posterior hippocampus observed in experienced taxi drivers (section 3.1). But it is important to note that there is an apparent cost to this specialization. Relative to London bus drivers, London taxi drivers performed significantly worse on tasks requiring acquisition or retrieval of new visuospatial information (Maguire et al. 2006). A new direction in research on growth of the mushroom body neuropil in the honeybee will be to look for evidence of specialization rather than enhancement. It should be noted that foraging-induced changes in the structure of the mushroom bodies, dependent upon the experience of the individual, will be a confound in studies of the cognitive effects of aging in the honeybee (Behrends et al. 2007; Rüppell et al. 2007).

We have focused our studies on the honeybee, but experience-dependent brain plasticity is not a rare phenomenon. If you are a behavioral neuroscientist, you can be confident that it is happening in your animal model—and, more to the point, happening in your own brain. Linking experience-dependent changes in brain structure to their functional consequences is important because doing so will give us insight into a powerful source of individual (experience-dependent) differences in animal behavior.

ACKNOWLEDGMENTS

The authors thank G. E. Robinson for thoughtful comments on this chapter.

PART II

Avian Cognition: Memory, Song, and Innovation

4 More on the Cognitive Ecology of Song Communication and Song Learning in the Song Sparrow

MICHAEL D. BEECHER & JOHN M. BURT

4.1. Introduction

In our chapter in the first volume of *Cognitive Ecology* (hereafter *CE I;* Beecher et al. 1998), we argued that insight into the function of song in territorial songbirds could be derived from viewing the songbird as a decision maker with respect to not only the songs he learns in his first year of life but also how he subsequently uses those songs in day-to-day interactions with his territorial neighbors. In this chapter, we continue to develop this idea in the context of a progress report on our research program on song learning, communication, and function in the song sparrow (*Melospiza melodia*).

Song is the first line of communication among territorial songbirds (Catchpole and Slater 1995). A bird announces his possession of a territory—"posts" it—by singing from perches throughout the territory; typically, his songs can be heard several territories away. In most temperate zone songbirds, the primary burden of territory defense falls on the male, and only the male sings. Our study species, the song sparrow, is typical in this respect. In tropical songbirds, it is common for the female to sing (for a full discussion of this case, see Riebel 2003), but in this chapter we will focus on the typical temperate zone case where just the male sings. In particular, we will focus on the way song functions in neighbor interactions beyond the purely defensive "posting" context.

Among animals that have complex acoustic communication signals, songbirds (oscines) are unusual in that the form of their song is influenced critically by early learning. Birds may also learn how to use their songs as well (discussed further below). The general functions of song learning are debated, but in this chapter we focus on the hypothesis that learning permits adaptation of the song repertoire to the species' local ecology, especially the social ecology, which we will call the social ecology hypothesis (Kroodsma 1983, 1988; Payne 1983; Baptista and Morton 1988; Slater 1989; D. Nelson and Marler 1994; Beecher et al. 1997). In many songbird species, birds sing songs distinctive of their local area. For example, in our study population of song sparrows, birds sing songs distinctive of their neighborhood of 6–12 birds. In this population, birds disperse from their birthplace when about a month old and remain in a new area for the rest of their lives. Song learning begins at 1 month of age, after they leave the birthplace, and ends at the beginning of the first breeding season approximately 9 months later (Beecher et al. 1994b; Nordby et al. 1999). Thus, the timing of song learning—beginning after the bird enters his new home area and before he becomes fully territorial—is consistent with the hypothesis that the bird needs to learn songs that are specific to his chosen neighborhood. Unfortunately, it is also consistent with the null hypothesis that the bird just happens to learn the songs that he hears during his song-learning period. To distinguish between these two hypotheses, we need to find some aspects of the song-learning strategy that depart from simply copying what is heard most often. We endeavor to do that in this chapter.

This chapter has two main themes. The first is that the social ecology hypothesis has strong implications for the study of song learning and song communication, suggesting that the rules of song learning reflect an evolved strategy designed to equip the bird with songs that are useful in his new neighborhood and, moreover, that in interactions with his neighbors the bird will use his songs in a way that capitalizes on their being local songs. The second major theme of the chapter is that the rules of song communication and song learning can be understood only when viewed in a cognitive context. As we argued in our chapter in *CE I,* a distinctly noncognitive view has prevailed in the study of bird song, a perspective derived from the classic mechanistic models of ethology, with song viewed as similar to a simple sign stimulus or releaser, and song learning as a type of imprinting. Even in field playback experiments, the implicit assumption has seemed often to be that the bird's response is determined entirely by the properties of the stimulus and that therefore the context could be ignored, as it might be if the experiment were being carried out in the laboratory. Our work has suggested a different, more cognitive perspective, in which the bird makes cognitive decisions about what songs to learn and how to use them.

4.2. Background

Our research has been carried out on a sedentary (nonmigratory) population of song sparrows in an undeveloped 540-acre park that borders Puget Sound in Seattle, Washington. Typically 100–150 song sparrow males are on territories in our study population in a given year. Song sparrows in our population usually maintain the same territory for life; sometimes they make small moves, but rarely do they move more than one or two territories away from their original territory. The birds live in small neighborhoods in territorial mated pairs. These neighborhoods are mixtures of longtime neighbors and first-year birds. Young birds generally establish their territories sometime between their first year (following dispersal from the birth area) and the following spring.

A song sparrow male has a repertoire of 6–12 (usually 8 or 9) distinct song types (Peters et al. 2000). About three-quarters of songbird species have song repertoires (Kroodsma 1982). The song types in a song sparrow's repertoire are as distinct from one another as are the song types of different song sparrows, and this, we will see, is because they are copied from a number of other birds in the song-learning phase. A song sparrow's songs do not have a distinctive vocal signature (Beecher et al. 1994a), which means that, to recognize his neighbors, a bird has to memorize all their individual songs, from about 50 to over 100 of them, and song sparrows seem to be quite good at this (Stoddard et al. 1992b). A song sparrow sings his songs with "eventual," rather than "immediate," variety; that is, he repeats a particular song type a number of times before switching to a new song type. Under conditions of free singing (i.e., singing at high rates, as most males do when unpaired or when the female is incubating), a bird usually does not return to a song type until after he has sung all or most of his other song types. In addition to the very large differences between song types, song sparrows also make small changes in successive renditions of a song type. Variation in a song type, however, is small compared with variation among songs, and song types are clearly defined by the eventual-variety style of singing (Stoddard et al. 1988; Podos et al. 1992; Nowicki et al. 1994).

Two males in a neighborhood in our population generally "share" some songs (i.e., have some very similar song types). Although the number of songs shared between two neighbors varies from zero to (rarely) all of them, the average percentage of shared songs is about 30 in our study population. An example of song sharing is shown in figure 4.1. This pattern of sharing arises because young birds learn the songs of the neighborhood they enter in their first year and in which they remain for the rest of their lives (Beecher et al. 1994b; Nordby et al. 1999). Song sparrows do not modify their song repertoire after their first breeding season (Nordby et al. 2002), but young birds moving into the area each year learn the songs of the older birds there, and thus song sharing is maintained in the neighborhood.

Male song sparrows use their songs in both intra- and intersexual contexts. We will focus on one context of song communication: counter-singing by territorial neighbors during the breeding season. Although song sparrows use song in intersexual contexts as well (Searcy 1984; O'Loghlen and Beecher 1997, 1999), we focus on the intrasexual context in this chapter.

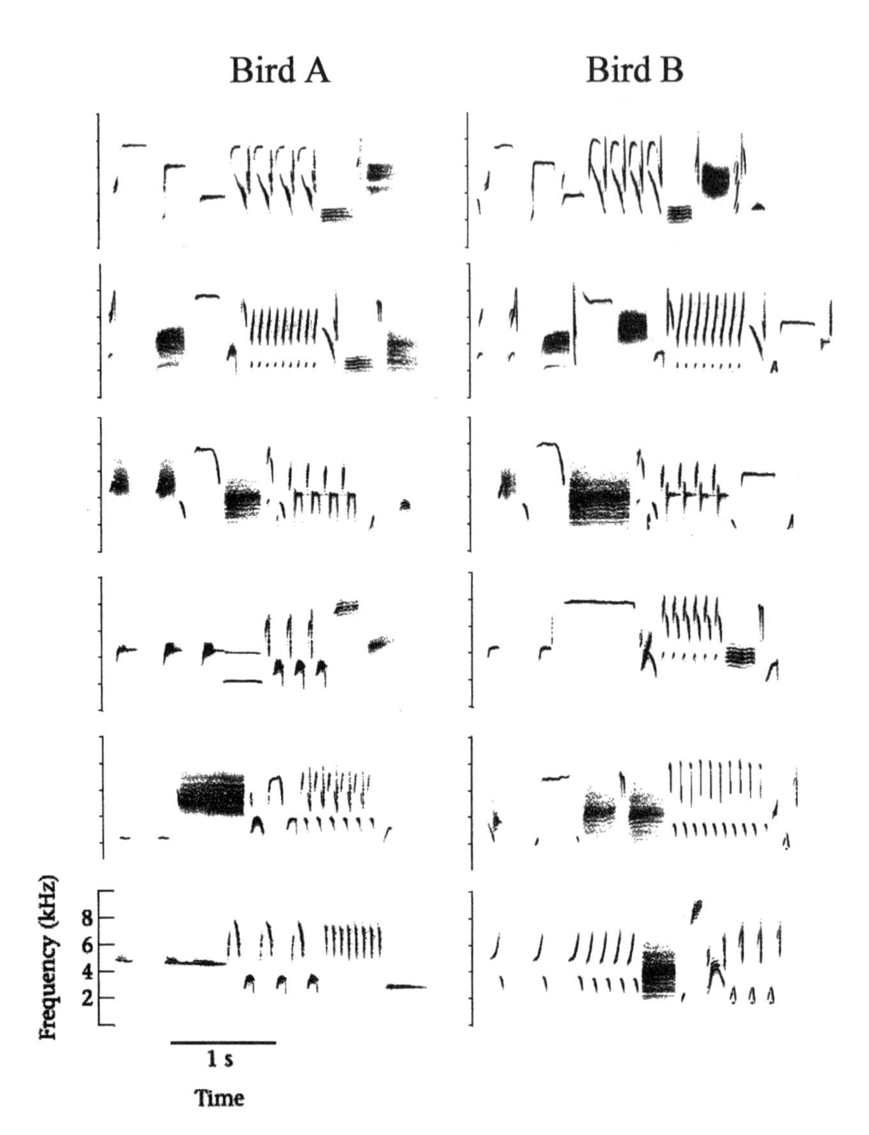

FIGURE 4.1. Partial song type repertoires of two song sparrow subjects. Birds A and B were neighbors and shared the first three songs in their 9-song repertoires (33% sharing). The shared songs of birds A and B are shown in the top three rows, while 6 of their remaining, unshared types are shown in the bottom three rows. Frequency (vertical) scale: 0–10 kHz, markers at 2 kHz intervals. Songs are 2–3 seconds long. From Beecher and Campbell 2005.

4.3. Song learning in the field

After a preliminary study with birds hatched in the years 1986–1990 (Beecher et al. 1994b), we color-banded all juvenile birds appearing in our population in 1992. Postdispersal learning is the typical pattern in songbirds (see review in Beecher et al. 1997), and in our population, birds we have banded in the nest and subsequently recaptured within the population postdispersal (typically having dispersed some distance from the nest) sing song types of their postdispersal area rather than their natal area. In the 1992 cohort, 41 birds survived to song crystallization the following spring (Nordby et al. 1999). We considered as possible song tutors all older birds in the study population who were on territory in the subject's hatching year. Because we had color-banded and recorded virtually all of the adult males in this population, we were able to identify the song tutors of the 41 young birds. We identified the older bird with the most similar rendition of the type (complete with idiosyncratic features not seen in other renditions of the type) as the young bird's "probable tutor" for that type. When two (or more) older birds had versions of a song that were highly similar to the young bird's—not unusual in this population, where neighbors share songs—they would both receive credit as the tutor for that song. In our analysis of song learning, we have relied on the judgments of multiple experienced human observers. These judgments have been informed by our extensive field observations of singing in this species, by the results of our field playback studies (Stoddard et al. 1988, 1990, 1991, 1992a; Beecher et al. 1996, 2000a; Burt et al. 2001, 2002), and by our laboratory perceptual experiments (Stoddard et al. 1992b; Horning et al. 1993; Beecher et al. 1994a). These issues are discussed in detail in Beecher 1996. In this section, we summarize the results of our field studies of song learning (Beecher et al. 1994b; Nordby et al. 1999, 2007).

A young song sparrow leaves his area of birth at around a month of age and begins learning songs during the second and third months of his life in the neighborhood where he may ultimately set up a territory. We have recently begun to radio-track male song sparrows in their first year (C. N. Templeton et al., unpublished data), and although their movement patterns are rather more complex that we originally thought, it is clear that the young bird is searching for a neighborhood where he can set up his territory. During this early stage, the young bird memorizes the song types of the adult birds in his new neighborhood. As discussed below, we do not know when the young bird ceases to memorize new song types, but we do know that he continues to be influenced by adult tutor-neighbors until the next spring and does not finalize his repertoire of 8 or so songs until around March. The results from our field studies are summarized below as "rules of song learning."

4.3.1. COPY SONG TYPES FAITHFULLY

Young song sparrows generally develop near-perfect copies of the songs of their older neighbors. These field results differ from the laboratory findings using tape tutors (Marler and Peters 1987, 1988). In the laboratory setting, song sparrows copy song elements quite precisely, but they frequently combine elements from different songs to form what we will call "hybrid" songs—songs made up of parts of different song types. That is, they often copy song elements but use them to improvise new song types.

4.3.2. LEARN THE SONGS OF MULTIPLE BIRDS

Although the minimum number of tutors needed to account for the young bird's full crystallized repertoire is easy to determine, the actual number of tutors that influenced the young bird's learning is probably larger than this minimum number, but it is difficult to estimate precisely because tutors typically share some songs, and as we will see below, young birds preferentially learn these tutor-shared songs. The criteria for tutor assignment are discussed in detail in Nordby et al. 1999. In any case, we estimate that the typical bird learns his songs from at least 3–5 older birds. Although it seems plausible that a bird might learn all his songs from a particular adult tutor, in Nordby et al. 1999 we found that only 1 of the 41 subjects appeared to be a song "clone" of a single older bird.

4.3.3. LEARN FROM YOUR NEIGHBORS

A bird's song tutors invariably turn out to have been neighbors in the young bird's hatching summer and, if they survive the winter, the following spring (the young bird's first breeding season) as well. The young bird usually establishes his territory within the territorial range of his song tutors, often replacing a tutor that died. The exceptions, where a young bird does not establish his territory among his tutor-neighbors, seem to occur when territorial vacancies do not open up because none or few of the tutor-neighbors died and/or other young birds took the available vacancies.

The young bird appears to commence song learning shortly after he has dispersed from his natal area. Because adult males (potential song tutors) in our population typically will remain on their territories from one year to the next unless they die in the interim, it is essentially impossible to determine from the field data when the young bird learns his songs. We originally thought that most or all of song memorization occurs in the traditional lab-determined sensitive period, roughly the second and third months of life (Marler and Peters 1987), but this was only a plausible guess and our lab studies have cast doubt on that assumption (Nordby et al. 2001).

4.3.4. PREFERENTIALLY RETAIN SONG TYPES OF TUTORS SURVIVING TO YOUR FIRST BREEDING SEASON

Birds often have song types that can be traced to tutors that were alive in the young bird's natal summer but died before the next breeding season. Nevertheless, they generally retain more songs of tutors that survive into the next breeding season than of tutors who do not. We refer to this late learning as "late influence" because it may not be de novo learning: these songs could have been memorized in the natal summer and retained because the bird continues to hear those songs the following autumn and/or spring. This would be the pattern hypothesized as typical by Nelson and Marler (D. Nelson 1992; D. Nelson and Marler 1994): the young bird memorizes songs during a sensitive period in the natal summer and the following spring and retains some of these songs and drops others on the basis of his social interactions with his territorial neighbors ("selective attrition"). Presumably, the bird retains those songs most similar to those of his spring neighbors, who will be his future competitors. We have recently compared the song repertoires of young song sparrows in the plastic song phase (late winter, early spring) and crystallized phases and found that they do indeed drop songs that are less similar to those of their springtime territorial neighbors (Nordby et al. 2007).

4.3.5. PREFERENTIALLY LEARN TUTOR-SHARED SONGS

As noted earlier, in our field population neighbors typically share a portion of their song repertoires, on average about 2–4 of their 8–9 song types. We have found that the young bird preferentially learns (or retains) song types shared by two or more of his tutors (Beecher et al. 1994b). There are several possible reasons that shared song types might be particularly salient, including the following: (*i*) they are heard more than unshared song types; (*ii*) "the same song" is being sung by several birds; (*iii*) they are heard more often in countersinging interactions than unshared songs. These possibilities are considered later in this chapter.

This preference for tutor-shared songs may indicate a "bet-hedging" strategy to guarantee that the young bird has song types he will share with his neighbors in his first breeding season. If instead the bird learned tutor-unique songs, he would have songs "specialized" for these particular tutor-neighbors (i.e., shared with that neighbor only). But these specialized songs would be good only until the tutor dies or moves, whereas a shared song is good until all the birds having it in the neighborhood die or move, and even beyond then because other young birds moving into the area also preferentially learn shared types.

4.3.6. PRESERVE TYPE AND TUTOR IN YOUR SONGS

As we noted in *CE I,* the young bird will sometimes break the perfect-copying rule and blend elements from different tutor songs. These exceptions fall into two categories and suggest an additional rule of song learning. First, the young bird will often blend two tutors' somewhat-different versions of the same song type rather than copying one or the other version. These are not true hybrids, because the song elements, although selected from two different tutors, are selected from the same or very similar song type. Second, the young bird will sometimes combine elements from two dissimilar song types of the same tutor. These two exceptions suggest the following rule: song elements of different songs are combined only if they are from (*i*) different tutors' versions of the same type or (*ii*) different song types sung by the same tutor. We summarize this principle as the student's "preserving type and/or tutor" in his songs.

4.3.7. INDIVIDUALIZE YOUR SONG REPERTOIRE

The rules so far can be interpreted to fit the overall rule: learn songs that you will share with your neighbors in your first breeding season. There is, however, an important exception to that rule. In the transition from plastic song to final, crystallized song, the young bird often modifies a song so that it actually becomes a poorer match to the model song of the putative tutor and to similar songs of his present neighbors (who may or may not include the tutor). Two examples are shown in figure 4.2. We interpret this as the bird's "individualizing" his repertoire. The song may still be perceived by the bird's neighbors as a shared song (although it may no longer meet the investigators' criteria for a "shared" song) while at the same time being perceived as that bird's particular version of that song type. If this is so, the bird gets to have his cake and eat it too: to have songs that are both shared with his neighbors and unique to him. It is not obvious, however, why the bird individualizes some of his songs but not others.

4.3.8. SONG-LEARNING RULES IMPLY COGNITIVE PRINCIPLES

As we noted in *CE I,* our field results differ from those of laboratory tape-tutor studies of song learning in song sparrows (Marler and Peters 1987, 1988). In the laboratory, young song sparrows copy elements from particular songs, but they commonly combine elements of the different songs to form hybrid songs. The avoidance by song sparrows in the field of this sort of improvisation was the basis for our conclusion in *CE I* that in the field a young

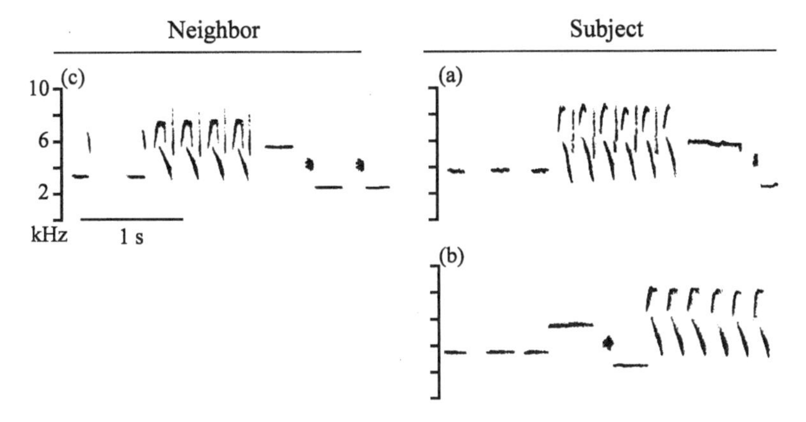

FIGURE 4.2. Plastic version (a) and crystallized version (b) of a song type that young bird GEMO shared with his neighbor (c). From Nordby et al. 2007.

song sparrow constructs his song repertoire in part on the basis of a cognitive classification of songs by type and singer identity. We suggest that a bird preserves a whole song type not simply because he has frequently heard those particular elements sung together but because (*i*) he has heard that song type contrasted with other distinct song types within the repertoire of a particular singer, and (*ii*) he has heard that song type sung by other birds as well. In other words, a song type is defined both by its being sung by multiple birds and its being contrasted with other song types within the singer's repertoire. In *CE I,* we discussed in some detail how the tape-tutor context—where both song type and singer identity are ill-defined—might interact with the bird's natural song-learning proclivities to spawn the hybrid songs that are rare in the field. (We discuss an alternative interpretation later in the present chapter: the hypothesis that eastern and western song sparrows might have very different song-learning strategies.) Apart from the tendency to individualize at least some of his songs, the young bird preserves type rather than recombining elements to make new song types, and thus he winds up sharing song types with his neighbors. As we will see in the next section, the adult bird uses his songs when interacting with neighbors according to parallel rules.

4.4. Communication by song in male-male interactions

As discussed in *CE I,* we have established that neighboring song sparrows in our population recognize one another by song (Stoddard et al. 1990, 1991). As we will see, this recognition is much more nuanced than we imagined in the beginning.

For a territorial species such as the song sparrow, the benefit of interaction by singing (counter-singing) is that it can substitute for more costly forms of negotiating territory boundaries, such as fighting. For example, if two male song sparrows cannot resolve the issue at a distance with song, they approach one another, cease singing, and switch to visual displays and "soft song," a low-volume, structurally very different type of vocalization (Searcy et al. 2006), and then often to fighting. Necessary for efficient negotiation is being able to recognize your neighbor, and it turns out that how the birds use the songs at their disposal is the key to the process.

In species with song repertoires (the most common case), the singer has a "choice" of which of its various song types to sing. How does the singer make these choices? We have developed a picture of the singing rules that are used by territorial male song sparrows and that seem to play an important role in maintaining neighbor relations and territorial boundaries. We have developed this picture through a combination of observational studies and experimental "playback" studies in which we simulate a neighbor singing near the subject's territorial boundary (Stoddard et al. 1988, 1990, 1991, 1992a; Beecher et al. 1996, 2000a; Burt et al. 2001, 2002; Beecher and Campbell 2005).

Several bird song researchers have argued that it is advantageous to be able to reply to a bird with the same song type he has just sung (e.g., Bremond 1968; Krebs et al. 1981), and many studies have shown that birds will, at least under some circumstances, reply to a shared song with the same song type: "song matching" or, more precisely, "type matching" (reviews in Krebs et al. 1981; McGregor 1991; W. Smith 1991). Most workers have viewed type matching as one of the ways in which the bird directs its response at the bird who has just sung. A puzzling finding, however, has been that higher rates of type matching are elicited when the stimulus song is the bird's own song or is a stranger's song that is similar to one of the bird's own songs than when it is a shared song of a neighbor (song sparrows, McArthur 1986 and Stoddard et al. 1992a; western meadowlarks, Falls 1985; great tits, Falls et al. 1982). Song sparrows, for example, match neighbor song only at about chance (<10%) level (Stoddard et al. 1992a).

We were able to shed some light on this puzzle in a study of established song sparrow neighbors, tested from mid- to late breeding season. While these birds, as expected, did not type-match the broadcast neighbor song, they did consistently reply with some other song they shared with that neighbor (Beecher et al. 1996). We dubbed this pattern of song selection "repertoire matching." In a subsequent experiment, we tested response to neighbor song by new neighbors twice during the breeding season: early, in April, and again a month and a half later (Beecher et al. 2000a). Early in the breeding season, new neighbors will have only recently established their territorial boundary, which may still be in dispute, and territorial skirmishes will have occurred recently or may still be occurring. A new neighbor singing at the boundary early in the season represents a more serious challenge than a well-established neighbor singing at that same boundary, and we predicted higher levels of type-matching on the earlier occasion. As predicted, early in the season birds usually replied to a shared neighbor song with a type match, whereas a month and a half later they usually replied with a repertoire match (in this experiment they never responded with unshared song either early or late). These results are consistent with the hypothesis that type matching is a more aggressive or escalated response than repertoire matching. In subsequent playback studies, we showed that song sparrows regard type matching as a threat and will respond to it more aggressively than they do to repertoire matching. If type-matched, they will either escalate by responding aggressively and staying on type or de-escalate by responding less aggressively while replying with an unshared song or stopping singing (Burt et al. 2001; Beecher and Campbell 2005). Repertoire matching appears to be intermediate.

Our results thus suggest that when neighbors share some song types and not others, the two neighbors may use their shared songs to modulate their social interactions. Song sparrows in our population typically share some but not all of their songs (most often they share 2–5 of their 8–9 songs), and they use the subset of shared types in a graded communication system. Some experience with one another is required for the two birds to learn the subset because a bird will share different subsets of his repertoire with different neighbors, but presumably they acquire this knowledge over the course of many counter-singing bouts. We summarize the results of our studies with the following "singing rules," diagrammed in figure 4.3. Consider two hypothetical neighbors sharing songs A, B, and C but no others. Bird 1 addresses bird 2 by singing one of their shared types A, B, or C (he sings in the direction of bird 2, since other neighbors may also share some of these types). Let us say bird 1 sings A. Bird 2 then can acknowledge the signal by replying with B or C (repertoire match), can escalate by replying with A (type match), can de-escalate by singing one of the unshared types, or can ignore bird 1 by not singing at all. If bird 2 type-matches bird 1 (sings A), bird 2 can escalate further by continuing to sing that song type, or he can de-escalate by switching to another shared song (repertoire matching) or de-escalate further by switching to an unshared type. Note that to type-match requires no prior experience with your opponent—the bird simply replies with his most similar song, and generally the "match" will be perceptually obvious—but to reply with a shared song, or with an unshared song for that matter, the bird needs to have

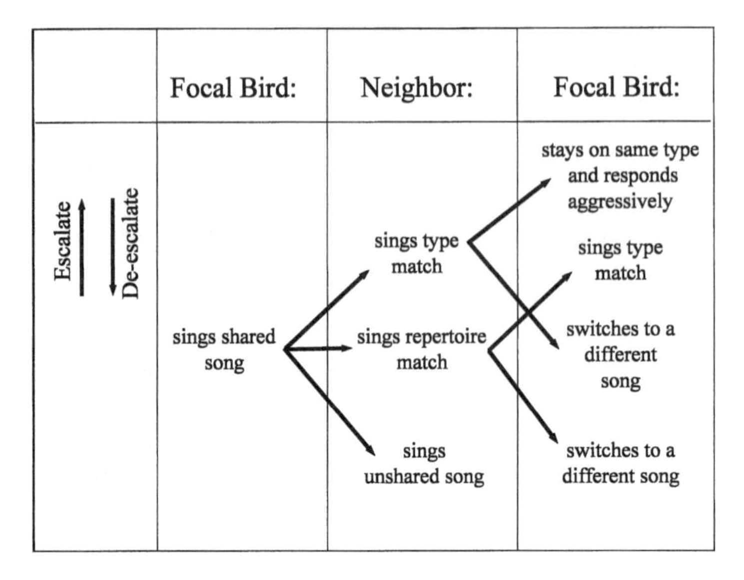

FIGURE 4.3. Diagram of singing interactions between a focal bird and his neighbor, who share song types. Escalation is indicated by behaviors higher in the diagram, and de-escalation by behaviors lower in the diagram. In this figure, the interaction begins when the focal bird sings a shared song type. The neighbor can then either type-match the focal bird (an escalation), repertoire-match (a directed but neutral signal), or sing an unshared song (a de-escalation). If the focal bird is type-matched, he may respond to the escalation by staying on the same type and responding aggressively (a further escalation) or de-escalate by switching to another song type and not responding strongly (the diagram does not show that this different song type may be either a repertoire match or an unshared song). "Aggressive response" refers to searching for, threatening, or attacking the singer. From Beecher and Campbell 2005.

had some experience with his neighbor: the bird needs to know which songs they share and which they do not.

A similar pattern of singing has been observed in the banded wren (Molles and Vehrencamp 2001), another territorial bird that exhibits song sharing with neighbors. The hypothesis that having shared songs facilitates communication between territorial neighbors is one hypothesis for the function of a song-learning program that leads to song sharing with neighbors. If the system of long-distance communication described here is beneficial in resolving most disputes with minimal cost, then we should see birds with shared songs faring better than those without shared songs. In a longitudinal study of 45 song sparrows followed from their first year on territory, we found that the number of songs a bird shares with his neighborhood group predicts how many years he will hold his territory, but his repertoire size does not (Beecher et al. 2000b). We also found that song sharing increases with repertoire size up to but not beyond 8–9 song types, which are the most common repertoire sizes in the population (the range in our sample was 5–13). This partial confounding of song sharing and repertoire size may account for an earlier finding of a correlation between territory tenure and repertoire size in song sparrows (Hiebert et al. 1989), but see below. A positive correlation between the number of songs a bird shares with his neighbors and the number of years he survives on territory has also been found in another western population, in California (P. Wilson et al. 2000). Moreover, it has also been shown in this California population that song sparrows are less aggressive toward neighbors with whom they share songs (P. Wilson and Vehrencamp 2001). In further studies of the Mandarte Island (British Columbia) song sparrow population studied by Hiebert et al. (1989), J. Reid et al. (2004) found that first-year males with larger repertoires were not more likely to acquire a territory, to acquire a larger territory, or to settle sooner. They were, however, more likely to mate, and their mates were more likely to begin laying earlier. Song sharing was not measured in this study, and because of the partial correlation of song sharing with repertoire size, we cannot disentangle these effects. One possible interpretation of the data on western song sparrows is that song sharing plays a key role in male-male competition, while repertoire size plays a key role in mate choice. Later in this chapter we discuss similar studies that have been carried out with eastern song sparrow populations.

4.5. Social eavesdropping hypothesis

The congruence of the singing rules and song-learning rules suggests a hypothesis about song learning that we have recently begun to study. According to the social eavesdropping hypothesis, the young bird learns his songs by eavesdropping on singing, but specifically on singing interactions between two or more birds rather than solo singing. Recent field experiments on songbirds have shown that males base their decisions on whom to challenge and females their decisions on whom to mate with on information about the dominance relationship of the singing males, information which they extract when eavesdropping on singing interactions (Naguib and Todt 1997; Naguib et al. 1999, 2004; Otter et al. 1999; Peake et al. 2001; Mennill et al. 2002, 2003). In the species studied to date, song overlapping, song type matching, and/or song leading seem to be the critical cues (Naguib 1999; Mennill and Ratcliffe 2004; Peake et al. 2005; Kunc et al. 2006). We have hypothesized that young birds too may use information they extract from singing interactions they overhear to decide which songs to learn or retain, and dominance would likely be one important dimension (Beecher and Burt 2004). The best example to date is the African village indigobird, *Vidua chalybeata:* younger birds copy the song of the dominant bird in the area (Payne 1985). Also relevant here is "social modeling" theory, as developed by Pepperberg, which suggests that the young bird's observation of communication interactions between individuals who have mastered the communication system may be critical for vocal learning (Pepperberg 1985).

There is a second type of information (besides dominance relationships) that a young bird could extract from the interactive singing of two adults but not from solo singing of these same birds: the singing rules that indicate the appropriate replies to particular songs in particular contexts. In the study of bird song learning, the focus has always been on the learning of particular songs rather than the learning of how to use them, but it may be that the two processes are intertwined. This is the case for human language learning, of course, and it might be true for songbirds as well. A key prediction of the social eavesdropping hypothesis—true regardless of whether dominance relationships or song reply rules are the key factor—is that a young bird who needs to interact with a new neighbor may select for his final repertoire not just the songs of that individual but songs the young bird has heard other birds singing to that individual as well. It also suggests a basis for the preferential learning of shared songs. As noted earlier, a song sparrow uses all of the songs in his repertoire equally when free-singing, but when he interacts with a particular neighbor, he preferentially uses the particular songs they share. This means that, if a young bird in the song-learning phase is particularly attuned to counter-singing, he will be exposed mostly to shared songs and therefore should tend to preferentially learn tutor-shared songs. Moreover, he should be learning the rules of song learning at the same time he is learning particular songs.

We have conducted one laboratory test of this hypothesis so far (Beecher et al. 2007). We compared two types of song tutoring: that resulting from direct interaction with the song tutor and that resulting from social eavesdropping (i.e., overhearing the singing interactions of other birds). Subjects were exposed to the songs of four live tutors during the early "memorization" phase (phase 1) of song learning and to just two of them again in the later "action-based" learning phase (phase 2). We have found in our field study (Nordby et al. 1999), as mentioned above, and in our lab studies as well (Nordby et al. 2000, 2001) that birds are more likely to retain songs for their adult repertoire that they heard in their natal summer if they are exposed to them again the following spring. Thus, in the present experiment we expected that birds would learn more from the two tutors present during both phases than from the two tutors present only in the early phase. Of the two tutors that returned in phase 2, one became a subject's interactive tutor,

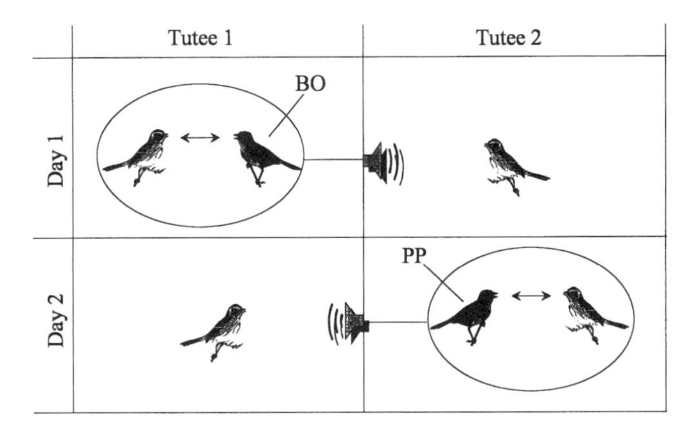

FIGURE 4.4. Schematic of yoked-subject design. In phase 2, a subject was exposed to one tutor live, the interactive tutor, on one day and overheard another tutor-subject pair on a second day. For one-half of the subjects, subject and interactive tutor were separated by a black cloth screen (not shown). On days 3 and 4 (not shown) the young bird was returned to his home cage in a closed chamber. Note that the diagram is not to scale and that the subject and interactive tutor were in their own separate cages within a larger sound-insulated chamber. From Beecher et al. 2007.

while the other became the subject's overheard tutor, that is, was overheard interacting with another, yoked subject. This yoked design is illustrated in figure 4.4. We found that subjects learned (retained) more songs from their overheard tutor than from their interactive tutor (about twice as many on average; fig. 4.5). The subject learned songs of the overheard tutor, not of the overheard yoked subject: the repertoire of a subject was no more similar to that of the yoked subject he heard than to that of the nonyoked subjects he never heard. Although many interpretations of this result are possible, we note here only that it is consistent with the social eavesdropping hypothesis. The birds may have learned more from the overheard interactions than from direct interaction with their tutors because the former were less threatening. The close, intense nature of the interactive tutor vis-à-vis the subject may have been intimidating and/or may have convinced the subject that he could not invade this bird's territory, and thus he directed his attention to the overheard, "more distant" songs. In this case he learned the songs of the tutor, the older and presumably more dominant member of the yoked pair, rather than of the subject, a younger bird.

We describe here a second experiment that shared some of the design features of the experiment above but utilized our new "virtual-tutor" methodology (Burt et al. 2007). The virtual tutor is basically a computer program that delivers digitized songs to subjects in a manner simulating one or

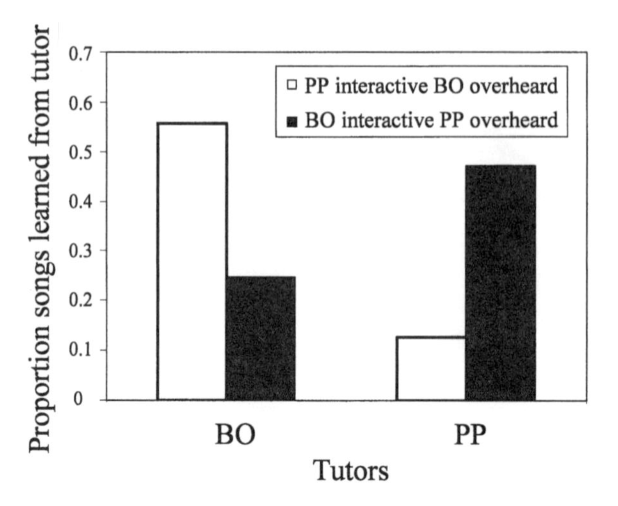

FIGURE 4.5. Relative effectiveness of late tutors BO and PP when in the two roles of interactive tutor and overheard tutor. From Beecher et al. 2007.

more live birds singing. The virtual-tutor method permits us to maintain the experimental rigor of the tape tutor paradigm while capturing at least some of the key features of natural singing, especially interactive singing, and it allows interaction with the subject (Beecher and Burt 2004). The virtual-tutor software, designed by John Burt (www.syrinxpc.com), can be programmed to (*i*) sing solo, noninteractively (i.e., "posting mode"), similar to a tape tutor, (*ii*) sing interactively with the subject, or (*iii*) sing interactively with another virtual tutor. In the interactive modes, it interacts as a live song sparrow would, using singing and song type selection rules that we have extracted from our field observations and playback studies, as discussed above. When interacting directly with the subject, the virtual tutor can identify what the subject is singing and reply appropriately. The repertoire of each virtual tutor is based on that of a real bird and consists of 8–10 song types, with 8–10 variations on each type (Stoddard et al. 1988; Podos et al. 1992). When more than one virtual tutor is used in an experiment, each virtual tutor's songs are played from different loudspeakers in the subject's chamber. Virtual tutors are fixed to particular loudspeakers to reinforce the impression that the subject is positioned between two neighbors.

In Burt et al. 2007, we used a hybrid design: in the early phase (early summer), young birds were exposed to two pairs of interacting live tutors on alternate days. In the second phase (January through March), the subjects were isolated and exposed to digitized songs from two of these tutors (i.e., they were now "virtual tutors"), one tutor from each of the two original pairs. On alternate days, a subject heard songs from each of the two tutors; for each subject one of the tutors was interactive, and the other was noninteractive (i.e., was just heard singing solo). The amount of song from each subject's interactive and noninteractive tutors was equated.

Subjects learned (or retained) more songs from the late interactive tutor than from the late noninteractive tutor, but the most interesting finding was that they also learned/retained more songs from the late interactive tutor's early-only singing partner. This implies that the young bird remembered the singing interactions he had heard the previous summer and selected songs for his final repertoire that he had heard sung to his present interactive tutor 6 months earlier. Thus, these results, like the results of the live-tutor experiment (Beecher et al. 2007) just discussed, also point to the importance of overheard singing interactions, though in this case the overheard interaction had occurred in the early phase rather than in the late phase of learning. The results of these two experiments taken together suggest that social interaction may indeed be critical for song learning, but in both cases it appears that the key social interaction was an overheard one.

4.6. Discussion

The song sparrows in our study population live in relatively stable neighborhoods, recognize one another, and use their songs according to specific rules to modulate territorial interactions. In this chapter, we developed two points. First, the key variables that determine how a song sparrow constructs his song repertoire in his first year and how he uses those songs subsequently when interacting with his neighbors pertain to the bird's social ecology. Second, song communication and song learning in song sparrows are largely organized with respect to two main cognitive categories: individual singer and song type. This two-way classification helps explain which songs birds learn and retain in their first year and the way in which birds use these songs as adults in counter-singing interactions. The adult bird knows which particular songs he shares with each of his neighbors, and he uses those shared songs preferentially and appropriately. Similarly, the young bird knows which songs his tutor-neighbors share, and he preferentially learns those songs. He could invent new song types by combining elements from the songs of different tutors, as a song sparrow will do in tape-tutor experiments, but he does not do so unless the songs represent different versions of the same type. Thus, the young bird typically winds up with a "useful" adult song repertoire: he will share at least some songs with each of his neighbors of his first breeding season, and these shared songs will permit him to communicate with them according to the rules summarized in figure 4.3.

In this chapter, we focused on our particular study species, the song sparrow. We hope that many of the general conclusions made apply more broadly, at the very least to many other songbird species. Considerable evidence supports the hypothesis that the function of song learning in many species is to equip the bird with songs he shares with his neighbors. A number of studies have shown that birds learn their new neighbors' songs after leaving their birthplace and/or later modify their song repertoires to increase song sharing with neighbors in the first or subsequent breeding seasons (Kroodsma 1974; Jenkins 1978; Payne 1981, 1982, 1983; Baptista and Morton 1988; McGregor and Krebs 1989; D. Nelson 1992; Lemon et al. 1994; Mountjoy and Lemon 1995; O'Loghlen and Rothstein 1995; W.-C. Liu and Kroodsma 2006).

There are two major difficulties with the hypothesis that the song sparrow learning program we have described here is an adaptive strategy. The first, mentioned above, is the competing "null hypothesis" that birds simply learn the songs that they happen to hear in the song-learning phase. Effects such as the preference for tutor-shared songs can be explained as simple dosage effects (these songs are heard more often). It will be difficult to eliminate this hypothesis until such time that we can separate out the various variables that are present when live birds are song tutors (whether in the field or in the lab). Earlier we described the virtual-tutor methodology that we think is capable of disentangling these variables. With this methodology we can remove dosage effects; for example, we can arrange for tutor-shared song types to be delivered to the song-learning young bird the same number of times as unique song types, or for some shared types never to be heard in counter-singing interactions while other shared types are heard in the typical counter-singing context.

The second problem with viewing the song sparrow learning program we have described here as an adaptive strategy is the fact that song sharing is the exception and not the rule in eastern populations of song sparrows (Hughes et al. 1998). Moreover, no advantage to song sharing has been found in this population, though that is not surprising since song sharing, when it occurs, occurs at very low levels (Hughes et al. 2007). Whereas birds in western populations in Washington, California, and British Columbia typically share 2–4 songs with a given immediate neighbor but none with birds just a few territories removed (Cassidy 1993; Beecher et al. 1994b; Nielsen and Vehrencamp 1995; C. Hill et al. 1999; P. Wilson et al. 2000; Reeves and Beecher, in preparation), birds in northeastern populations (Pennsylvania, Maine, Ontario) rarely share songs with neighbors and share no more with neighbors than with nonneighbors (Borror 1965; Harris and Lemon 1972, 1974; Kramer and

Lemon 1983; Hughes et al. 1998; but see Foote and Barber 2007). Hughes et al. concluded that the difference in song sharing is the result of a difference in the underlying song-learning program wherein "Washington birds copy whole songs, while Pennsylvania birds appear to copy and recombine song segments, as has been found in laboratory studies of song learning. Thus both song learning and the function of song repertoires differ between populations of song sparrows" (Hughes et al. 1998, 437). This hypothesis challenges the assumption that differences in social learning between individuals of two populations are usually due to differences in the learning environment and suggests instead that they may be due to differences in the underlying genetic-developmental programs that affect the social learning in question (Beecher and Brenowitz 2005).

Most studies that have examined neighbors for song sharing in repertoire species, or song similarity in single-song species, have found greater similarity between neighbors than between nonneighbors (Verner 1975; Avery and Oring 1977; Payne 1982, 1985; McGregor and Krebs 1982; Schroeder and Wiley 1983; D. Nelson 1992; Briskie 1999; Beebee 2002; W.-C. Liu and Kroodsma 2006). There are other populations, however, where birds share no more with neighbors than with other birds in the population: for example, chaffinches, *Fringilla coelebs* (Slater and Ince 1982); western meadowlarks, *Sturnella neglecta* (Horn and Falls 1988); Kentucky warblers, *Oporornis formosus* (Tsipoura and Morton 1988); Gambel's white-crowned sparrows, *Zonotrichia leucophrys gambelii* (D. Nelson 1999); and, as just noted, most eastern song sparrow populations. Nelson and Marler (1994) have argued that in many populations, especially migratory populations, birds may memorize songs from one set of birds in their natal summer and then, in the following spring, select from this large pool of songs those songs that best match the songs of the totally different birds who are now their neighbors. This process would produce populations where neighbor song sharing might not pass the song-sharing threshold applied by our group or by most song researchers but might be detectable by someone looking for these subtle similarities. For example, Nelson (1992) showed that field sparrow males return to the breeding area with two or three songs and then retain the one that most closely resembles the song of the most actively singing neighbor. Still, the degree of song similarity between neighbors will surely be greater in populations where birds learn their songs directly from their neighbors than in cases where they follow the more indirect Nelson-Marler process.

A second kind of exception to the neighbor-sharing rule may be exceptions that prove the rule, cases where the animal's social-ecological circumstances rule out the possibility and/or the benefit of song sharing with neighbors. If, in general, a song-learning strategy that equips the birds with songs that he shares with his neighbors is favored by selection, then what is the best song-learning strategy in populations where neighbors change within as well as between breeding seasons? Kroodsma (1996) has argued that for birds without long-term neighbors, there is no advantage to shared songs, and so the development of generalized species-typical songs will be favored. The sedge wren (*Cistothorus platensis*) provides a test of this prediction. Northern populations of sedge wrens are migratory, and during the breeding season they are seminomadic as well. Thus, even in the breeding season, they have a constantly changing set of neighbors. These sedge wrens show a unique pattern of song learning in tape-tutor experiments: they do not imitate tutor songs but rather improvise songs (different from but derived from the tutor songs) or invent songs (totally new), all of them normal species songs (Kroodsma and Verner 1978; Kroodsma et al. 1999a). Consequently, each bird winds up with a repertoire of unique songs, and two neighbors in the field (who probably will not be neighbors for long) will share no song types. In contrast, the closely related but sedentary marsh wrens faithfully copy tutor songs in comparable experiments, and in the field they share songs with their neighbors (Verner 1975; Kroodsma and Pickert 1984). Furthermore, tropical populations of sedge wrens, which, unlike the seminomadic northern populations, are sedentary, show the common pattern of song sharing with neighbors that is generally taken to imply song learning from neighbors (Kroodsma et al. 1999b).

An Ontario song sparrow population studied by Weatherhead and Boak (1986) appears to have much in common with the northern sedge wren. For example, Weatherhead and Boak report that approximately two-thirds of the males holding territories at the beginning of the breeding season abandoned those territories at some point during the breeding season; moreover, fewer than 20% of those males returned to the population in subsequent years. Hence, song sharing with neighbors may be neither possible nor beneficial in populations like this.

The existence of populations with little or no song sharing among neighbors does not necessarily weaken the hypothesis that song sharing with the neighbors is the "goal" of the song-learning program. For one thing, the hypothesis suggests no prediction concerning what songs the bird should learn if he is unlikely to have any permanent or semipermanent neighbors. Second, in such populations it is still possible that birds follow similar singing rules. For example, even in populations with low song sharing, neighbors may still have songs that are similar—or at least that the birds regard as similar. This can happen in two ways. First, as noted above, birds may prune their song repertoire in their first spring, or even after, keeping songs that are similar to those of their neighbors and dropping songs that are not (e.g., D. Nelson 1992). Indeed, the Nelson-Marler model of song learning suggests that this is a common pattern of song learning (D. Nelson and Marler 1994). If that is true, demonstrations that birds in a population share no more with their neighbors than with distant birds (Hughes et al. 1998) are perplexing. The second way neighbors can have similar songs is essentially by chance. Two neighbors will have some songs in their repertoires that are more similar than others. Long-term neighbors should come to recognize which songs these are, and they could then treat these as "shared" songs. We have shown that song sparrows will song-match using songs that we do not classify as the same type but that are similar in some general way (e.g., beginning with an accelerating trill; Burt et al. 2002). R. Anderson et al. (2005) have shown that in eastern song sparrows—who as noted share no more songs with neighbors than with nonneighbors—a bird will nevertheless song-match songs that are partially similar to one of his own, specifically in the initial song element. Falls (1985) has shown that, in a population of western meadowlarks where neighbors are said not to share songs, birds still recognize their neighbors and show the typical patterns of song matching. For that matter, many studies have shown that birds will song-match stranger songs, and probably in many cases (usually not documented in the papers) the "matching" songs are similar yet would not be described as "shared" by most investigators.

Finally, in this chapter we have focused on the aspect of the song sparrow's social ecology that pertains to social interactions among neighboring males. We have neglected male-female interactions, yet these undoubtedly are also of great importance in shaping song learning and song use in this species and other songbirds (Searcy 1984; Catchpole et al. 1984; Catchpole 1986; West and King 1988; Nowicki et al. 1998, 2001). Our own efforts at evaluating female influences on song in our study species have so far provided evidence both for (O'Loghlen and Beecher 1997, 1999) and against (C. Hill et al., in preparation) the hypothesis that female sexual preferences shape song repertoires. As mentioned above, there is some evidence that female preferences shape song repertoires in another western population of song sparrows (J. Reid et al. 2004). As we noted in our earlier chapter, it will be interesting to see if the study of intra- and intersexual selection provides reinforcing or opposing arguments for the cognitive ecology of song learning and song communication in this species.

4.7. Summary

In song sparrows, the strategy of song learning in young birds and the mechanisms of song communication between neighboring adult males are shaped by two major sets of variables: cognitive factors at the proximate level and variables in the species' social ecology at the ultimate level. At the proximate level, song learning and adult singing rules appear to be shaped and guided by the bird's concepts of song type and singer identity. At the ultimate level, the function of song learning appears to be the acquisition of songs that will be shared with territorial neighbors. We argue that a cognitive ecology perspective—which focuses on both cognitive factors and social-ecological variables—captures the key features of song learning and song communication in this species and will probably do so as well in other species of songbirds. Finally, we suggest that parallels between song-learning rules in song sparrows in their first year and song usage rules in their subsequent years may relate to the nature of the song-learning process. According to the social eavesdropping hypothesis, the young bird is particularly influenced by singing interactions (counter-singing) between adults that the young bird hears in his first year. Our laboratory song-learning experiments have been consistent with this hypothesis so far, and we are collecting data in the field on radio-tracked young birds in their first year that we suspect will also be consistent with this hypothesis. Understanding the song-learning process may ultimately be the key to understanding why it is that song learning in one life stage parallels song communication in later life stages.

ACKNOWLEDGMENTS

We thank the National Science Foundation for supporting our research and Liz Campbell, Chris Hill, Cindy Horning, Adrian O'Loghlen, Cully Nordby, Phil Stoddard, and Chris Templeton for collaborating in this work.

5 Consequences of Brain Development for Sexual Signaling in Songbirds

WILLIAM A. SEARCY & STEPHEN NOWICKI

5.1. Introduction

Song learning in songbirds is a cognitive task of great complexity, in which young males copy many of the features of their songs from models they hear during early development. The adult songs that result function integrally in male-male interactions (see Beecher and Burt, chapter 4 in this volume) and in attracting females and stimulating them to mate. In the latter function especially, issues of signal reliability become crucial. As in any system of sexual advertisement, some degree of signal reliability is needed for the signaling system to be evolutionarily stable (Searcy and Nowicki 2005). Whereas the signalers presumably are always under selection to exaggerate their displays, the signaling system can be stable only if exaggeration is kept within bounds, so that the relationship between the signal and signaler quality—that is the signal's reliability—is maintained. To keep the system functioning, then, mechanisms must exist to limit exaggeration, and it is logical to expect such mechanisms to depend on signal costs (Zahavi 1975; Grafen 1990). This logic focuses our attention on the costs of song, and more specifically on the costs of song development.

In this chapter, we review evidence on the developmental costs of song and their relevance to signal reliability. Developmental costs are costs that are paid at the time that a signal develops rather than at the time it is produced (Searcy and Nowicki 2005). Large, anatomical structures that function as signals, such as the antlers of deer or the trains of peacocks, obviously must impose costs of this nature. It is less obvious that important developmental costs can be imposed by a signal that, like song, is a behavior rather than an anatomical structure. Song learning and song production, however, do require investment in a complex and specialized neural system, and it is investment in this system that has been suggested to maintain the reliability of many of the song attributes to which females attend (Nowicki et al. 1998, 2002a). The cost of the "song system" is important not only because of the magnitude of the cost but also because of its timing: much of the song system develops during the period soon after hatching, when a songbird is growing rapidly and is potentially under stress from poor nutrition, parasites, and other external factors, and when many other physiological and anatomical systems are competing for investment, including other systems in the brain.

The idea that the developmental costs of song are crucial to maintaining its reliability has been termed the "developmental stress hypothesis" (Nowicki et al. 1998, 2002a; Buchanan et al. 2003). Stress enters into the hypothesis because juvenile songbirds in their period of rapid growth are exposed to many sources of stress, such as poor nutrition, parasitism, and temperature extremes, all of which can affect the overall investment an individual can afford to make in growth. Those individuals that happen to escape stress, or whose genotypes buffer them against stresses that they do encounter, are able to afford adequate investment in both their song and their general phenotype. Those individuals who are exposed to stresses and whose genotypes leave them vulnerable may sacrifice investment in song development in preference to sacrificing systems crucial to survival rather than to advertisement (Andersson 1986; Nowicki and Searcy 2005a). Song then becomes a reliable indicator of both phenotypic and genetic quality.

The developmental stress hypothesis predicts that early stresses will negatively affect the development of the song system in the brain and hence the quality of adult song. Although the hypothesis was first proposed just 10 years ago (Nowicki et al. 1998), a number of experiments have already been carried out to test these predictions. Our primary goal in this chapter is to review the results of these experimental tests, paying particular attention to the kinds of stresses employed and the effects recorded on the brain and on song. We will also review evidence on the effects that the same types of stress have on the development of other aspects of the phenotype, so that we can draw inferences about the aspects of male quality that may be signaled by song. First, however, we need to provide some background on song neurobiology and on female preferences for song, to give the context necessary for the interpretation of tests of the developmental stress hypothesis.

5.2. The song system

Song in songbirds is controlled by a "song system" consisting of two main pathways of brain nuclei linked by projecting neurons (Nottebohm et al. 1976; Nottebohm 2004; Reiner et al. 2004). The descending motor pathway starts in the dorsal part of the forebrain with the HVC, which projects to the robust nucleus of the arcopallium (RA), which in turn projects to the tracheosyringeal part of the hypoglossal nucleus (nXIIts). Motor neurons in the nXIIts innervate the syrinx, the organ responsible for generating vocal sounds in songbirds. The anterior forebrain pathway also begins with HVC, which projects to area X in the anterior forebrain, then to the dorsolateral thalamic nucleus (DLM), to the lateral magnocellular nucleus of the anterior nidiopallium (LMAN), and finally to the RA.

As its name implies, the descending motor pathway is important to song production. Lesions in either the HVC or the RA disrupt singing in adults (Nottebohm et al. 1976), and neurons in the pathway are active in synchrony with singing behavior (McCasland 1987; Fee et al. 2004). Recent work suggests that motor patterns at the level of note structure are stored in the RA, while patterns at the level of the sequence and timing of notes are stored in the HVC (Leonardo and Fee 2005; Kozhevnikov and Fee 2007). The anterior forebrain pathway apparently functions in song learning rather than song production. Lesioning the nuclei exclusive to this pathway (area X, DLM, and LMAN) does not disrupt the crystallized song of adults but does disrupt normal song development in juveniles (Bottjer et al. 1984; Sohrabji et al. 1990).

The sizes of certain of the song nuclei are correlated with song attributes important to females, notably with measures of song complexity such as syllable repertoire size and song type repertoire size. Such an association was first demonstrated by Nottebohm et al. (1981), who showed that across individual canaries *Serinus canaria* syllable repertoire size was positively associated with the volume of both the HVC and the RA. Subsequent studies found similar relationships in some other songbird species (e.g., Brenowitz et al. 1995; Airey et al. 2000), but not all (Kirn et al. 1989; Brenowitz et al. 1991). A recent meta-analysis (Garamszegi and Eens 2004a) concluded that existing studies as a whole demonstrate a positive association between song complexity and both HVC and RA volume within the songbird species that have been investigated. Song length is also correlated with the size of these nuclei overall. Negative results of individual studies were attributed to small sample sizes rather than to any heterogeneity of associations across species. Comparative studies have also shown that HVC volumes covary with repertoire sizes across songbird species as well as within species (DeVoogd et al. 1993; Székely et al. 1996).

As might be expected from their functions, the anterior forebrain pathway develops before the descending motor pathway (Brenowitz and Kroodsma 1996). In the zebra finch *Taeniopygia guttata*, for example, functional connections between the nuclei in the anterior forebrain pathway are present by 15 days (F. Johnson and Bottjer 1992; Nordeen et al. 1992), whereas connections from the HVC to the RA in the motor pathway do not begin to develop until 25–35 days posthatching (Konishi and Akutagawa 1985). The HVC itself increases in size in zebra finches between 10 and 50 days (Bottjer et al. 1985; Nordeen and Nordeen 1988). The RA also increases in volume during the same period (Bottjer et al. 1985; Konishi and Akutagawa 1985). In canaries, most HVC neurons that project to the learning pathway are born when the bird is still in the egg, whereas most HVC neurons that project to the RA, thus forming part of the motor pathway, are born between 10 and 240 days posthatching (Alvarez-Buylla et al. 1988).

Learning of external song models tends to be concentrated in songbirds in a limited time period early in life. This "critical period" or "sensitive phase" overlaps with the period in which the song system, and in particular the motor pathway, develops. Thus, in zebra finches, the sensitive phase for song learning extends from 25 to 65 days posthatching (Immelmann 1969; Eales 1985; A. Roper and Zann 2006), while, as we have seen, the motor pathway develops in large part 10–50 days posthatching. In canaries, the sensitive phase is 40–240 days posthatching (Nottebohm et al. 1986), while the motor pathway develops 10–240 days posthatching.

5.3. Female preferences for song attributes

Female mating preferences in songbirds are influenced by a variety of male song attributes (reviewed in Searcy and Andersson 1986; Catchpole and Slater 1995; Searcy and Yasukawa 1996). Only a subset of the preferred attributes, however, seems likely to have developmental costs and thus to be relevant to the developmental stress hypothesis (Nowicki and Searcy 2005b). This subset includes song complexity, quality of learning, and vocal performance. We review evidence for these preferences below, emphasizing results on species that have been used in tests of the developmental stress hypothesis.

Song complexity is usually measured as the number of syllable types or song types in a male's repertoire. A preference for larger song repertoires is quite common across songbirds (Searcy and Yasukawa 1996). In song sparrows (*Melospiza melodia*), for example, females tested in the laboratory show a greater courtship response to large song type repertoires than to small (Searcy and Marler 1981; Searcy 1984) and in the field are more likely to mate with a first-year male the larger is his repertoire (J. Reid et al. 2004). In European starlings (*Sturnus vulgaris*), males with large repertoires obtain a female earlier than males with smaller repertoires both in the field and in the laboratory (Eens et al. 1991). In perhaps the most convincing experimental demonstration of any song preference, Lampe and Saetre (1995) showed that female pied flycatchers (*Ficedula hypoleuca*) prefer to build nests in boxes from which large-syllable repertoires are broadcast over building nests in boxes associated with smaller repertoires. In other species, male song repertoire size is positively correlated with extrapair mating success (Hasselquist et al. 1996).

Quality of learning is difficult to study in the field, because one cannot know with certainty which model songs young males are attempting to copy, making judgments of copy quality difficult. Accordingly, the one study that has looked directly at female preferences based on learning quality used songs from laboratory-raised males (Nowicki et al. 2002b). Hand-reared male song sparrows were tutored with songs from a specific population. The songs these males produced as adults were then rated on two measures of learning quality: the proportion of notes identified as having been copied from tutor songs and the accuracy of the copies as assessed by spectrogram cross correlation. Adult females taken from the same population gave a significantly greater copulatory response to songs that rated well on both measures of quality of learning than to songs that rated poorly on these measures (fig. 5.1). Although this study provides the only direct evidence that we know of that female birds prefer well-learned songs, the commonly observed preference of females for local songs over foreign (Baker 1983; Baker et al. 1987) may reflect the same underlying preference (Searcy et al. 2002).

Vocal performance measures the ability of individuals to sing elements that are physically difficult to produce. At present, the best-understood example of vocal performance involves the trade-off between bandwidth and syllable repetition rate in trills. Because of limitations on how fast the bill can move, male songbirds find it difficult to produce songs that combine syllables with

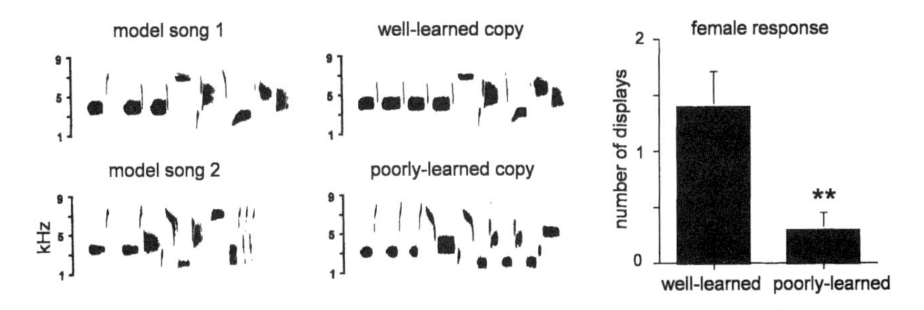

FIGURE 5.1. Female song sparrows respond preferentially to well-learned songs. The two copy songs are examples of songs learned by young males raised in the laboratory and tutored with tape-recorded songs. The well-learned copy matches model song 1 better than any other model song and is a good match in both the proportion of notes copied and the accuracy of note copying. The poorly learned song best matches model song 2 but is a poor match in both the proportion of notes copied and the accuracy of note copying. The female response bar graph shows the mean number (±SE) of copulatory displays given by adult female song sparrows to a larger sample of well-learned and poorly learned songs. ** indicates a significant difference at $P < 0.01$. Spectrograms reprinted from Nowicki et al. 2002b.

a wide frequency range (high bandwidth) and a rapid trill rate (M. Westneat et al. 1993). This limitation produces a negative upper-bound regression line between bandwidth and trill rate that can be seen both within and between species (Podos 1997). In swamp sparrows (*Melospiza georgiana*), females have been shown to respond preferentially to songs that closely approach this upper performance limit (Ballentine et al. 2004). Canaries have similar preferences, responding most strongly in laboratory tests to song phrases with high bandwidths and rapid syllable repetition rates (Vallet and Kreutzer 1995; Vallet et al. 1998; Draganoiu et al. 2002).

Theory suggests that display attributes can be honest indicators of an aspect of signaler quality only if they are costly along a dimension relevant to the kind of information being signaled (Grafen 1990). Thus, one way to identify the information conveyed by a preferred signal attribute is to identify its costs. Some categories of costs seem very unlikely for the three song attributes reviewed above. Energy costs, for example, are very low for bird song in general (Oberweger and Goller 2001; Ward et al. 2003, 2004) and do not seem likely to vary appreciably with song complexity or learning quality. Costs through increased risk of predation also seem unlikely for any of these song parameters—that is, it seems unlikely that predators would be differentially attracted to songs of high performance over those of low performance, to large repertoires over small repertoires, or to well-learned songs over poorly learned songs. In most songbirds, parameters such as repertoire size and the acoustic structure of song are set when adulthood is reached and thereafter cannot be varied; thus, these attributes are unlikely to be conventional signals whose reliability is enforced by receiver responses, as signals of this sort require that individuals have a free choice of which signal to make (Guilford and Dawkins 1995). A pattern of developing early and then not changing thereafter is, by contrast, what ought to be expected in signals with a pronounced developmental cost.

5.4. Experimental tests of the developmental stress hypothesis

The developmental stress hypothesis assumes that some of the song attributes preferred by females have important developmental costs, creating trade-offs between song development and the development of other aspects of the phenotype. Only males who do not encounter major stresses during development, or whose genotypes are especially well buffered against stress, can invest adequately in song development and other aspects of growth. The hypothesis thus predicts that stresses encountered early in development will have lasting effects on the song system in the brain and on adult song features. Experimental tests of this hypothesis have now been carried out in several species of songbirds.

5.4.1. SWAMP SPARROWS AND SONG SPARROWS

Nowicki et al. (2002a) subjected young swamp sparrows to nutritional stress during a period starting 3–7 days posthatching and ending at 14 days. Both control and experimental subjects were fed by hand with a standard hand-rearing mix. Control subjects were given as much food as they would take, while experimental birds were limited to 70% of the average intake of the control group. All birds were tape-tutored with the same set of tutor songs between 20 and 80 days, which spans the critical period for song memorization in this species (Marler and Peters 1987). The songs produced by the subjects were recorded starting at 8 months and continuing through their first spring as adults. At the age of 14 months, all subjects were euthanized and their brains were removed for anatomical measurement.

The experimental treatment had significant effects on the volume of two song system nuclei—HVC and RA. HVC volumes were about 25% smaller and RA volumes about 30% smaller in nutritionally stressed birds than in control birds. Part of the effect could be ascribed to an effect on overall telencephalon size, which was about 13% smaller in the nutritionally stressed group. Even as a proportion of telencephalon size, however, the RA was significantly smaller in the experimental birds than in controls.

MacDonald et al. (2006) manipulated nutrition in a similar way in song sparrows, which are a close relative of swamp sparrows. In this experiment, the manipulation began at 3 days posthatching, and experimental young were limited to 65% of the food intake of controls. The subjects were euthanized at 23–26 days posthatching to see whether the manipulation had an immediate effect on the song system. Experimental birds did not differ from controls in the volume of the RA or of area X, but they did have significantly smaller HVC volumes.

Effects of nutrition on song could not be assessed in the song sparrow experiment, because the subjects were sacrificed before singing began. In swamp sparrows, nutritional stress had no effect on song repertoire sizes, which were virtually identical in control and experimental birds (Nowicki et al. 2002a). One measure of the quality of learning was affected, however: note copy accuracy. Note copy accuracy was measured with spectrogram cross correlation, which provides an objective way of measuring the correspondence between original and copy in frequency and time profile. Cross correlation scores were significantly higher in the well-fed control birds than in the nutritionally restricted experimental birds.

5.4.2. EUROPEAN STARLINGS

Buchanan et al. (2003) stressed young European starlings by making their access to food unpredictable. The experimental group was deprived of food each day for a randomly chosen four-hour period, or about 25% of daylight hours. Controls were given constant access to food. The manipulation began when the birds were 35–50 days of age and ended when they were 115–130 days, thus spanning the usual critical period for song memorization.

The experiment tested for effects on song but not for effects on the song system. During the subjects' first breeding season, experimental males showed lower performance than controls on a number of measures of singing behavior: experimentals spent significantly less time singing, they produced significantly fewer bouts of song, and they produced song bouts of significantly shorter duration. Additional song analysis by K. Spencer et al. (2004) showed that experimentals had significantly smaller repertoires: repertoire size averaged 42 syllables in controls but only 36 syllables in stressed birds, a reduction of about 15%.

5.4.3. ZEBRA FINCHES

Young zebra finches have been subjected to three forms of stress in experimental tests of the developmental stress hypothesis. K. Spencer et al. (2003) and Buchanan et al. (2004) studied the effects of nutritional stress. Here parent birds in the experimental group were given access to seeds mixed with husks, in a ratio of one part seeds to three parts husks, while control parents were given seeds with no husks. The rationale for this treatment is that mixing nutritionally valuable seeds with valueless husks ought to make it harder for parents to deliver adequate nutrition to their young. In addition, experimental young were given only half as much of a high-protein dietary supplement as were the controls. The two nutritional treatments were applied from 5 to 30 days posthatching. K. Spencer et al. (2003) and Buchanan et al. (2004) also studied treatment with corticosterone. Corticosterone is a hormone that is secreted in response to external stresses and promotes effects such as mobilization of energy stores and immunosuppression (Buchanan 2000). In nestling songbirds, nutritional deprivation is one of the stresses that elicit elevated corticosterone production (Saino et al. 2003; Pravosudov and Kitaysky 2006). Zebra finch young in the experimental group were given corticosterone orally mixed in peanut oil, and controls were given the same amount of peanut oil without corticosterone. The hormone treatment was applied from 5 to 30 days posthatching. Gil et al. (2006) studied the effects of stress by manipulating brood size. Chicks were moved between broods at 2 days posthatching to produce brood sizes that were large (5–6 nestlings), medium (4 nestlings), or small (2–3 nestlings). Mean brood size before manipulation was 4.5 nestlings. Brood size manipulation ought to be somewhat equivalent to manipulation of nutrition, as parents with enlarged broods may have difficulty providing sufficient food per chick.

Buchanan et al. (2004) found that both nutritional restriction and corticosterone treatment reduced the size of the HVC of males in the experimental groups relative to HVC size in controls (fig. 5.2). Sizes of the RA and the LMAN were not affected by either treatment. Brain volumes were measured when subjects were about 215 days old, well beyond the age at which song crystallizes in this species. Effects on the HVC were fairly dramatic, with the volume of the HVC in food-restricted birds being only about 53% of the volume in controls, and in corticosterone-treated birds only about 44% (fig. 5.2). Gil et al. (2006), by contrast, found that brood size manipulation had no effect on the volume of the HVC, the RA, or the LMAN.

Both nutritional restriction and corticosterone treatment affected song development, and in similar ways (K. Spencer et al. 2003). Male zebra finches exposed to either one of these stresses developed shorter songs with fewer syllables and lower peak frequencies than controls (fig. 5.3). The number of syllable types per song was not affected. Results of brood size manipulations were quite different: no effects were seen on any song parameter, including measures of the quality of song learning as well as song length and syllable repertoire size (Gil et al. 2006).

K. Spencer et al. (2005b) tested directly for a female preference for songs of unstressed males. Females were tested in an operant-conditioning chamber, set up so that when a bird landed on a specific perch, playback of a particular category of songs was triggered. Test songs were taken from those produced by the subjects of the K. Spencer et al. (2003) study of the effects of nutritional stress and corticosterone treatment on song development. The female subjects showed a significant preference for hearing songs of unstressed control males over the songs of the two categories of stressed males.

5.4.4. CANARIES

K. Spencer et al. (2005a) subjected young canaries to stress in the form of exposure to *Plasmodium relictum,* the parasite that causes avian malaria. Experimental birds received two injections of blood from a pigeon (*Columba livia*) previously infected with the malarial parasite. Injections were administered when the subjects were between 20 and 45 days posthatching. Controls were given two injections of saline solution during the same period. Transmission of the parasite to the experimental birds was confirmed using blood smears. Songs

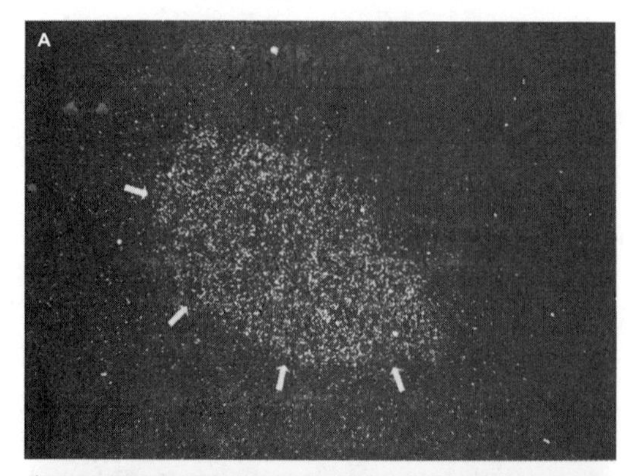

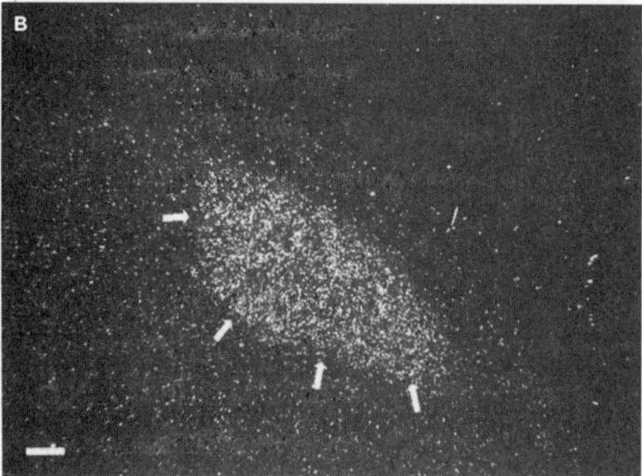

FIGURE 5.2. The size of the brain nucleus HVC in stressed and unstressed individuals in two species of songbirds. a. A zebra finch from an unstressed control group. b. A zebra finch from a group stressed by corticosterone treatment. The zebra finch pictures (a, b) are darkfield photomicrographs showing androgen receptor mRNA expression, reprinted with permission from Buchanan et al. 2004.

were recorded when the subjects were 240–300 days old, and the subjects were then euthanized so that their song system nuclei could be measured.

The experimental treatment had a significant effect on size of the HVC (fig. 5.2) but not on size of the RA. HVC volume was 36% lower in the stressed birds than in the controls. The treatments also had a significant effect on syllable repertoire size, with stressed birds developing a repertoire that was on average 12% smaller than the repertoires of controls. No effect was seen on the types of complex syllables that previous work has indicated are especially preferred by female canaries (Vallet and Kreutzer 1995; Vallet et al. 1998).

5.4.5. CONCLUSIONS

A variety of forms of stress have been shown to affect either adult song or the song system or both (table 5.1), including nutritional limitation, corticosterone treatment, and parasite infection. The one exception has been brood size manipulation, which had no effect on either song or the song system in zebra finches (Gil et al. 2006). Brood size enlargement ought in general to have effects that parallel those of food limitation, as young in enlarged broods

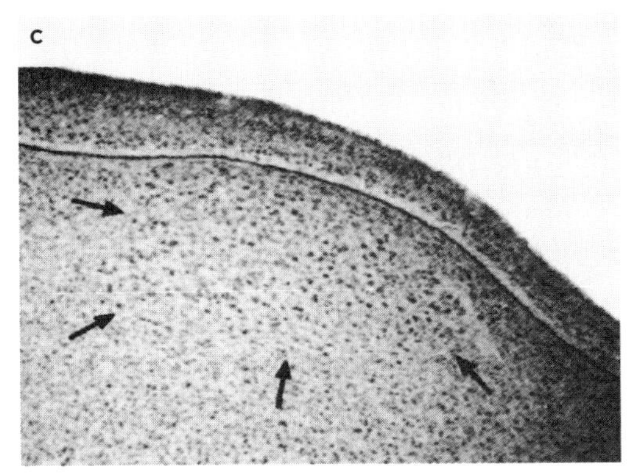

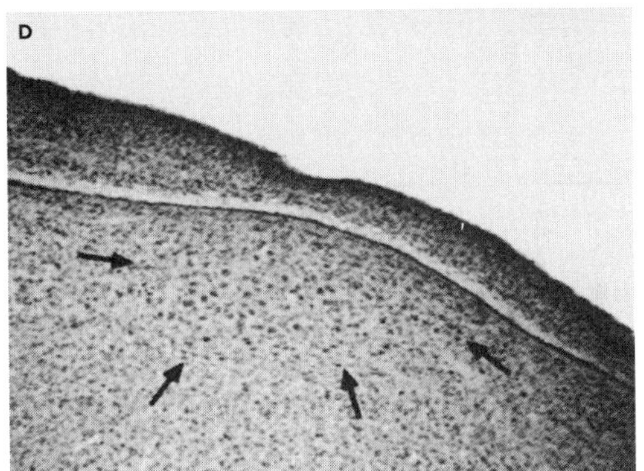

FIGURE 5.2. (*continued*)
c. A canary from an unstressed control group. d. A canary from a group stressed by malaria parasites. The canary pictures (c, d) are brightfield photomicrographs, reprinted with permission from K. Spencer et al. 2005a.

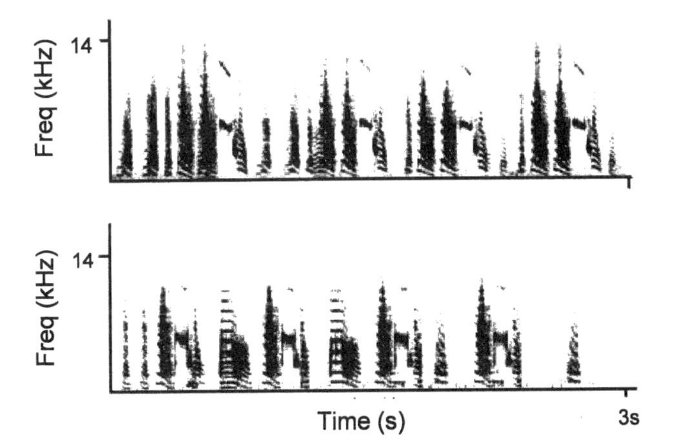

FIGURE 5.3. Songs from (top) a control male zebra finch and (bottom) a male zebra finch stressed early in development. The two songs were learned from the same tutor, but the song of the stressed bird is shorter, with fewer syllables. Reprinted with permission from K. Spencer et al. 2005b.

TABLE 5.1. Effects of stress on the song system and adult song

Species	Swamp sparrow	Song sparrow	European Starling	Zebra finch	Zebra finch	Zebra finch	Canary
Stress	Nutritional limitation	Nutritional limitation	Unpredictable nutrition	Nutritional limitation	Corticosterone	Brood size manipulation	Parasite infection
Effects on song system	Smaller RA, smaller HVC	Smaller HVC, no effect on RA	Not tested	Smaller HVC, no effect on RA	Smaller HVC, no effect on RA	No effects on HVC or RA	Smaller HVC, no effect on RA
Effects on adult song	Lower note copy accuracy, no effect on repertoire size	Not tested	Smaller repertoires, fewer song bouts, shorter songs	Shorter songs, lower peak frequencies, less preferred by females	Shorter songs, lower peak frequencies, less preferred by females	No effect on song learning, song duration, repertoire size	Smaller repertoire size, no effect on complex syllables
References	Nowicki et al. 2002a	MacDonald et al. 2006	Buchanan et al. 2003	K. Spencer et al. 2003, 2004; Buchanan et al. 2004	K. Spencer et al. 2003, 2004; Buchanan et al. 2004	Gil et al. 2006	K. Spencer et al. 2005a

typically obtain less food per chick and experience lower rates of mass increase (Pettifor et al. 2001). Thus, the negative results of brood size enlargement in zebra finches are difficult to reconcile with the positive results of nutritional limitation studies, especially those of the studies done with the same species (Buchanan et al. 2004; K. Spencer et al. 2003). The most parsimonious explanation for the difference in results is that the two experimental stresses differed in severity and timing. Leaving the one brood size manipulation experiment aside, experimental tests of the developmental stress hypothesis have otherwise been strongly positive.

5.5. Effects of developmental stress on phenotypic quality

If developmental stresses such as nutritional limitation, elevated corticosterone levels, and parasite infection affect attributes of adult song such as repertoire size and note copy accuracy, then those attributes become a signal to females indicating how stressful a male's development has been. If the same stresses affect other aspects of the singer's developing phenotype, then females can gain information about those aspects by attending to song. It thus becomes important to ask what other aspects of the phenotype are known to be affected by the same stresses that affect song.

5.5.1. ADULT BODY SIZE

Nutritional limitation early in life generally has negative effects on the growth rates of young birds and consequently on their body sizes (Skagen 1988; Konarzewski et al. 1996; Searcy et al. 2004), but whether these negative effects last into adulthood is variable. Young birds can counter the effects of poor early nutrition either by accelerating the rate of growth once the stress has ended ("compensatory growth") or by extending the period of growth (Schew and Ricklefs 1998). Nevertheless, negative effects on body size that last until adulthood have been demonstrated in some songbird studies, for example, in zebra finches (Birkhead et al. 1999) and song sparrows (Searcy et al. 2004). Furthermore, there is evidence for other songbird species that females prefer larger males as mates (Weatherhead and Boag 1995; Kempenaers et al. 1997), though such a preference is not universal (Schluter and Smith 1986). Why females should sometimes prefer to mate with larger males is not obvious, but one possibility is that large males tend to be dominant and that there are advantages to being mated to a dominant male—for example, in gaining preferential access to resources.

Individuals who are able to compensate completely for a poor early start in growth may nevertheless pay long-term fitness costs for having to undergo a period of accelerated, compensatory growth (Metcalfe and Monaghan 2001). In mammals, long-term costs of compensatory growth often stem from metabolic disorders, leading to problems in glucose tolerance and adult obesity (Metcalfe and Monaghan 2001).

5.5.2. DOMINANCE AND COMPETITIVE ABILITY

K. Spencer and Verhulst (2007) tested the effects of early corticosterone treatment on adult dominance in zebra finches. As in K. Spencer et al. 2003, experimental subjects were treated with corticosterone in peanut oil from day 7 to day 18 posthatching, while control subjects were given peanut oil only. Then at 50–60 days, experimental subjects were tested against controls matched for body mass in competition for access to a single perch. Birds that had been exposed to corticosterone early in life showed reduced competitive ability, spending significantly less time on the perch and being supplanted significantly more often. Thus, early postnatal corticosterone treatment affects adult dominance as well as adult song.

Richner et al. (1989) restricted growth rates of hand-reared carrion crows (*Corvus corone*) by limiting the amount of food they were given. After growth had ended, the experimental birds were significantly smaller in both mass and linear dimensions than were control birds that had been fed ad libitum. As juveniles, the nutritionally restricted birds ranked significantly lower in dominance than same-sex control birds. Thus, again a developmental stress that affects song also has long-lasting effects on dominance.

In carrion crows, the effects of early nutrition on dominance could be attributed to long-lasting effects on size, but effects of nutrition on dominance can also be seen in cases in which compensatory growth is sufficient to eliminate size differences between treatment groups. In an experiment with green swordtail fish (*Xiphophorus helleri;* Royle et al. 2005), one group of subjects was fed three times a week from 2 to 6 months of age while a second group was fed daily. The poorly fed group grew more slowly during the treatment period but caught up in body mass after the treatments had ended. As adults, individuals from the well-fed treatment were nevertheless dominant during aggressive interactions with size-matched individuals from the poorly fed treatment in 12 of 13 cases. Early nutritional stress thus had a negative effect on dominance even after size differences had been eliminated by compensatory growth.

5.5.3. IMMUNOCOMPETENCE

Developmental stress has been shown to have short-term negative effects on immunocompetence in songbirds. In European starlings, for example, birds 35–130 days old were stressed with an unpredictable food supply and then tested during the period of stress for humoral response to injections of sheep red blood cells (Buchanan et al. 2003). Individuals in the stressed group had a significantly lower secondary humoral response than did control birds. The two groups did not differ in their cell-mediated response to injection of a plant protein. In a second study, nestling sand martins (*Riparia riparia*) subjected to nutritional restriction produced a lower humoral response than did control birds when tested with injections of a plant protein (Brzek and Konarzewski 2007). Other types of stress have also been shown experimentally to depress measures of immune response in nestling songbirds (Soler et al. 2003; Snoeijs et al. 2005). All these studies, however, have looked at immediate effects of stress on young birds during the period of stress rather than examining the later effects of early stress on immunocompetence as adults. We know of only one study of the latter type, by Birkhead et al. (1999) on zebra finches. Nestlings were reared on a low-protein or a high-protein diet and then tested as adults for both a cell-mediated immune response and an antibody response. Neither response differed between the treatment groups.

Although this one test of the effects of developmental stress on adult immunocompetence in songbirds was negative, long-term effects of early stress on immune function have been demonstrated in mammals. In particular, nutritional stress in the early postnatal period has been shown to have effects on immune function that last up to several months in rodents and up to several years in humans (Jose et al. 1973; Gershwin et al. 1985; Myrvik 1994). Given that short-term effects of nutrition on immunocompetence have been demonstrated in songbirds, and that long-term effects are known in mammals, long-term effects in songbirds also seem possible and even probable.

5.5.4. LONGEVITY

In the Birkhead et al. (1999) study in which nestling zebra finches were stressed with a protein-poor diet, the stressed birds as adults were significantly smaller than control birds in body mass but otherwise showed no obvious phenotypic differences from control birds. Nonetheless, the experimental birds showed significantly lower survival than controls after the treatment had ended. The effect was considerable: stressed birds had a mean life span of 352 days, compared with 461 days for the controls, a difference of 31%. Birkhead et al. (1999) suggested that the longevity effect may have been due to subtle differences in immune function, which they were unable to measure, or to effects on dominance and competitive ability.

5.5.5. OTHER COGNITIVE ABILITIES

The same types of developmental stress that affect song learning can also affect other cognitive abilities. The best evidence for songbirds comes from the work of Pravosudov and colleagues on spatial memory in the western scrub jay, *Aphelocoma californica* (see Pravosudov, chapter 6 in this volume). Pravosudov et al. (2005) manipulated nutrition in young scrub jays by limiting the intake of an experimental group to 65% of the intake of control birds. The manipulation lasted from 7 days posthatching until the birds were independent of hand-feeding, at about 11 weeks for experimental birds and 8 weeks for controls. At the age of 6.5 months, the birds were tested for their ability to remember locations where they had cached food. The nutritionally deprived birds inspected significantly more locations before finding their caches than did controls at both 1 day and 10 days after caching. At 8 months, the nutritionally stressed birds performed significantly worse on a second spatial memory test, one that required remembering where the researchers had hidden food. Stressed birds did not perform worse than controls on learning tasks that did not involve spatial memory. At 1 year of age, the birds were sacrificed so that their brain anatomy could be examined. The one brain measurement that differed between experimental and control birds was the volume of the hippocampus, a region thought to function in spatial memory. Hippocampal volumes were significantly smaller in stressed birds.

The brains of rats are also vulnerable to early nutritional stress, with the hippocampus seeming to be particularly at risk. Bedi (2003) reviewed studies of the effects of undernutrition imposed soon after birth on total neuron numbers in various areas of the brain of rats and found clear evidence of a detriment only for the dentate gyrus of the hippocampus. Early postnatal nutritional stress has also affected later performance of rats in spatial memory tests in some studies (Fukuda et al. 2002; Valadares and Almeida 2005), though not in others (L. Campbell and Bedi 1989).

M. Fisher et al. (2006) stressed young zebra finches during the early posthatching period by providing the parents of experimental birds with diets lacking protein, vitamin, and mineral supplements that were provided to parents of control birds. The stressed young grew more slowly in mass during the period of deprivation but then caught up with controls after deprivation ended. When tested as adults on a task the required learning an association between a color and food, the stressed birds did not show significantly lower performance than controls. Performance on this cognitive task, however, was negatively related to the rate of compensatory growth among the stressed group, suggesting that the compensatory growth in this case has a cost in impairment of cognitive abilities.

5.6. Conclusions and prospects

Experimental studies have shown that developmental stresses such as nutritional restriction, corticosterone treatment, and parasite infection have detrimental effects on the neural structures that control the learning and development of song in songbirds. Associated with these negative effects on brain anatomy are detrimental effects on attributes of adult song that are known to be important to female songbirds in mate choice, such as song repertoire size, quality of learning, and vocal performance. Not surprisingly, the same types of developmental stress also have negative effects on other aspects of the adult phenotype, aspects that we might think of as contributing to phenotypic "quality." Thus, as we have shown, early nutritional stress in particular leads in some cases to reduced adult body size, lower competitive ability, reduced longevity, possibly reduced immunocompetence, and lower ability to master spatial learning tasks. All of these aspects of phenotypic quality may be important to female songbirds in choosing a mate, in part because these traits affect the level of resources a male is able to provide to a female and to her offspring. Thus, a male that has low competitive dominance may be unable to acquire a high-quality territory, and a male with poor spatial memory may have low success in finding food when provisioning offspring. Certainly, a male at risk of becoming ill or even dying will be less likely to be able to complete a breeding attempt successfully. In addition, individuals must vary in their susceptibility to stress, in the sense that some individuals will be more affected by a given amount of stress than are others. If susceptibility to stress has a genetic component, then females choosing a male whose song shows little evidence of effects of stress may on average acquire a mate whose genotype is resistant to stress. Thus, female songbirds may obtain both direct and indirect benefits from using song as an indicator of male quality.

Study of developmental stress in birds has begun only recently, so our knowledge of the effects of developmental stress on song and the song system is very incomplete. Only a handful of songbird species have been studied, and those with regard to only a subset of relevant stresses and song attributes. Although most of the stresses that have been studied have produced some effects on song, it is puzzling that the one study of brood size manipulation found no effects (Gil et al. 2006); this negative result ought to be followed up. Even more incomplete is our knowledge of the effects of developmental stress on other aspects of the adult phenotype in birds. Further studies of the effects of early stress on adult social dominance, immunocompetence, and cognitive abilities would be particularly interesting.

ACKNOWLEDGMENTS

We thank Reuven Dukas for his advice and encouragement, Susan Peters for long-term collaboration, and the National Science Foundation for support through grants IBN-0315566 (to Searcy) and IBN-0315377 (to Nowicki).

6 Development of Spatial Memory and the Hippocampus under Nutritional Stress: Adaptive Priorities or Developmental Constraints in Brain Development?

VLADIMIR V. PRAVOSUDOV

6.1. Introduction

During development, abundant nutrition is necessary for successful growth. Even though most animals reproduce when food supplies are relatively plentiful and predictable, great variance in foraging success may still exist. As a result of such variance, many species may exhibit significant variation in growth rates both within and between the broods (Ricklefs 1983; O'Connor 1984). Postnatal development appears to be especially vulnerable to perturbations in food supply because during that period young depend on their parents to gather enough food for both self-maintenance and developing offspring. While mammals may potentially compensate for variation in available food supplies by using the mother's body reserves, posthatching development in birds depends entirely on availability of food and, most importantly, on parents' ability to find and deliver it to their offspring (Ricklefs 1983). Most animals that survive food shortages during postnatal development appear to compensate for growth deficiencies when they become nutritionally independent (Ricklefs 1983; O'Connor 1984), but from an evolutionary standpoint it is crucial to understand whether variation in development rates caused by insufficient nutrition might be associated with fitness consequences. Some available data suggest that such an association does exist because slower-growing individuals often show higher mortality than faster-growing ones (Ricklefs 1983; O'Connor 1984; Merilla and Wiggins 1995). The next question, then, concerns the mechanisms leading to reduced fitness in undernourished animals and how natural selection might act to reduce fitness consequences of undernourishment during postnatal development. One possible path for selection might be to act on parental quality to select the highest-quality parents that are capable of providing sufficient nutrition to their growing offspring regardless of environmental conditions (Hoelzer 1989). Another path might be to act on mechanisms of resistance to malnutrition during development and on the way available nutrition is partitioned by the growing organism (Schew and Ricklefs 1998).

In particular, it has been hypothesized that when nutrition during development is limited, natural selection may favor selective investment of available nutrition into the morphological and physiological structures that are most crucial for survival, while reducing nutrition allocation to less important structures (Schew and Ricklefs 1998). Neural structures appear to be of high priority during development, and it has been suggested that learning abilities of young may contribute significantly to their probability of survival and that learning abilities might be under intense selection pressure (Weathers and Sullivan 1989; Dukas 1998c). Learning is dependent on underlying neural mechanisms, and natural selection might favor evolution of mechanisms that give high nutritional priorities to development of neural tissue. The brain consists of many areas responsible for somewhat-different functions, and hypothetically, some of these functions might be more critical for survival than others. In such a case, if there is not enough nutrition for developing the entire brain normally, more important brain areas might be given nutritional priority at a cost of underdeveloping less important brain areas, as stated by the adaptive priorities in brain development hypothesis (Nowicki and Searcy 2005a). Some support for the adaptive priorities in brain development hypothesis comes from studies of bird song learning and the song nuclei system in the brain (Nowicki and Searcy 2005a). Bird song is a sexual display trait that does not seem to be critical for survival, and thus structures responsible for song learning and production might be of lower priority when nutrition is in short supply during development (Andersson 1986; Nowicki and Searcy 2005a). Within the avian brain, song nuclei are responsible for song learning, and development of these nuclei is therefore likely to be compromised first if there are not enough resources to invest in all brain areas equally (Nowicki and Searcy 2005a). Supporting this view, many studies have reported that several song nuclei, most notably the high vocal center (HVC), are smaller in birds that were subjected to nutritional deprivation during posthatching development and that these birds also showed inferior song-learning abilities (Nowicki et al. 1998, 2002a; Buchanan et al. 2003, 2004; K. Spencer et al. 2003, 2004; MacDonald et al. 2006; Pfaff et al. 2007).

The data on avian song nuclei do not provide a complete test of the adaptive priorities in brain development hypothesis, however, because they test only one component of this hypothesis. Additional tests are needed to show that brain areas crucial for survival are spared when resources are limited. Here, I review data on the development of the hippocampus, a brain area that is involved in spatial learning and appears to be important for survival in most animals, and how it responds to nutritional stress.

6.2. Spatial memory and the hippocampus in birds

Avian spatial memory and the hippocampus have been under fairly intensive investigation for quite some time (Shettleworth 1998; Macphail 2002), and most work has focused on homing pigeons (Bingman et al. 2003, 2005, 2006), food-caching birds (Shettleworth 1995, 2003; Pravosudov 2007), and migratory songbirds (Healy et al. 1996; Cristol et al. 2003; Pravosudov et al. 2006). In all these groups, spatial memory appears to be a crucial component of fitness: homing pigeons should be able to find their home loft, food-caching birds appear to use spatial memory to find hundreds to thousands of individually hidden food caches, and migratory birds appear to use memory during long-distance migration.

Similar to mammals, spatial memory in birds appears to be dependent on the hippocampus in the brain, and experimental lesions of the hippocampus resulted specifically in spatial memory impairments (Sherry and Vaccarino 1989; Hampton and Shettleworth 1996) while memory for color was not affected (Hampton and Shettleworth 1996). If spatial memory is largely dependent on the hippocampus, it may be further hypothesized that intense selection pressure on spatial memory stemming from specific ecological conditions should result in evolution of enhanced spatial memory and an enlarged hippocampus—this is the adaptive specialization hypothesis (Krebs et al. 1989, 1996; Sherry et al. 1989). For example, fitness of food-caching birds depends on their ability to successfully recover their previously hidden caches, and cache recovery has been shown to rely, at least in part, on spatial memory, making it an important fitness component (Shettleworth 1995). As a result, if the adaptive specialization hypothesis were true, it might be expected that food-caching birds have better spatial memory and a relatively larger hippocampus than noncaching species (Krebs et al. 1989, 1996; Sherry et al. 1989). Initial comparative studies of the hippocampal volume in food-caching and noncaching avian species supported the adaptive specialization hypothesis by showing that caching species indeed have larger hippocampi (Krebs et al. 1989; Sherry et al. 1989; Healy and Krebs 1992, 1996; Hampton et al. 1995). However, when Brodin and Lundberg (2003) reanalyzed all available data on corvids and parids, no obvious relationship between food-caching propensity and hippocampal volume was discovered. Further analysis of the same data by Lucas et al. (2004) revealed that there are significant differences in the hippocampal volume between Eurasian and North American parids and corvids, and when the differences between the continents were statistically controlled, more food caching was positively related to larger hippocampi. The latest large-scale species comparison included more data (mostly on noncaching species) and also seemed to support the adaptive specialization hypothesis by showing that food-caching species have larger hippocampi than noncaching species (Garamszegi and Eens 2004b).

Most of the comparative research of the hippocampus focused only on hippocampal volume and did not consider the mechanisms behind volumetric changes. It is likely that volumetric changes between species result from changes in neuron numbers (Pravosudov and Clayton 2002; Pravosudov et al. 2006). Recently, it has also been shown that, compared with noncaching species, food-caching species have greater hippocampal neuronal recruitment rates (Hoshooley and Sherry 2007). Density of hippocampal neurons might also change without volumetric changes (Cristol et al. 2003), which suggests that it is critical to expand comparative analyses of the hippocampus beyond volumetric measurements.

Not all data, however, provide support for the idea that strong selection pressure on spatial memory should result in an enlarged hippocampus. For example, the pinyon jay *Gymnorhinus cyanocephalus,* which relies heavily on cached food for survival and reproduction, has a relatively smaller hippocampus than the western scrub jay *Aphelocoma californica,* which is a much less specialized cacher (Basil et al. 1996), even though this conclusion might still be controversial (Pravosudov and de Kort 2006). Among four compared woodpecker species, the two noncaching species actually had larger hippocampi than the two food-caching species, but other factors, such as differences in territory use among these species, may have contributed to these findings (Volman et al. 1997). The data on differences in spatial memory performance between food-caching and noncaching species were also not entirely consistent with the adaptive specialization hypothesis, and such inconsistencies resulted in vocal criticism of the entire approach to understanding the relationship between memory and the hippocampus (Bolhuis and Macphail 2001; Macphail and Bolhuis 2001). Whereas this critique had some valid points, it failed to recognize that inconsistencies in available data may be due to incomplete information on numerous potential differences in memory-dependent behaviors unrelated to food caching (Pravosudov and Clayton 2002; Pravosudov et al. 2006), as well as incomplete information on phylogeny of caching behavior in different lineages (de Kort and Clayton 2006).

If we want to understand how selection pressures related to food-caching behavior might have affected memory and the hippocampus, it might be better to compare species that differ only in food-caching behavior and not in other ecological parameters. Comparing populations of the same species, for example, might meet this criterion fairly well. Analysis of two populations of black-capped chickadees (*Poecile atricapillus*) revealed that birds from the more northern population (which appears to depend on cached food more heavily than the more southern population because of significantly shorter day length, lower ambient temperature, and more variable food supply in the winter) cached more food, showed significantly better spatial memory performance, and had significantly larger hippocampi with more neurons even after several months of being kept in identical laboratory conditions with the chickadees from the southern population (Pravosudov and Clayton 2002). Although the existence of population differences that are unrelated to caching behavior cannot be excluded, such differences are likely to be much smaller than those between different species.

A similar within-species comparison was also successful in testing the adaptive specialization hypothesis in migratory birds (Cristol et al. 2003; Pravosudov et al. 2006). Even though migratory birds are known to rely on various mechanisms for their orientation, it appears that spatial memory might be an important part of this arsenal, especially in older birds (Mettke-Hofmann and Gwinner 2003). Migratory birds appear to reuse the same stopover and wintering sites, and thus, they may have to remember more spatial information than species that use the same territories year-round. Supporting this view and the adaptive specialization hypothesis, Cristol et al. (2003) reported that migratory dark-eyed juncos (*Junco hyemalis*) appear to have better spatial memory and a higher density of hippocampal neurons than nonmigratory dark-eyed juncos. Pravosudov et al. (2006) compared two populations of the white-crowned sparrow *Zonotrichia leucophrys*—migratory *Z. l. gambelii* and fully resident *Z. l. nuttalli*—and reported that, compared with nonmigratory *Z. l. nuttalli*, migratory sparrows showed better spatial memory performance and had relatively larger hippocampi with more neurons. Although the issue is still far from being fully resolved, data on migratory species, together with most of the data on food-caching birds, seem to support the adaptive specialization hypothesis, suggesting that increased selection pressure on spatial memory appears to result in evolution of an enhanced spatial memory and an enlarged hippocampus with more neurons.

While much research on avian spatial memory and the hippocampus has been focused on comparative studies and on the adaptive specialization hypothesis, much less attention has been devoted to investigating the development of the hippocampus and whether spatial memory and the hippocampal structure in adults might be affected by environmental conditions during early development. Most existing work on the development of the hippocampus and spatial memory considered the period after young birds had fledged and become nutritionally independent from their parents (Clayton and Krebs 1994a; Clayton 1996, 2001). Such work produced groundbreaking results showing that birds need to use their memory (e.g., in caching and retrieving caches) to initiate the final stages of hippocampal growth and to maintain the enlarged hippocampus commonly found in food-caching birds (Clayton and Krebs 1994a; Clayton 1996, 2001). Little attention, however, has been devoted to development of the brain and the hippocampus during the earlier developmental stages when young altricial birds are entirely dependent on their parents for food. Most of the brain growth occurs before birds become nutritionally independent, and it is crucial to understand whether conditions during early posthatching development can produce strong long-lasting effects on the brain, the hippocampus, and memory.

6.3. Nutritional deficits during posthatching development, spatial memory, and the hippocampus in western scrub jays

In both mammals and birds, the hippocampus is a brain area involved in spatial learning (Sherry and Vaccarino 1989; Hampton and Shettleworth 1996). Spatial learning and spatial memory are likely to be important fitness components in most animal species because they are necessary for remembering large amounts of crucial spatial information such as location of food patches, shelters, nests, and so on. The adaptive priorities in brain development hypothesis predicts that, when there are limited nutritional resources during postnatal development, the hippocampus should be preserved at the expense of less crucial brain areas such as avian song nuclei (Nowicki and Searcy 2005a). Such a pattern should be especially pronounced in food-caching animals that rely on spatial memory to find previously made food caches. Compared with noncaching species, food-caching animals appear to have significantly larger hippocampi relative to the rest of the brain, which reflects the importance of spatial learning and the hippocampus for their survival (Krebs et al. 1989; Garamszegi and Eens 2004b; Lucas et al. 2004). Thus, food-caching species present an especially interesting model to test the adaptive priorities in brain development hypothesis.

Pravosudov et al. (2005) investigated the effects of nutritional deprivation during the posthatching period on brain development in food-caching western scrub jays (*Aphelocoma californica*). Experimental scrub jays were fed 65% of the food fed to control birds until both groups became nutritionally independent and started eating completely on their own at about 2 months of age (Pravosudov et al. 2005). All birds were then provided with unlimited food until they were sacrificed for the brain analyses at about 1 year of age. Food deprivations provided in this experiment resulted in approximately

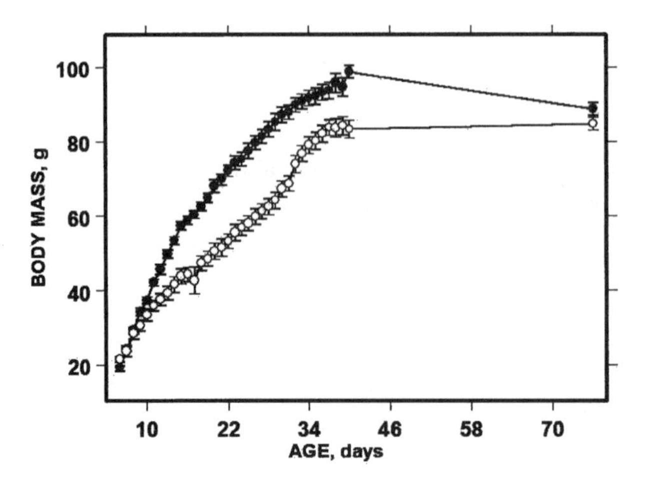

FIGURE 6.1. Body mass of nutritionally deprived (open circles) and non-deprived western scrub jays (filled circles) during the posthatching period. Redrawn from Pravosudov et al. 2005.

23% lower body mass in experimental chicks between beginning of treatment and 30 days of age (fig. 6.1), which is similar to the variation regularly occurring within and between broods in western scrub jays (Ritter 1984). Based on the similarity in body mass variation between the experimental and natural conditions, it might be assumed that the consequences of the experimental reduction in nutrition resemble those in natural nests.

6.3.1. SPATIAL MEMORY

When scrub jays were about 6.5 months old (about 4 months after unlimited food became available to nutritionally deprived chicks), they were tested on a cache recovery test in which the birds were allowed to cache food and retrieve it either 1 or 10 days after caching (Pravosudov et al. 2005). Compared with control birds, scrub jays that were nutritionally deprived during posthatching development performed significantly worse on cache recovery tasks during the two trials with a 1-day retention interval and the one trial with a 10-day retention interval, even though there were no differences between control and experimental scrub jays in the amount of food caching (Pravosudov et al. 2005).

When scrub jays were 8 months old (5.5 months after unlimited food became available to experimental birds), they were also tested on a spatial version of the associative learning task in which the birds were required to memorize 1 of the 38 available locations containing food over five trials. During each trial, birds were first allowed to find a location containing visible food and then search for hidden food in the same location after a 2-hour retention interval. During all five trials and a following probe that contained no food in any of the locations, scrub jays that had been food deprived during the posthatching period showed inferior spatial memory compared with control birds that experienced unlimited nutrition (fig. 6.2; Pravosudov et al. 2005).

Taken together, both the cache recovery and the associative learning tests indicated that nutritional deprivations during posthatching development resulted in impaired spatial memory even after 4–6 months of nutritional rehabilitation with unlimited food. These results suggest that spatial memory deficits must have originated during the posthatching period and were unaffected by unlimited nutritional resources after a certain critical period.

6.3.2. THE HIPPOCAMPUS

One-year-old scrub jays that experienced nutritional shortages during the posthatching period had significantly smaller relative hippocampal volume containing fewer neurons than control birds (fig. 6.3; Pravosudov et al. 2005). The fact that these differences were present 9.5 months after unlimited nutrition was restored suggests that the posthatching period when young scrub jays are dependent on their parents for feeding represents a critical period for brain development and lack of required nutrition may result in long-term

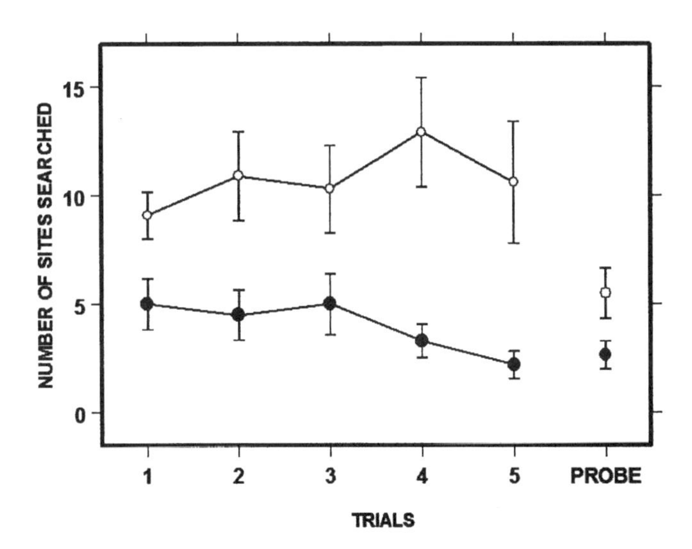

FIGURE 6.2. Memory performance in a spatial version of an associative learning task by western scrub jays that were nutritionally deprived during posthatching development (open circles) and control scrub jays (filled circles). During the probe, no food was available in any of the sites. Fewer number of sites searched indicates better memory performance. Redrawn from Pravosudov et al. 2005.

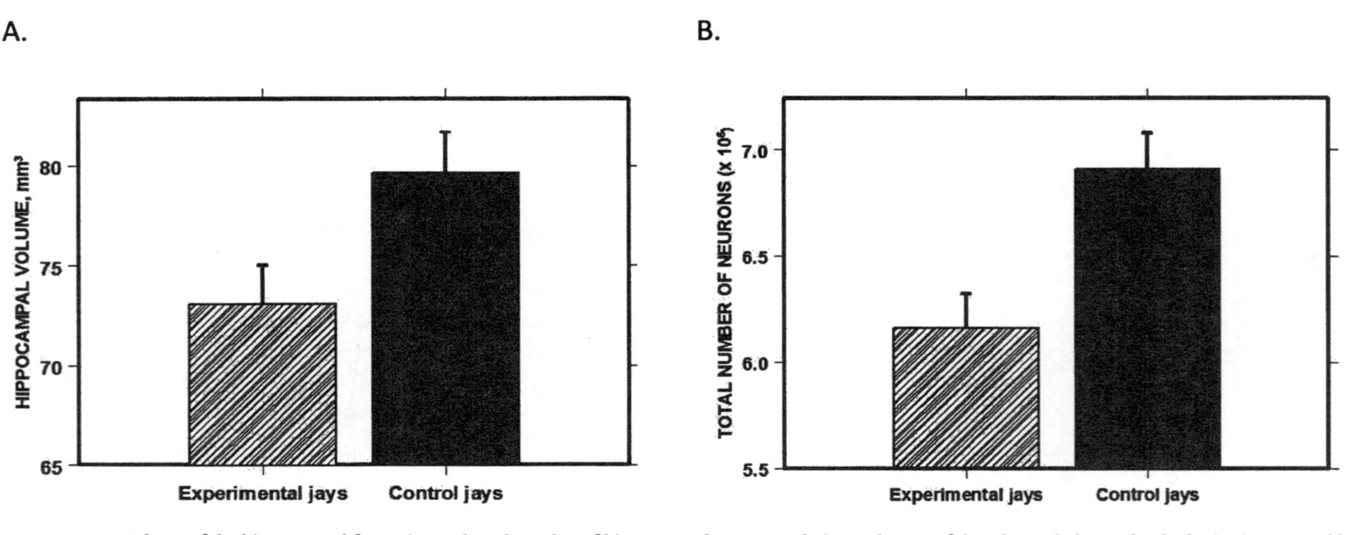

FIGURE 6.3. Volume of the hippocampal formation and total number of hippocampal neurons relative to the rest of the telencephalon and to body size in 1-year-old western scrub jays that experienced nutritional deprivation during the posthatching period (hatched bars) and in 1-year-old control scub jays (black bars). Redrawn from Pravosudov et al. 2005.

structural changes in the hippocampus. Differences in the hippocampal volume and neuron numbers between nutritionally deprived and control birds also explain the differences in spatial memory performance, as spatial memory appears to be dependent on the hippocampus (Sherry and Vaccarino 1989; Hampton and Shettleworth 1996).

Overall, the data on the effect of nutritional stress on memory and the hippocampus in food-caching western scrub jays do not seem to support the adaptive priorities in brain development hypothesis, because these data showed that the hippocampus, a brain area involved in function crucial for survival, was not spared when nutrition was in short supply. One question here is whether the nutritional deprivation in the Pravosudov et al. 2005 study was so strong that it affected the entire brain. If that were the case, other brain areas, as well as behaviors independent of the hippocampus, should also be affected by such treatment. However, analyses of nonspatial learning and other brain areas showed no significant differences between nutritionally deprived and control scrub jays, suggesting that the effects were not general to the entire brain.

6.3.3. NONSPATIAL MEMORY

To test whether nutritional shortages during posthatching development affected nonspatial memory, Pravosudov et al. (2005) used a nonspatial, color version of the associative learning task in which 8-month-old scrub jays were supposed to learn to associate a unique color with a single location containing food among the 38 available locations. Memory for color appears to be hippocampus independent (Hampton and Shettleworth 1996), and therefore, this nonspatial task presents a good test for nonhippocampal brain functions. Scrub jays were given five repeated trials to learn color association and there were no significant differences between nutritionally deprived and control scrub jays in nonspatial memory performance (fig. 6.4; Pravosudov et al. 2005). These results also do not fit well with the adaptive priorities in brain development hypothesis, because they indicated that color memory, which is likely to be less important than spatial memory in food-caching birds, was spared during the nutritional deprivation period, whereas spatial memory was negatively impacted.

6.3.4. OTHER BRAIN AREAS

Pravosudov et al. (2005) also reported that neither total brain mass nor the telencephalon minus the hippocampus volume in 1-year-old birds was significantly different between nutritionally deprived and control western scrub jays. In addition, Pravosudov (unpublished) also measured the volume of the

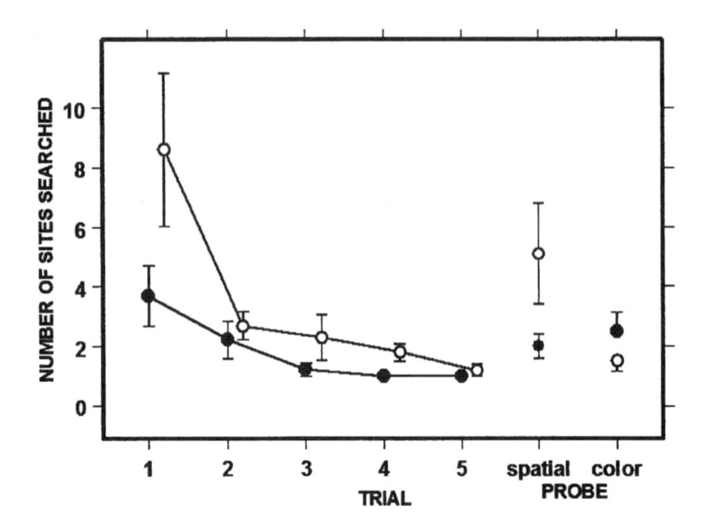

FIGURE 6.4. Memory performance in a nonspatial (color) version of an associative learning task by western scrub jays that were nutritionally deprived during posthatching development (open circles) and control scrub jays (filled circles). The probe with no food in any of the sites was a color dissociation test in which color associated with the rewarded site was moved to a new location. Fewer number of sites searched indicates better memory performance. Redrawn from Pravosudov et al. 2005.

septum, a brain area that is located under the hippocampus and shares some connections with the hippocampus (Shiflett et al. 2002). In mammals, the septum appears to play a role in spatial memory function (Shiflett et al. 2002; H. Smith and Pang 2005), but it is not known whether the septum is similarly involved in spatial memory function in birds. Shiflett et al. (2002) reported that food-caching black-capped chickadees (*Poecile atricapillus*) have relatively larger septums than two noncaching parid species. Shiflett et al. (2002) also reported that septum volume in black-capped chickadees varied seasonally and concluded that the avian septum may be involved in spatial memory. These correlative comparisons, however, do not allow unambiguous conclusions about the septum's involvement in spatial memory function. In western scrub jays, septum volume in 1-year-old birds was unaffected by nutritional deprivation during the posthatching period, whereas spatial memory performance and the hippocampus were negatively affected (Pravosudov et al. 2005; Pravosudov, unpublished data). Statistical power to detect at least a 10% differences in septum volume between experimental and control groups was more than 90%, suggesting that the sample size was sufficient to detect significant differences if they were indeed present. These results have two important implications. First, these data suggest that the role of the septum in spatial memory in birds may be minimal or, at least, not as crucial as that of the hippocampus, because birds with the same septum volume showed significant differences in spatial memory performance (Pravosudov et al. 2005). Second, these data show that, unlike the hippocampus, septum volume is not affected by nutritional shortages during posthatching development.

Overall, these data suggest that the effects of malnutrition during early posthatching development were specific to the hippocampus and spatial memory, whereas malnutrition did not affect nonspatial memory and the entire brain. Pravosudov et al. (2005) did not measure any of the song nuclei in western scrub jays, although based on previous studies of other passerine species (Nowicki et al. 2002a; Buchanan et al. 2004; MacDonald et al. 2006), it is likely that these areas were also affected. It would be important to detect the effects of malnutrition during posthatching development on the hippocampus and avian song nuclei in the same species, but such data are currently not available. The data available on western scrub jays, nonetheless, do not appear consistent with the adaptive priorities in brain development hypothesis, because important brain areas (the hippocampus) were affected by nutritional deprivation during early development while seemingly less important brain areas (septum) were not affected.

Nowicki and Searcy (2005a) argued that western scrub jays might not rely on spatial memory as much as some other food-caching species because they cache relatively little food. I would argue, though, that spatial memory and spatial learning abilities are important for most animals, not just food-caching animals. While western scrub jays do not cache as much as pinyon jays (*Gymnorhinus cyanocephalus*) or Clark's nutcrackers (*Nucifraga columbiana*), they cache food all year round, they cache both animal (perishable) and vegetable (not perishable) food items, and they appear to be capable of hippocampus-dependent episodic-like memory by remembering timing and content of caches in addition to their location (see review in Pravosudov and de Kort 2006). In addition, western scrub jays have one of the largest hippocampi among corvids (Pravosudov and de Kort 2006), which, along with behavioral data, suggests that this species is dependent on spatial memory and the hippocampus. The question therefore remains why such an important brain area is selectively vulnerable to nutritional stress during development.

6.4. Nutritional deficits during postnatal development and the hippocampus in mammals

The effect of malnutrition on cognitive abilities and the brain has been studied quite intensively in mammals, and mammalian studies have identified the hippocampus and spatial learning as especially vulnerable to malnutrition during postnatal development (Dauncey and Bicknell 1999; Bedi 2003). Malnutrition has been reported to have many effects on the hippocampal structure, including the total number of neurons and synapses (Dauncey and Bicknell 1999), but here I will focus on research dealing only with neuron numbers and volumes of brain structures. Whereas numerous negative effects of malnutrition may exist beyond these levels of analyses, changes in neuron numbers and volume may indicate especially severe consequences of malnutrition.

Bedi (2003) reviewed studies investigating the effect of early malnutrition on the number of neurons in different brain areas and concluded that "it is difficult to cause a loss in the total number of neurons by undernutrition during early life" (149). Bedi (2003) also pointed out that, since the studies he reviewed provided quite severe malnutrition, the fact that the number of neurons in most brain areas was not affected was likely not caused by insufficient levels of malnutrition. Most interestingly, Bedi (2003) concluded that only one mammalian brain structure shows a marked response in neuron numbers to malnutrition during postnatal development—the hippocampus. More specifically, Bedi (2003) reported that the granule cells in the dentate gyrus of the adult rat hippocampus showed a marked decrease in total number as a result of reduced nutrition during early postnatal development. Similarly to the western scrub jays (Pravosudov et al. 2005), rats that experienced poor nutrition during early development had a reduced number of hippocampal (dentate gyrus) neurons despite a period of nutritional rehabilitation (Bedi 2003). Also similarly to scrub jays, rats undernourished during early life showed long-term deficits in spatial memory performance (L. Campbell and Bedi 1989; Bedi 1992).

It is interesting that, even though other brain areas may also show deficits in development as a result of malnutrition during early postnatal development, these areas compensate for such deficits once full nutrition is restored (Bedi 2003). In contrast, nutritional rehabilitation does not appear to reverse the negative changes in the hippocampus (Bedi 2003; Pravosudov et al. 2005). Thus, of all the brain structures, the hippocampus seems to be especially vulnerable to malnutrition in mammals (Bedi 2003), while in birds the vulnerability to malnutrition also seems to extend to song nuclei, most notably the HVC (Nowicki et al. 2002a; Buchanan et al. 2004; MacDonald et al. 2006).

6.5. Hippocampus and song nuclei in birds

All available data suggest that both the hippocampus and avian song nuclei are affected by nutritional deprivations during postnatal development despite the fact that these structures appear to have vastly different effects on survival. Song nuclei are unlikely to have a negative effect on survival, whereas the hippocampus is likely to have a strong negative impact on survival. Neither structure appears to compensate in size and/or neuron numbers when unlimited nutrition is restored. The effects of nutritional stress seem to be long-lasting, as nutritional rehabilitation after birds become independent does not reverse these changes. What makes both the hippocampus and song nuclei vulnerable during postnatal development? It appears that these two structures share at least two processes that may make them susceptible to malnutrition during early life: timing of development and reliance on neurogenesis for growth and maintenance.

6.5.1. TIMING OF DEVELOPMENT

Unlike most of the other brain areas, both the hippocampus and song nuclei continue their development for several months posthatching when birds are already nutritionally independent (Clayton and Krebs 1994a; Clayton 1996; Nowicki et al. 2002a; Brenowitz 2004). In food-caching birds, the hippocampus continues to grow after the birds become nutritionally independent and start caching food, and in fact, such late hippocampal growth appears to be dependent on memory-based food-caching experience (Clayton and Krebs 1994a; Clayton 1996). Under laboratory conditions, only birds allowed to experience food-caching and retrieval continue hippocampal growth, whereas birds deprived of such memory-based experiences halt further hippocampal growth (Clayton and Krebs 1994a; Clayton 1996). In migratory white-crowned sparrows (*Zonotrichia leucophrys gambelii*), adults have significantly larger absolute hippocampal volume than juveniles, suggesting that the hippocampus continues to grow in these birds until they are at least 1 year old and that such growth may depend on memory-based migratory experience (fig. 6.5; Pravosudov et al. 2006). Interestingly, the rest of the brain in white-crowned sparrows actually reduces in size between the juvenile and adult age (fig. 6.5; Pravosudov et al. 2006).

Avian brain song nuclei also continue to grow after birds become nutritionally independent, and most development occurs within several months after fledging (Nowicki et al. 2002a). The fact that both the hippocampus and song nuclei develop rather late in life might make them vulnerable to variation in nutrition during the posthatching period.

6.5.2. NEUROGENESIS

Both the hippocampus and song nuclei appear to depend on neurogenesis for their development and maintenance. In food-caching birds, intense

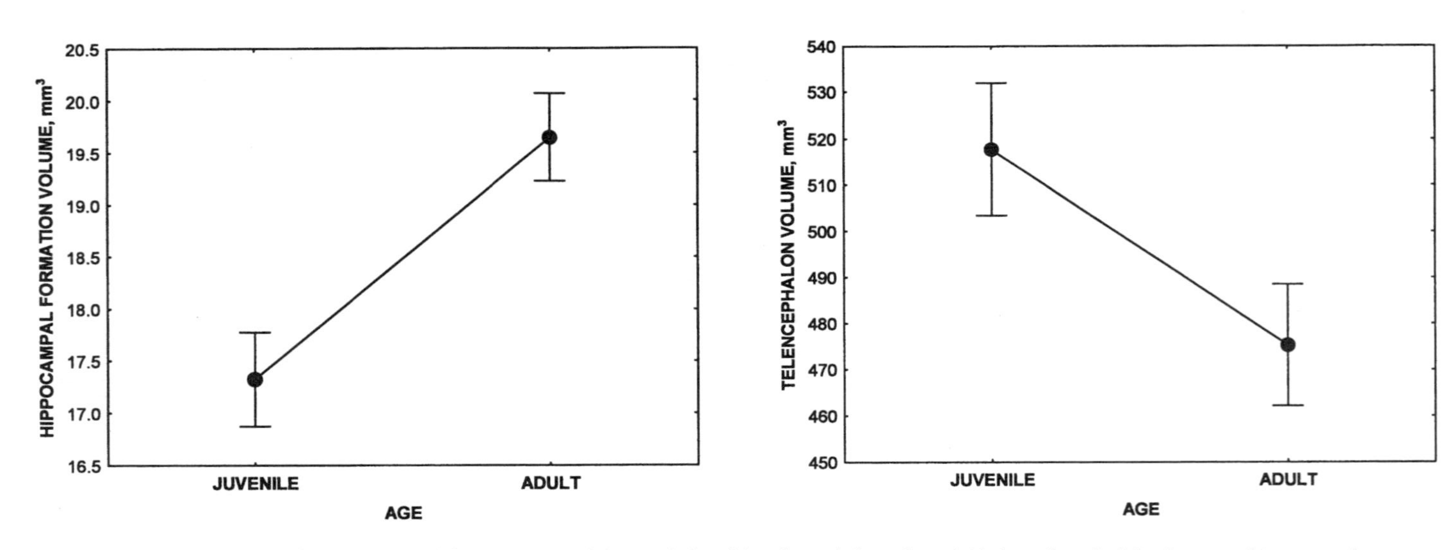

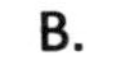

FIGURE 6.5. Absolute volume of the hippocampal formation (A) and the remainder of the telencephalon volume (B) in juvenile and adult migratory white-crowned sparrows (*Zonotrichia leucophrys gambelii*). Juvenile sparrows were sampled after their first migration to the wintering grounds in October, and they were about 4 months old, assuming a hatch date in early June. Data from Pravosudov et al. 2006.

neurogenesis is responsible for postfledging memory-experience-dependent hippocampal growth (Patel et al. 1997). Adult neurogenesis, consisting of cell proliferation and neuron survival, regularly occurs in the hippocampus of both mammals and birds, and it has been linked to memory function (Barnea and Nottebohm 1994; Kempermann 2002; Nottebohm 2002a, 2002b; Leuner et al. 2006). It also appears that the hippocampus in adult animals requires neurogenesis for its normal functioning and maintenance (Kempermann 2002; Nottebohm 2002a, 2002b; Leuner et al. 2006), and reduced neurogenesis and/or increased cell death in the hippocampus may potentially lead to reduction in hippocampal size and/or the number of hippocampal neurons.

Avian song nuclei continue to develop past postfledging by increasing in size and by adding new neurons (Alvarez-Buylla and Kirn 1997; Nottebohm 2002a, 2002b). The HVC experiences especially dramatic growth (Brenowitz 2004). In some bird species, the HVC also experiences large seasonal variation in volume and neuron numbers; the number of neurons may almost double from fall to breeding season (Brenowitz 2004). Such large changes in neuron numbers have been attributed to seasonal changes in neurogenesis and cell death (Brenowitz 2004).

Thus, it appears that for both the hippocampus and the HVC neurogenesis may be essential for initial growth and for further maintenance and normal functioning (memory, song learning). Additionally, new neurons for the hippocampus and the HVC seem to be produced in approximately the same general brain area—the ventricular zone (Alvarez-Buylla et al. 1990b; Barnea and Nottebohm 1994; Scott and Lois 2007). It appears that neurons born on the hippocampal side of the ventricular zone migrate into the hippocampus, whereas neurons born on the opposite side of the ventricular zone (the mesopallium side) migrate into the HVC and possibly other song nuclei. Because these neurons are born in the same area, it is possible that the same population of stem cells residing in the ventricular zone gives rise to both hippocampal and HVC neurons. Because the HVC and the hippocampus both depend on adult neurogenesis to reach and maintain their target sizes, both structures may depend on the same environmental factors that control cell proliferation rates in the ventricular zone and further neuron survival, even though these structures have vastly different effects on survival.

For example, dominance status in mountain chickadees (*Poecile gambeli*) affects cell proliferation rate (the first stage of neurogenesis) in the ventricular zone adjacent to both the hippocampus and the mesopallium (fig. 6.6; Pravosudov and Omanska 2005), and socially subordinate chickadees have reduced cell proliferation rates in both areas.

Another example suggesting that neurogenesis in the hippocampus and the HVC may be linked concerns the findings that testosterone has a positive

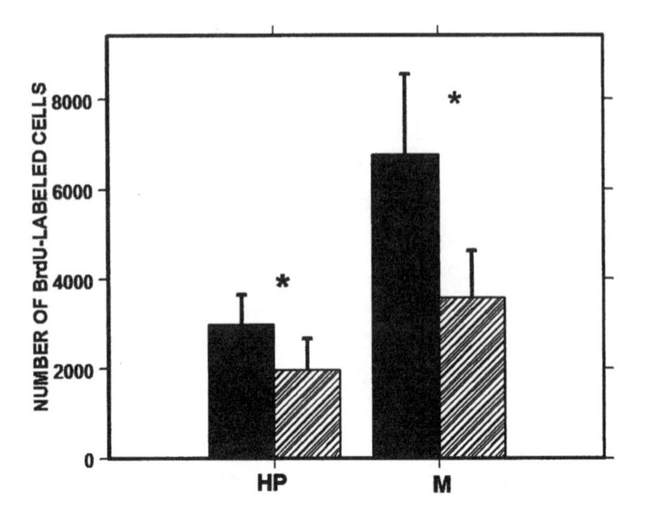

FIGURE 6.6. Cell proliferation rates in the ventricular zone adjacent to the hippocampus (HP) and adjacent to the mesopallium (M) in dominant (black bars) and subordinate (hatched bars) mountain chickadees. Redrawn from Pravosudov and Omanska 2005.

effect on neuron survival in both the hippocampus and the HVC (G. Smith et al. 1997; Louissaint et al. 2002; Spitzer and Galea 2007).

6.6. Does lack of nutrition directly cause changes in the brain?

The adaptive priorities in brain development hypothesis assumes that lack of nutrition directly affects development of the brain and that available resources should be routed first to the more important brain areas (Nowicki and Searcy 2005a). It is not clear, however, whether the effects of malnutrition on the hippocampus and song nuclei system are caused directly by malnutrition or whether the effects of malnutrition are mediated by stress, in which case the stress and mechanisms associated with stress might be directly responsible for the observed effects.

In mammals, it has been shown that almost any kind of developmental stress may negatively impact spatial learning and the hippocampus (Dauncey and Bicknell 1999; Lemaire et al. 2000; Coe et al. 2003; Mirescu et al. 2004; Mirescu and Gould 2006). Examples of such stressful events include maternal separation (Mirescu et al. 2004), malnutrition (Bedi 2003), prenatal stress (Lemaire et al. 2000; Coe et al. 2003). The fact that different kinds of stress may result in similar long-term effects on the hippocampus suggests that the mechanism causing such effects may be the same in all cases. In particular, it has been suggested that such mechanism may be related to glucocorticoid stress hormones, such as cortisol and corticosterone (Mirescu et al. 2004; Mirescu and Gould 2006). It is well known that many stressful events, including food deprivation, result in elevation of glucocorticoids (Harvey et al. 1984; Wingfield et al. 1998; Kitaysky et al. 1999; Lynn et al. 2003). It has been suggested that stressful events during early development may lead to long-term changes in the hypothalamus-pituitary axis (HPA) resulting in significantly higher HPA activity in adulthood (Lemaire et al. 2000; Coe et al. 2003; Mirescu et al. 2004). There is experimental evidence that animals stressed during early development show stronger HPA response to stress in adulthood (Meaney et al. 1996; D. Liu et al. 1997; Anisman et al. 1998; Vallee et al. 1999; Penke et al. 2001; Coe et al. 2003), and it has been suggested that increased HPA activity may suppress adult hippocampal neurogenesis (Gould and Tanapat 1999; Gould et al. 2000; Coe et al. 2003; Mirescu and Gould 2006; Mirescu et al. 2006), which appears to be crucial for hippocampal maintenance.

In western scrub jays, nutritional deprivations during posthatching development resulted in significant plasma corticosterone elevation within at least 10 days after the beginning of treatment (fig. 6.7; Pravosudov and Kitaysky 2006). One-year-old scrub jays that experienced nutritional stress during early development also showed stronger adrenocortical response to a standardized stress test (fig. 6.7; Pravosudov and Kitaysky 2006). Thus, it is possible that the reduction in spatial memory performance and reduced hippocampal volume with fewer neurons were caused by elevated corticosterone levels rather than by lack of nutrition.

Lemaire et al. (2000) exposed pregnant rats to stress by restraining them for 45 minutes three times a day and discovered that such stress to unborn rats resulted in a lifelong reduction in hippocampal cell proliferation rates and in impaired spatial memory performance. Clearly, such an effect of prenatal stress on the hippocampus and spatial memory was not caused by insufficient nutrition.

Maternally separated rat pups showed significantly reduced hippocampal cell proliferation rates in adulthood, but lowering corticosterone in these animals reversed cell proliferation suppression (Mirescu et al. 2004). In rhesus macaques (*Macaca mulatta*), prenatal stress resulted in reduced hippocampal volume, suppressed neurogenesis, and increased HPA activity in 2- to 3-year-old animals (Coe et al. 2003).

Existing evidence suggests that, rather than nutrition itself, stress and adrenocortical response to stress associated either with limited nutrition or with a number of other stressors may produce a direct effect on the brain in general, and the hippocampus and spatial learning in particular. Kitaysky et

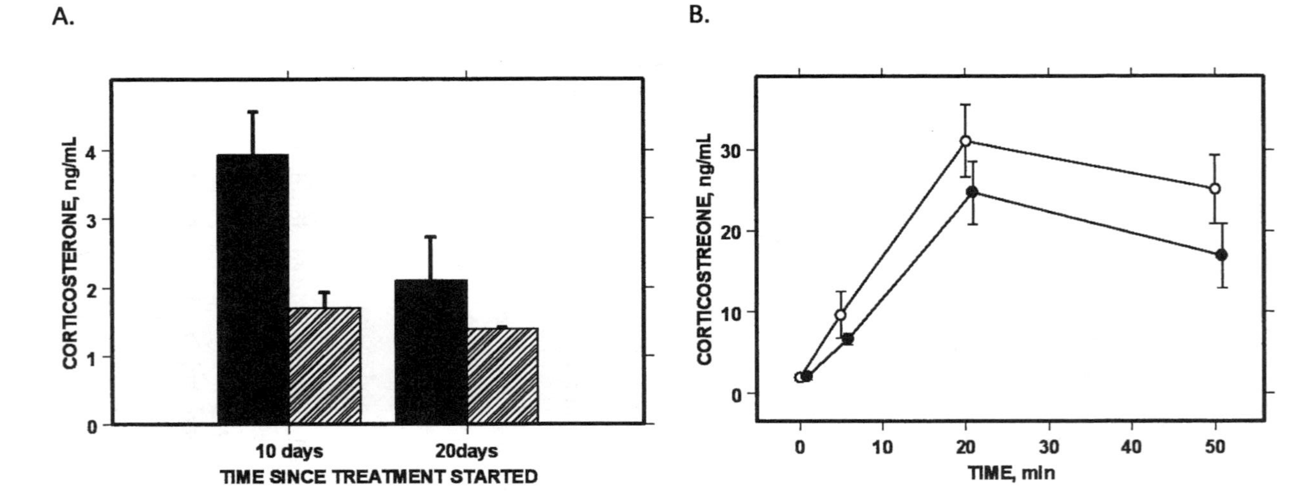

FIGURE 6.7. A. Baseline plasma corticosterone levels in nutritionally deprived (black bars) and control (hatched bars) western scrub jays. B. Corticosterone response to acute stress of handling and restraint in 1-year-old western scrub jays that were nutritionally deprived during posthatching development (open circles) and in controls (filled circles). Redrawn from Pravosudov and Kitaysky 2006.

al. (2003) reported that experimentally elevated corticosterone in posthatching black-legged kittiwakes (*Rissa tridactyla*) resulted in deteriorated cognitive performance in adulthood independently of nutrition. In zebra finches (*Taeniopygia guttata*) experimentally elevated corticosterone levels during early development resulted in reduced song complexity, suggesting that such treatment also affected brain song nuclei (K. Spencer et al. 2003). K. Spencer et al. (2005a) also showed that experimental infection with parasites during posthatching development resulted in simpler songs and reduced HVC volume in adult canaries (*Serinus canaria*). K. Spencer et al. (2005a) reported that parasite infection did not affect baseline corticosterone levels, but they sampled birds only once and it is possible that their negative results were due to the timing of sampling. It is possible that birds elevated their corticosterone levels immediately after the infection but then reduced these levels within a few days (Pravosudov and Kitaysky 2006). Alternatively, parasite infection may have affected free but not total corticosterone levels (Breuner et al. 2006).

It is possible that negative effects of malnutrition on both the hippocampus and song nuclei are a by-product of the trade-offs associated with HPA activity during development. When growing chicks experience a shortage of food, their corticosterone levels are significantly elevated (Kitaysky et al. 1999, 2006). Such elevations might be adaptive, as they cause chicks to intensify their begging calls as a signal to the parents that more food is needed (Kitaysky et al. 2001, 2003). Chicks with elevated corticosterone levels also increase their aggressiveness, which may help them to compete better with their siblings (Kitaysky et al. 2003). When begging is intensified, parents should respond by bringing more food, which would result in lowered corticosterone levels. However, when parents are incapable of bringing more food, corticosterone elevation in hungry young would become chronic, which may result in higher HPA activity and reduced neurogenesis during adulthood (Gould et al. 2000; Coe et al. 2003). If neurogenesis controls growth and maintenance of both the hippocampus and the HVC, both brain structures would be affected simultaneously irrespective of their different significance for survival.

Thus, increased HPA activity caused by stress during early development may be at least one of the mechanisms responsible for suppressed hippocampal neurogenesis in adulthood (Lemaire et al. 2000; Coe et al. 2003; Mirescu et al. 2004). Although some studies provide support to this hypothesis by reporting that stress in early development might result in heightened HPA activity in adulthood in both mammals and birds (Meaney et al. 1996; D. Liu et al. 1997; Anisman et al. 1998; Vallee et al. 1999; Penke et al. 2001; Coe et al. 2003; Pravosudov et al. 2005), other studies did not detect such effects (Mirescu et al. 2004). One explanation might be that most studies looked at the levels of total glucocorticoids without taking into account the concentration of glucocorticoid-binding globulins (Breuner and Orchinik 2002; Breuner et al. 2006), but the negative effects on neurogenesis might be caused by free, unbound glucocorticoids (Mirescu et al. 2004). For example, Lynn et al. (2003) showed that elevation of free but not total corticosterone levels persisted longer after short-term food deprivation in birds. Alternatively, it is possible that reduction in adult neurogenesis in animals that experienced stress during early development is unrelated to HPA activity in adulthood but rather is caused by changes in the numbers and/or properties of stem cells producing neurons (Lemaire et al. 2000).

6.7. Stem cells

It has been hypothesized that stress during early development may result in reduction of the total number of stem cells that produce hippocampal neurons (Lemaire et al. 2000). This hypothesis may potentially explain age-related decline in hippocampal neurogenesis detected in mammals (Lemaire et al. 2000). When rats were subjected to prenatal stress, their neurogenesis showed a decline with age similar to that in control rats with the exception that, at each tested age, rats stressed during early development had significantly lower neurogenesis rates (Lemaire et al. 2000). If neurogenesis links growth and maintenance of both the hippocampus and the HVC in birds, the hypothesis about a reduction in number of neuronal progenitors/stem cells provides a clear and testable prediction that both hippocampal and HVC neurogenesis should show a decline with age and that it should also be reduced by stress (including nutritional stress) during posthatching development.

Another alternative hypothesis is that stress during early development somehow changes properties but not the numbers of neuronal progenitors/stem cells in the ventricular zone, which results in reduced cell proliferation rates and/or reduced neuronal life cycle (Lemaire et al. 2000). Different experiences appear to stimulate more intense neurogenesis either in the hippocampus (memory experience; Patel et al. 1997) or in the HVC (song-learning experience; Alvarez-Buylla et al. 1990a; Brenowitz 2004), and that is why we may see differences between hippocampal and HVC neurogenesis during different seasons and/or life stages. Nonetheless, the properties of the stem cells residing in the ventricular zone may affect neurogenesis rates in both of these areas equally. If stress during early development affects the ability of stem cells to produce neurons, the responses to specific experiences (spatial memory or song learning) may be uniformly reduced.

6.8. Conclusions

The hippocampus is a crucially important brain area that is involved in spatial learning. Spatial learning is most likely critical for survival in most vertebrate species, especially mammals and birds, and it has been implicated as an important component of juvenile mortality (Weathers and Sullivan 1989; Dukas 1998c). According to the adaptive priorities in brain development hypothesis, brain areas critical for survival should be preserved during development at the expense of less important brain areas, such as those involved in sexual displays (Nowicki and Searcy 2005a). Nevertheless, all available data suggest that the hippocampus is one of the brain areas most vulnerable to stress during early development (Bedi 2003; Coe et al. 2003), which argues against the adaptive priorities hypothesis. What is especially intriguing is that many brain areas may show negative responses to malnutrition during early development, but most of them seem to compensate once required nutrition is restored (Bedi 2003). The hippocampus in mammals and the hippocampus and song nuclei, most notably the HVC, in birds, on the other hand, seem to suffer unrecoverable consequences in volume and neuron numbers when animals are exposed to nutritional stress during early development. Importantly, these negative effects appear to be caused by numerous types of stress and not just nutritional stress, which suggests a unifying cause resulting from different stressors.

The fact that brain areas with drastically different consequences for survival such as the hippocampus and the HVC are jointly affected by developmental stress suggests that these brain areas may be dependent on the same mechanism, which may constrain the ability of natural selection to decouple the hippocampus and the HVC (and possibly other song nuclei) during development. Thus, developmental constraints might provide a better explanation for the fact that brain areas both crucial (the hippocampus) and noncrucial (song nuclei) for survival are simultaneously affected by developmental stress. While it is likely that natural selection pressure on spatial learning and the hippocampus is high, it appears that such selection may be focused on parental quality (Hoelzer 1989) and optimal brood size, which would allow optimal nutrition for developing young. In addition, song-learning and spatial-learning abilities appear to be linked during development, and sexual selection for song learning appears to be based on honest signaling, as poor song learning indicates more serious problems, such as deficits in spatial learning, which may be directly linked to survival. Therefore, favoring animals with better song/spatial learning might provide an easier path for natural selection to select for better parents that are able to produce high-quality offspring, rather than resolving the potential developmental constraint issue.

I hope that the ideas presented here provide testable predictions for future research, which should be able to enhance our understanding of brain development and its evolution.

ACKNOWLEDGMENTS

I would like to thank Pierre Lavenex, Alicia Omanska, and Alexander Kitaysky, who collaborated with me on the project investigating the effects of nutritional deficits on memory and the hippocampus in western scrub jays, and Tim Roth for numerous discussions about this chapter. I am also grateful to Reuven Dukas for critical comments. I was supported by grants from the National Science Foundation (IOB-0615021) and from the National Institutes of Health (MH079892 and MH076797).

7 The Cognitive-Buffer Hypothesis for the Evolution of Large Brains

DANIEL SOL

> No one, I presume, doubts that the large proportion which the size of man's brain bears to his body, compared to the same proportion in the gorilla or orang, is closely connected with his mental powers.
>
> CHARLES DARWIN, *The Descent of Man* (1871)

> Brains exist because the distribution of resources necessary for survival and the hazards that threaten survival vary in space and time.
>
> ALLMAN, *Evolving Brains* (2000)

7.1. Introduction

The brain confers to animals the ability to integrate and store information about their environment and use this information for the regulation of behavioral responses. Because these faculties have obvious benefits for the control of critical body functions, it is easy to understand why the brain has evolved (Allman 2000; Dunbar and Shultz 2007a). However, since Darwin (1871) it has been known that some animals have brains that are much larger than appear to be required for usual body functioning (see Jerison 1973). Humans, for example, have brains that are ten times larger than those of wolves of similar body size. The evolution of such large brains is puzzling because growing a disproportionately large brain requires more time for development. Thus, in addition to being energetically expensive (Isler and van Schaik 2006), larger brains incur important fitness costs through delayed reproduction (Deaner et al. 2003; Iwaniuk and Nelson 2003) and lower offspring survival to maturity (Ricklefs 2004). If the development of a larger brain is so costly, why do many animals possess brains than are substantially larger than expected given their body size?

Natural selection should have favored the evolution of large brains only if they provide some sort of selective advantage that counterbalances the costs of their production and maintenance. A priori, these advantages are expected to mostly arise later in life, when the brain is fully functional, and should help compensate for the fitness lost through delayed maturity. One obvious way animals can compensate for the costs of delayed maturity is by lengthening their reproductive life. Thus, it has been suggested that a major benefit of a large brain might be to prolong life span. There is some evidence that species with larger brains do live longer. In mammals, Sacher (1978) found a strong correlation between life span and brain mass, and Allman et al. (1993) and Deaner et al. (2003) confirmed that the brain–life span relationship held when the effect of body size was statistically removed. In birds, Ricklefs and Scheuerlein (2001) used multiple-regression models to show that brain mass is related to rate of aging when included in a model with body mass. While these studies support the association between brain size and life span, the question remains as to why larger brains should be linked with a longer life.

A number of hypotheses have been proposed to explain the brain size–life span relationship (reviewed in Deaner et al. 2003), including a role of large brains in improving precision of large physiological regulation or in governing the growth of the whole body. As noted by Deaner et al. (2003), however, most of these hypotheses involve developmental constraints, implying that large brains can evolve only in ancestral lineages with long-lived life histories. Despite the central importance of constraints in the brain size–life history association, it seems clear that making the brain increase more rapidly than the whole body in spite of substantial costs must require more than the mere possibility of being able to do so. Yet, what are the benefits of developing and maintaining larger brains?

Perhaps the most popular adaptive hypothesis is that a large brain contributes to prolonging life span by providing a "behavioral buffer" against the vagaries of the external world (Allman et al. 1993; Allman 2000), the so-called cognitive-buffer hypothesis. The hypothesis is based on two primary assumptions. First, a big brain confers flexibility in both the utilization of information and the production of behavioral responses (Jerison 1973). Second, this behavioral flexibility allows the possibility of constructing an adapted response to unusual or novel socioecological challenges (Ricklefs 2004). If a large brain buffers the animal against vagaries in the environment, this should reduce extrinsic mortality and prolong life span, providing a mechanism for compensating the costs of delayed maturity.

While the cognitive-buffer hypothesis is widely accepted by many students of brain evolution, the logic of the theory has been questioned by others (Healy and Rowe 2007). Why should a large brain enhance behavioral flexibility? Why should behavioral flexibility provide survival advantages? What evidence do we have that large-brained animals survive better in nature? Critics also highlight findings that do not support the hypothesis. In primates, Reader and MacDonald (2003) found no evidence that large-brained species inhabit environments with more variable climatic conditions, regardless of the temporal scale at which variation was measured. In birds, Nicolakakis et al. (2003) failed to show that species with larger brains are less vulnerable to extinction risk, despite the fact that environmental alterations are considered a major cause of extinction. The existence of some negative results does not invalidate a hypothesis but highlights lack of generality and/or problems in its formulation or testing procedure.

In this chapter, I discuss the general validity of the cognitive-buffer hypothesis, trying to integrate old ideas with new findings into a coherent framework. I begin by reviewing current evidence for the two primary assumptions of the hypothesis: that any increase in brain size must provide some increase in the capacity for adaptive, flexible behaviors, and that this capacity provides individuals with some protection from the vagaries of the environment. After having shown that these assumptions are reasonably well supported, I then focus my attention on describing the current evidence that the brain protects individuals against environmental vagaries and how this buffer effect translates to enhanced survival and life span. My final goal is to highlight some important issues that remain to be integrated into the cognitive-buffer hypothesis and that are likely to be important avenues for future research.

7.2. Assumptions of the cognitive-buffer hypothesis

7.2.1. WHAT IS BEHAVIORAL FLEXIBILITY?

Behavior flexibility can be defined as the general cognitive capacity of an animal to adaptively modify its behavior. Cognition refers to the neuronal processes concerned with the acquisition, retention, and use of information (Dukas 2004a). Because of their cognitive basis, new or altered behaviors do not generally express a prewired genetic program but a plastic reaction norm (B. Robinson and Dukas 1999). Unlike other plastic responses, however, these flexible behavioral responses are generally not context specific, are reversible, and can be generalized to situations that differ partially or totally from those in which they were initially acquired (B. Robinson and Dukas 1999). Thus, behavioral flexibility provides a great adaptive potential to deal with a variety of problems in a variety of ecological contexts.

Novel or modified behaviors arise from a combination of different cognitive processes, including exploration, innovation, learning, memory, and decision making. Although our understanding of the cognitive components of behavioral flexibility is still poor, some of these cognitive processes are known to co-vary in ways consistent both within and across species. The studies of Lefebvre et al. (2002) in birds and those of Reader and Laland (2002) in primates have revealed a consistent positive association between behavioral innovation, learning, and tool use, suggesting that these cognitive abilities are the result of the same evolutionary process. Deaner et al. (2006) used a meta-analysis to show that some primate genera consistently perform better than others across a range of nine cognitive paradigms. Within species, exploration and flexible decision making have been shown to positively co-vary in great tits (*Parus major;* Marchetti and Drent 2000), whereas in pigeons (*Columba livia*) a positive relationship exists between innovative problem solving and social learning (Bouchard et al. 2007). These and other studies provide support for the long-standing belief that animals differ in some kind of general behavioral flexibility.

7.2.2. DO LARGER BRAINS CONFER HIGHER BEHAVIORAL FLEXIBILITY?

7.2.2.1. *Why should larger brains enhance behavioral flexibility?*

Large brains are believed to confer higher behavioral flexibility by enhancing the cognitive capacities underlying behavioral changes (Aboitiz 2001). The rationale of this view is that organs that show the most growth over the course of evolution are likely to perform functions that confer a selective advantage (Healy and Rowe 2007). Because brain size is known to be a heritable character (Jensen 1979; Cheverud et al. 1990; G. Miller and Penke 2007), the assumption is that selection favoring the cognitive demands of enhanced behavioral flexibility drives increases in brain size.

Animals are capable of a variety of sophisticated cognitive skills, such as the use and manufacture of tools, counting objects, remembering specific past events, and reasoning about the states of individuals, many of which were in the past considered to be unique to humans. The ability to perform such apparently complex cognitive acts has been described only in the animals with the largest brains, perhaps reflecting the need of more advanced neural processing (Byrne and Whiten 1988; Lefebvre et al. 2006). However, if large brains are the result of differential increases in parts of the brain (Barton and Harvey 2000; Iwaniuk et al. 2004), why should we expect the whole brain to be associated with behavioral flexibility? There are several reasons. First, behavioral flexibility involves a range of processes, including perception, motor, and cognitive processing (Changizi 2003), and this processing is unlikely to be localized in one brain area (J. Lewis 2006). Second, many specific brain areas are responsible for a number of functions, and at the same time several brain areas may be involved in a given, specific function (Allman 2000). Third, larger brains, relative to body size, tend to contain more neurons than smaller brains, at least in rodents (Herculano-Houzel et al. 2006) and primates (Herculano-Houzel et al. 2007). Finally, many brain component volumes are tightly correlated with whole-brain size, particularly the large parts of the brain, such as the mammalian neocortex and the avian forebrain, that are involved in higher-order and multimodal integration (Timmermans et al. 2001; Iwaniuk et al. 2004). In primates, for example, correlation coefficients between overall brain size and the sizes of the neocortex, cerebellum, and corpus striatum range from 0.96 to 0.99, implying that the sizes of these higher processing centers can essentially be predicted from overall brain size (K. Gibson 2002).

7.2.2.2. *Evidence that larger brains enhance behavioral flexibility*

Whether or not the whole brain correlates with general behavioral flexibility is in any case an empirical issue. However, assembling evidence for a correlation between brain size and direct measures of behavioral flexibility has proven to be problematic, primarily because of the difficulties in defining and measuring behavioral flexibility in a variety of species with a wide range of lifestyles. As a way to overcome this problem, some authors (reviewed in Lefebvre and Bolhuis 2003) suggested quantifying the degree of behavioral flexibility of a species as the frequency of field observations of some of its components (e.g., innovation, deception, learning, and tool use) reported in field naturalists' journals. Using this approach, several studies have found support for an association between brain size and different measures of behavioral flexibility, such as innovation, tool use, tactical deception, and learning (Madden 2001; Timmermans et al. 2001; Reader and Laland 2002; Byrne and Corp 2004; Lefebvre et al. 2004).

Healy and Rowe (2007) have recently criticized the comparative analyses of brain size, highlighting a number of problems with both data collection and the ways in which the data are compiled for the subsequent meta-analyses. While it is clear that there are areas in need of improvement for the comparative study of the brain and behavior, it should be noted that the described studies linking brain size and field measures of behavioral flexibility have been based on reliable measures of brain size (Garamszegi et al. 2002; Iwaniuk and Nelson 2002) and have yielded results that are robust against a number of potential confounding effects and are consistent for a variety of taxonomic groups (Lefebvre et al. 2004). Moreover, the validity of the measures used to quantify behavioral flexibility has been confirmed using the methods of experimental psychology (reviewed in Lefebvre et al. 2004), in which components of behavioral flexibility are assessed by means of performance tests in the laboratory.

Recent human studies provide additional evidence that brain size limits the capacity for developing flexible behaviors. Evidence is accumulating that brain size correlates with a variety of intelligence tests. Witelson et al. (2005) estimated the intelligence of 100 neurologically normal, terminally ill volunteers and found that intelligence correlated with cerebral volume, although the relationship depended on the realm of intelligence studied. Evidence is also accumulating that disease-causing alleles associated with unusually small brains result in deficits in normal cognitive functioning. For example, microcephalin is a gene linked to the evolution of large brains in humans (P. Evans et al. 2005). Mutations in this gene cause primary microcephaly, a disease defined as severe reduction in size of an otherwise-normal brain. This brain reduction is coupled with mental retardation.

7.2.3. WHY SHOULD BEHAVIORAL FLEXIBILITY FACILITATE SURVIVAL IN NATURE?

7.2.3.1. *Behavioral flexibility and response to environmental changes*

While it may be that the payoff for increasing brain size is greater general behavioral flexibility, it is not obvious why a large brain and enhanced behavioral flexibility should facilitate better survivorship of animals in the wild. Animals that must decide for themselves how best to avoid predators, which foods are beneficial and which potentially toxic, and if a habitat is suitable for breeding—decisions that take time and may be risky—should often be more vulnerable than those that do so by means of instinctive behaviors (Reader 2004). Indeed, in nature, many animals continue to thrive with small brains and limited behavioral flexibility. Then, why is there any need to invest in larger brains and enhanced behavioral flexibility?

One important reason is that in nature animals often have to deal with problems for which they must devise novel or flexible solutions (Ricklefs 2004). Behavioral flexibility permits the possibility of producing an appropriate response to unusual or novel ecological challenges (Ricklefs 2004). These adaptive responses may well serve a variety of purposes in a broad array of behavioral contexts (table 7.1), allowing better adjustment to conditions that are novel or likely to change (Plotkin and Odling-Smee 1979).

Consider a population with two genotypes, one that always expresses the same behavior regardless of the context and one that can express different behaviors in different contexts (fig. 7.1). Because organisms often evolve specialized behaviors to adapt to local environmental conditions, we assume that the stereotyped genotype is better adapted (i.e., attains higher fitness) than the plastic one in their ancestral environment, where the behavior matches the environment in which it has been selected. However, if we move the individuals to a novel environment, then the stereotyped phenotype will be displaced

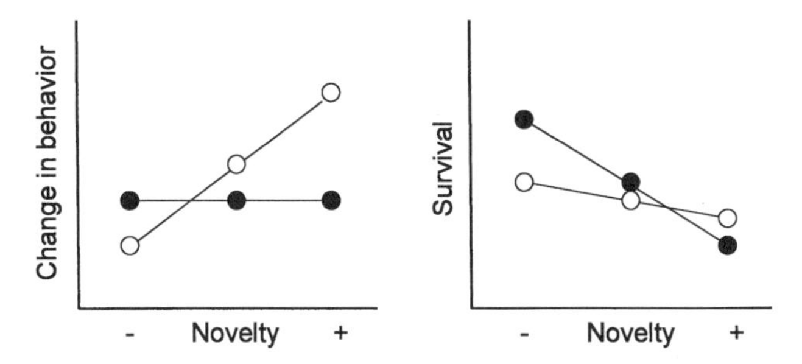

FIGURE 7.1. Schematic representation of the costs and benefits of stereotyped (black dots) vs. flexible (white dots) behavioral responses (see text for details).

from their adaptive peak for proper function in the wild, thereby decreasing fitness. In contrast, individuals that can adjust their behavior to the new situation will experience a less dramatic decrease in fitness. In such circumstances, the strategy to behave flexibly would be favored over the stereotyped one.

The streamside salamander *Ambystoma barbouri*, studied by Andrew Sih and coworkers, illustrates well the disadvantages of stereotypy in changing environments. In this species, individuals exhibit similar high levels of activity in the presence and absence of the predatory sunfish *Lepomis cyanellus* (Sih et al. 2003). A high foraging rate in the absence of predators provides important fitness advantages, as it considerably increases growth rate, yet in the presence of predators the same strategy increases the risk of being killed. Thus, while a stereotyped strategy is advantageous in the absence of predators, a flexible strategy (i.e., adjust activity rates to current levels of predation) is more advantageous when individuals are exposed to changing predatory pressures.

7.2.3.2. *Gathering information*

From the above discussion, it is tempting to conclude that behavioral flexibility is an adaptation that is always useful when the distribution of resources necessary for survival and the hazards that threaten survival vary in space and time. However, it seems clear that not all variable environmental conditions select for such flexibility. Resources that are clumped in space or that change very rapidly often can be better exploited with other strategies, such as high aggressiveness or high mobility (e.g., Goldberg et al. 2001). If an animal does not need to substantially change its behavior to efficiently track and deal with changes in its environment, then it likely does not require a large brain. Consequently, it is important to clarify which features of environmental variation should select for enhanced behavioral flexibility. The theoretical development in this area is currently limited, but it seems clear that one feature particularly important is that variation is in some way predictable. If there are cues that give information on the state of the present or future environment, then we will expect that individuals will evolve means of using such cues to adopt the appropriate behavior (Roff 2002; Warkentin and Caldwell, chapter 10 in this volume). The implication is that environmental information acquisition is fundamental in the development of flexible behavioral responses (Allman 2000; Barton 2004) and hence in the selection for enlarged brains. Selection for improved sensorial function has been suggested to explain the initial increase in the mammalian brain (Jerison 1973), as survival in their crepuscular world demanded improved senses of smell and hearing and thus bigger and more sophisticated olfactory and auditory systems. Barton (2004) and Garamszegi et al. (2002) have also emphasized the central role of the sensory function to explain brain evolution in, respectively, primates and birds.

7.2.3.3. *Producing behavioral responses*

The acquisition and use of information contribute importantly to the ability of animals to gain essential resources or reduce the risk posed by enemies and predators (Balda et al. 1996; Mettke-Hofmann 2000). However, the ability to integrate information to change the behavior is useless if one is unable to express that change (Dukas 1998c). The cognitive processes that appear to be more important for this purpose are innovation, learning, and decision making.

Innovation defines a behavior that significantly departs from the usual repertoire of the animal and that can thus be considered a novel behavioral pattern (Lefebvre 2000; Reader and Laland 2003). The invention of new behavioral patterns is a common way in which animals devise solutions to new problems, as, for example, the incorporation of novel foods into an individual's and population's diet. A well-known example is the opening of milk bottles by English blue tits (*Parus caeruleu*). Innovations can also facilitate development of novel solutions to old problems (Reader and Laland 2003). The potato washing by Japanese macaques (*Macaca fuscata*) is a behavior that was first observed in a young individual in 1953. Three years later, this same macaque developed another novel solution: separating wheat from sand by throwing mixed handfuls into water and scooping out the floating grains (Reader and Laland 2003; see also Kendal et al., chapter 13 in this volume).

Learning can be broadly defined as the capacity of animals to modify or fix new behaviors through experience and is considered an advanced form of reversible plasticity (Dukas 2004a; Dukas and Ratcliffe, chapter 1 in this volume). The fitness benefits of learning have been quantified in the laboratory (Dukas and Bernays 2000) but rarely in the field. However, these fitness advantages are apparent in a number of case studies (table 7.1). For the oys-

FIGURE 7.2. Oystercatchers can adapt and adjust their behaviors according to what type of food is available, thus effectively increasing their food intake rate. Picture courtesy of Xavier Ferrer.

tercatcher *Haematopus ostralegus* (fig. 7.2), learning allows for the adoption of a range of foraging techniques as well as the ability to better differentiate tiny differences in the prey and environments (Goss-Custard 1996; Nagarajan et al. 2002). In southeastern England, wintering oystercatchers mainly feed on mussels (*Mytilus edulis*) either by hammering a hole through the shell on the ventral or dorsal side or by stabbing between the intact valves. An individual oystercatcher will either hammer mussels or stab them but will not do both. This is because individuals must learn the technique from their parents' when they are young, and they appear to have limited capacity to learn more than one technique. By learning mechanisms, however, oystercatchers can adjust their behaviors to the type of mussel most common in their environment (see below), thus effectively increasing their feeding intake rate.

Decision making involves the determination of action based on the known states of relevant environmental features and experience (Dukas 2004a). Animals can improve their fitness by making decisions on how, where, and when to search for prey and by assessing the relative value of different prey types encountered based on nutritional value and handling time. Oystercatchers are good at fine-scale discrimination and decision making (Nagarajan et al. 2002). Individuals that ventrally hammer mussels select ventrally thin mussels that support few barnacles and have brown-colored and ventrally flat shells, which are easier to break. The improvement in the overall net energy intake rate that could be achieved by such a valve thickness discrimination strategy is around 4%. The fitness benefits of decision making are supported in a number of other case studies (table 7.1).

It is important to make the distinction between innovation, learning, and decision making, as they likely involve different cognitive processes and have different ecological consequences, but in practice it is likely that these mechanisms act together in many behavioral responses along with other cognitive processes (e.g., memory) and temperament traits (e.g., neophobia). In the bottle-

TABLE 7.1. Examples of studies reporting fitness benefits of behavioral flexibility

Behavioral domain	Example	Source
	Foraging behaviors	
Adopt a novel food	Killer whales (*Orcinus orca*) have shifted to preying on sea otters after the decline in their traditional preys.	Estes et al. 1998
Develop a novel foraging technique	Black rats (*Rattus rattus*) have learned to open pine cones of Jerusalem pines after most natural forests were cut down.	Terkel 1996
Use tools	Woodpecker finches (*Cactospiza pallida*) have learned to use twigs and cactus spines to pry out food from tree holes and crevices.	Tebbich and Bshary 2004
Hoard food	Food hoarding improves winter survival in tits (*Parus* spp.).	Jansson et al. 1981
Discriminate high-quality food	Grasshoppers (*Schistocerca americana*) that employ associative learning for diet choice experience higher growth rates than individuals that do not.	Dukas and Bernays 2000
Improve handling performance	Chipping sparrows (*Spizella passerine*) learn the relative values of seeds that differ in handling time.	Pulliam 1981
Avoid toxic preys	Blue tits (*Parus major*) learn to avoid toxic conspicuous prey.	Lindstrom et al. 2001
	Antipredator behavior	
Recognize unfamiliar predators	Wallabies (*Macropus eugenii*) can recognize predators by social learning.	A. Griffin and Evans 2003
Avoid novel predators	Moose (*Alces alces*) mothers who lost juveniles to recolonizing wolves in North America's Yellowstone region developed hypersensitivity to wolf howls.	Berger et al. 2001
Respond to novel predator cues	European minnows (*Phoexinus phoexinus*) learn to respond to fish alarm substances recognized as dangerous.	Magurran 1989

	Mate choice	
Flexible mating	In the spotted woodpecker (*Picoides minor*) behavioral flexibility in the mating system is predicted to increase reproductive success when the sex-ratio varies.	Beardsley 2006
Divorce	Seabirds change their mates after a reproductive failure.	Johnston and Ryder 1987
Mate recognition	Females of Japanese quail (*Coturnix japonica*) that have seen a conspecific male mating subsequently show an enhanced tendency to affiliate not only with that particular male but also with males that share his characteristics.	D. White and Galef 2000
	Reproduction	
Decide the best time for reproduction	In blue tits (*Parus caeruleus*) egg-laying date is causally linked to experience in the previous year.	Grieco et al. 2002
Choose the breeding host	Associative learning during oviposition in *Crataegus* or apple hosts can significantly influence the propensity of apple maggot flies (*Rhagoletis pomonella*) to accept or reject these hosts in future encounters.	Prokopy et al. 1982
Respond to nest parasites	Reed warblers (*Acrocephalus scirpaceus*) modify their level of defenses in response to the intensity of cuckoo parasitism.	Brooker et al. 1998
	Social interactions	
Kin discrimination	Long-tailed tits (*Aegithalos caudatus*), cooperative breeders, discriminate between true and fostered siblings through learning processes.	Hatchwell et al. 2001
Resource patch choice	Feral pigeons (*Columba livia*) distribute across food patches according to food availability and competitor number.	Lefebvre 1983
Respond to competitors	Feral pigeons that learn faster may enjoy a competitive advantage in terms of reduced delay in responding to the presence of a conspecific.	Plowright and Landry 2000

opening behavior observed in blue tits, the new behavior is likely to have been initiated by the decision of some individuals to approach and explore the bottle in search of food, followed by innovation and learning processes that should have first allowed and then improved the efficiency of opening the bottle.

7.2.3.4. *Copying behavioral responses*

Although the acquisition of new behavioral patterns is risky and takes some time and energy, these costs can be mitigated by observing or interacting with companions. Animals can learn important things socially. Galef (1991) has shown that the Norway rat (*Rattus norvegicus*) is able to tell what food another rat has eaten recently and to learn from the other rat's experience. It may have obvious advantages for an individual to adopt a new food into its diet when a member of its species has already consumed it and ostensibly will benefit from its ingestion.

Because new or modified behaviors are often developed by one or a few individuals, social learning is critical to spread the fitness benefits of novel behavioral patterns to other members of the population (Lefebvre 2000). The rapid spread of the milk-bottle-opening behavior in blue tits and the potato-washing behavior in macaques has been attributed to a process of cultural transmission (for further discussion of social learning in other animals, see Kendal et al., chapter 13 in this volume). The vertical transmission of learned behaviors from parents (or other relatives) to offspring and the use of public information in habitat choice are two other examples of the advantages of interacting with other individuals in the development of flexible behavioral responses.

7.2.3.5. *Adaptive function of behavioral flexibility*

We still know little about the kind of problems for which large brains and behavioral flexibility are most relevant. In birds and primates, most examples of novel and learned behaviors are found in the foraging domain, yet there is also abundant evidence for the importance of flexible behaviors in predator avoidance and mate selection (table 7.1). It can be argued that if the brain provides some kind of general behavioral flexibility, perhaps the exact ecological context in which it is useful is not so important. Still, it is likely that each of innovation, learning, and decision making is more important in some ecological contexts than in others (Nicolakakis and Lefebvre 2001; Pollen et al. 2007). Identifying the ecological contexts in which behavioral flexibility is more relevant is important because it will inform us what environmental factors may have selected for larger brains in the past, as well as help assess whether selection for larger brains differs between sexes (Lindenfors et al. 2007).

While the adaptive function of behavioral flexibility is well supported by evidence, it is also clear that we need more direct field evidence that changes in behaviors increase fitness of individuals when exposed to novel or changing conditions. Yeh and Price (2004) provide an example of the type of study required. These authors investigated the role of phenotypic plasticity in the establishment of a small population of dark-eyed juncos (*Junco hyemalis*) in a Mediterranean climate in coastal southern California. The breeding season of coastal juncos was more than twice as long as that of the ancestral population, and they fledged approximately twice as many young. This plastic response in the population contributed substantially to the persistence of this population, counterbalancing the effect of the higher mortality that coastal juncos experienced during their first year. The behavioral change studied by Yeh and Price (2004) is unlikely to require high cognitive skills, but the same approach can be used to study more complex behaviors.

The alternative of investigating the fitness benefits of flexible behaviors using field experiments has never been attempted, yet population translocations seem an obvious option (Reznick et al. 1997; Losos et al. 2004). Losos et al. (2004) used a field experiment to study the behavioral response of *Anolis* lizards toward an introduced predator in small islands in the Bahamas. They found that *Anolis* lizards become increasingly arboreal in the presence of an introduced predator. Although this study did not specifically examine the fitness benefits of behavioral changes, it is expected that individuals less exposed to the predator reduce the probability of being killed.

7.3. Predictions of the cognitive-buffer hypothesis

7.3.1. HOW CAN WE TEST WHETHER THE BRAIN BUFFERS INDIVIDUALS AGAINST ENVIRONMENTAL VAGARIES?

The evidence that brain size enhances behavioral flexibility and that this behavioral flexibility is advantageous in the face of novel challenges gives credence to the proposed mechanism underlying the cognitive-buffer hypothesis. However, to demonstrate the validity of the hypothesis it is also necessary to show that animals with larger brains are more successful in dealing with novel or changing ecological challenges than are smaller-brained species. One problem in trying to validate this crucial prediction is that of defining what we mean by novel or altered environmental conditions. First, simple categorizations such as stable versus fluctuating environments are of little use if the environmental factors that vary are not clearly specified. Is habitat unpredictable with respect to food supply? Predators? Both? Second, if the brain functions to solve a wide array of ecological problems, perhaps we should look for situations that simultaneously confront a species with a variety of ecological challenges. Finally, we should look for ecological challenges with obvious fitness consequences, the solution of which will require the development of flexible behavioral responses. We can imagine three situations in which animals confront ecological challenges that meet the above criteria: (*i*) the invasion of novel environments, (*ii*) exposure to habitat alterations associated with human activities, and (*iii*) living in seasonal habitats. Below I present studies that have examined whether large brains are associated with higher success in the face of such changes in the environment.

7.3.1.1. *Novel environments*

The most obvious way to study the function of the brain in dealing with novel circumstances is by examining the survival of species varying in brain size when introduced to new regions (Sol and Lefebvre 2000; Sol 2003; Sol et al. 2005a). A species that is exposed to a novel environment will generally face many novel environmental challenges, and success will depend on whether the invader can rapidly gather information on the environment and develop behavioral responses to these new challenges. If the brain buffers individuals against novel environmental challenges, we should expect a higher success in establishment in large-brained species than in small-brained species. This prediction has been tested with information on human-aimed species introductions. In birds, large-brained species are more likely to be successful in novel environments than are small-brained species, a pattern first described in New Zealand (Sol and Lefebvre 2000) and then globally (Sol et al. 2005a). These patterns are robust against the effect of other factors also known to influence establishment success, including introduction effort and the habitats themselves. Successful invaders are also characterized by a high propensity of innovative behavior in their native ranges, supporting the view that larger brains help birds respond to novel conditions by enhancing their innovation propensity rather than indirectly through noncognitive mechanisms.

Nevertheless, critical to developing a theory is establishing common, repeated patterns in different animal taxa (Marino 2005). In an attempt to test the universality of the brain-invasion association, Drake (2007) explored the role of brain size in the likelihood of establishment in introduced fishes. There was no evidence of a correlation between relative brain size and establishment rate across fish species. Drake (2007) proposed two alternatives to explain this finding. First, perhaps fishes do not exhibit the same degree of cognitive variation as birds, such that differential cognitive abilities among species have relatively little impact on colonizing ability. Alternatively, it might be that the behaviors more relevant during colonization in fishes are those involved in reproduction rather than in feeding, behaviors that perhaps are poorly represented by differences in brain size.

To study the role of the brain in response to novel environments, mammals appear to be more suitable than fish, as they contain the species with the largest brains to ever have evolved. Moreover, the neural basis of their flexible responses is better understood (see section 7.2.2). Thus, mammals provide a crucial challenge to the cognitive-buffer hypothesis. Using a global database documenting the outcome of over 400 introduction events, we have recently shown that mammal species with larger brains tend to be more successful at establishing themselves in novel environments, an effect that cannot be attributed to other factors also known to influence establishment success (Sol et al. 2008). The finding that brain size, relative to body mass, is statistically associated with establishment success in mammals and birds, the animals with the largest brains, provides important support for the cognitive-buffer hypothesis.

7.3.1.2. *Altered environments*

If large-brained animals are more capable of dealing with novel environments, they should also be less vulnerable to changes in their current environment. One way to validate the general importance of brain size and behavioral flexibility for the adaptive response to habitat alterations is to examine demographic fluctuations in populations facing large-scale habitat transformations. The prediction is that populations of big-brained species will be less negatively affected by habitat alternations. Shultz et al. (2005) used long-term demographic data on common farmland birds from the United Kingdom to test this prediction. In the United Kingdom, agricultural intensification, associated with a loss of habitat diversity and simplification of farming systems, has been a major cause of avian population declines, providing a unique opportunity to test the response of animals to dramatic changes in habitat. Two factors, resource specialization and brain size, were significantly associated with species-specific population trends; that is, species using atypical resources and with relatively small brains were most likely to have experienced overall declines. Further analyses of specific brain components indicated that the relative size of the telencephalon, the part of the brain associated with problem solving and complex behaviors, and the brain stem might be better predictors of population trends than overall brain size. These results suggest that flexibility in resource use and behavior are the most important characteristics for determining a species' ability to cope with large-scale habitat changes.

This same study (Shultz et al. 2005) may explain why Nicolakakis et al. (2003) found no relationship between brain size and extinction risk. In the latter study, neither the cause of extinction nor the possibility that the cause might vary geographically was explicitly considered. Behaviorally flexible species might be expected to be more tolerant to habitat loss than less flexible species, but not necessarily more tolerant to human persecution. Owens and Bennett (2000) showed that, in birds, ecological generalism (measured as the number of different habitats used by the species) reduces extinction risk associated with habitat loss but not that associated with persecution and introduced predators. Thus, future tests of the role of behavioral flexibility in extinction risk will need to take into account the specific causes of extinction.

7.3.1.3. *Changing environments*

Temporal and spatial variation in resource availability is often cited as a key ecological factor selecting for the evolution of larger brains. The spatiotemporal-mapping hypothesis argues that the need to exploit ecological resources dispersed in time and space should select for higher cognitive skills and larger brains (Allman 1977; Clutton-Brock and Harvey 1980). In primates, the larger brains of fruit-eating species compared with those of leaf-eating species might result from the higher cognitive demands of exploiting fruits (Clutton-Brock and Harvey 1980), which tend to be distributed more patchily in space and time than are leaves, although other explanations are also possible (Aiello and Wheeler 1995).

Fluctuations in resources are a major feature of seasonal environments, providing a good opportunity to test the role of the brain in dealing with changes. One way to do so is by examining migratory patterns in birds. In temperate regions, winter is particularly hard for birds, as food supplies become more variable and are more likely to fall below threshold levels for survival, forcing many birds to migrate to warmer regions. However, there are species that confront the seasonal changes and manage to remain in the same region for the whole year. A large brain and enhanced behavioral flexibility might alleviate the adverse conditions during cold, harsh seasons and hence could be one of the adaptations facilitating year-round residency in birds (Sol 2003). One well-known cognitive strategy is food caching in corvids and tits, which reduces the variance in food supply between seasons. Another cognitive strategy is exploration and learning, which facilitate an animal's ability to exploit local resources and persist during periods of scarcity. Although migratory birds excel in some specialized cognitive skills, such as long-term memory (Mettke-Hofmann and Gwinner 2003), recent studies show that resident species tend to have relatively larger brains and to show a higher propensity for feeding innovations (Winkler et al. 2004; Sol et al. 2005b). Evidence is also available suggesting selection favoring larger brains in sedentary populations.

In the white-crowned sparrow *Zonotrichia leucophrys*, an increase in brain size has been observed in a subspecies that recently changed from a migratory to a residential strategy (Pravosudov et al. 2007; Pravosudov, chapter 6 in this volume).

The above evidence is correlational, and hence the possibility that the results can be interpreted in a different way must be addressed (Sol et al. 2005b). An alternative explanation is that differences in flexibility between migratory and resident species result from directional selection for reduced behavioral flexibility in migrants (Berthold and Terrill 1991). Such selection is still predicted by the cognitive-buffer theory. The information gathered by migrants as they travel through novel environments may be useful only for short periods, and information relevant to one environment may also expose individuals to risks (e.g., novel predators) in another. Thus, in migratory species learning and innovation may have greater costs than benefits, and genetic programs may be favored over flexible behaviors. However, there is also the possibility that selection for smaller brains in migratory species has nothing to do with cognition. One possibility is that a large brain is energetically too expensive to produce, maintain, and carry by migratory species, some of which travel intercontinental distances (Winkler et al. 2004).

7.3.2. DO LARGE-BRAINED ANIMALS SURVIVE BETTER IN NATURE?

Evidence for a function of larger brains in dealing with the vagaries of the environment provides support for the cognitive-buffer hypothesis. However, to demonstrate the validity of the hypothesis we also need evidence for two additional key predictions. First, if a large brain buffers the animal against vagaries in the environment, this should reduce extrinsic mortality (figs. 7.3b and 7.4). Although the cognitive-buffer hypothesis was first proposed some time ago (Allman et al. 1993), this crucial prediction has received empirical support only recently. The main difficulty has been the lack of good estimates of mortality rates in wild populations. Nevertheless, in the past years an unparalleled amount of data on mortality has been assembled for many wild populations, notably for bird species (Liker and Székely 2005), making it possible to conduct such a comparative test. Using data from birds, it has become clear that there are consistent differences in adult mortality rates across species and that these differences are in part explained by differences in brain size (Sol et al. 2007). As predicted, avian species with larger brains, relative to body size, consistently show lower rates of adult mortality in the wild than species with smaller brains (fig. 7.3a).

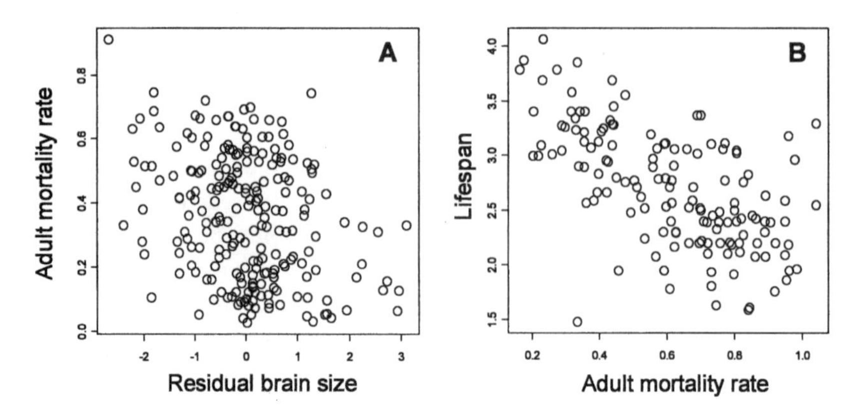

FIGURE 7.3. Relationship between annual adult mortality and (A) brain size, relative to body size, and (B) life span in birds. From Sol et al. 2007.

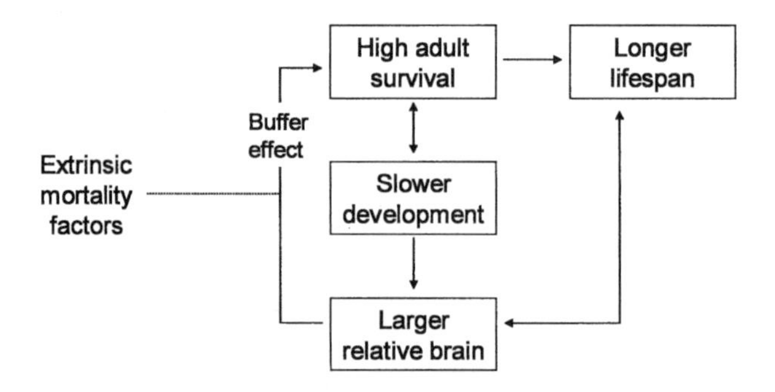

FIGURE 7.4. Schematic representation of the potential role of large brains in the life history strategy of long-lived species.

Second, if a large brain reduces extrinsic mortality (figs. 7.3b and 7.4), this should prolong the reproductive life of individuals. As noted in the introduction, the prediction that species with larger brains live longer is well supported in birds, primates, and carnivores (Sacher 1978; Allman et al. 1993; Ricklefs and Scheuerlein 2001; Deaner et al. 2003). However, it must be noted that the statistical effect is relatively weak in all these groups and nonsignificant within insectivores, bats, and cetaceans (Deaner et al. 2003). One possible explanation is that the cognitive-buffer hypothesis is not as general in explaining the evolution of large brains as previously thought. There are other, alternative explanations, however. First, life span is difficult to quantify, and thus estimates are subject to substantial error, which might detract from our ability to resolve the actual strength of an association between brain size and life span. Second, the relationship between brain and life span is not direct but is, more accurately, the indirect effect of the specific cognition capabilities that enhance behavioral flexibility. Finally, not all long-lived species necessarily require enhanced cognition to properly function in nature (Deaner et al. 2003). Lefebvre et al. (2006), for example, noted that the Columbiformes (pigeons and doves) and Caprimulgiformes (nightjars) are, in relative terms, not much more encephalized than lineages with much faster life history strategies such as the Galliformes and Ratites.

7.4. Synthesis

The cognitive-buffer hypothesis provides a mechanism for understanding how large-brained animals compensate for the costs associated with delaying sexual maturation. When a species has a long-life strategy, individuals should gain greater fitness benefits by investing in adult survivorship rather than in current reproduction (Ricklefs and Wikelski 2002), implying that selection will favor the evolution of adaptations that increase survival over early reproduction. The cognitive-buffer hypothesis suggests that a large brain should play such an adaptive role. Several studies have now found support for the assumptions and predictions of the hypothesis (sections 7.2 and 7.3). These studies show that a large brain confers advantage to individuals in the form of flexibility in the utilization of information and that the resultant behavioral flexibility enables them to deal with novel environmental challenges more successfully. If a large brain plays such a buffer function, this should increase survival rates and favor a longer reproductive life, which may well compensate for the costs of delayed reproduction. Although still limited, there is evidence that a large brain is associated with reduced adult mortality (Sol et al. 2007) and a longer reproductive life (Deaner et al. 2003), suggesting that the evolution of large brains in long-lived species might be part of an adaptive life history strategy to maximize individual fitness.

One of the most attractive qualities of the cognitive-buffer hypothesis is that it builds a bridge between the common viewpoint that larger brains enhance the cognitive capacities underlying behavioral flexibility (Darwin 1871; Jerison 1973) and the view that behavioral flexibility is a useful strategy to deal with a variety of ecological problems (Klopfer 1962; Mayr 1965; Morse 1980). In doing so, the hypothesis explicitly recognizes that the selective pressures favoring larger brains are likely to be diverse and to vary according to the cognitive level of the animal, helping integrate many of the hypotheses proposed for the evolution of large brains (fig. 7.5). Indeed, the clearest evidence available to date for the role of a large brain in response to change involves an animal's adaptability to a broad diversity of changes (see section 7.3.1).

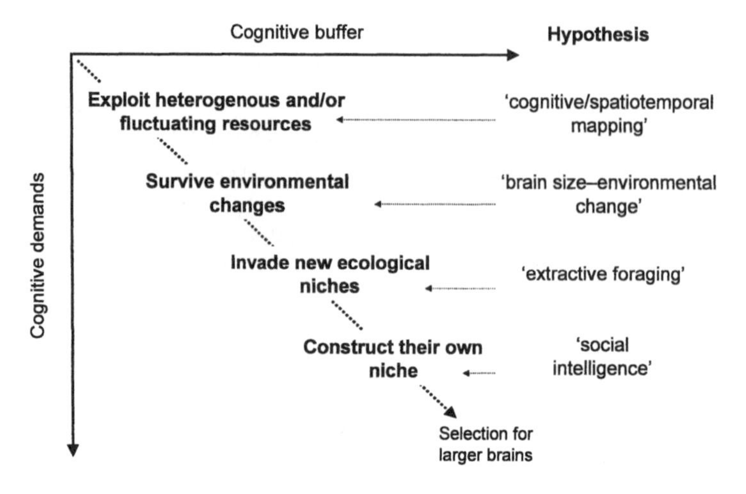

FIGURE 7.5. Scheme describing several general processes favoring the evolution of larger brains through an increase in cognitive skills and buffer mechanisms against environmental challenges. Several classic hypotheses on brain evolution are represented on the right side, to illustrate the generality of the cognitive-buffer hypothesis.

Another attractive quality of the cognitive-buffer hypothesis is that it integrates brain evolution within life history theory. The evolution of large brains and complex behaviors is rarely considered within a life history framework (but see Deaner et al. 2003; Ricklefs 2004). However, the existence of conflicting benefits and costs resulting from limited time and energy is clearly a life history issue (Ricklefs 2004). A life history perspective is essential to understanding brain evolution because it forces us to consider large brains as part of a more general adaptive strategy that allows some long-lived species to maximize individual fitness.

7.5. Avenues for future research

With at least some components now well developed, the cognitive-buffer hypothesis offers a good framework from which to build a more general theory of brain evolution. Below I highlight four important aspects of the theory that remain poorly understood and that can be addressed within the framework of cognitive ecology.

7.5.1. THE ROLE OF BODY SIZE

Body size is expected to play a central role in the cognitive-buffer theory, both because of its allometric relationship with absolute brain size (Deaner and Nunn 1999) and because body size itself is another factor that buffers individuals against environmental vagaries (J. Brown and Sibly 2006). Despite much effort devoted in the past, we still do not know why the brain correlates with body size. The most popular explanations are the "switchboard model" and "traffic maintenance hypothesis" (Byrne and Corp 2004). Both hold that at any given level of cognitive processing larger animals require larger nervous systems to coordinate their larger bodies. Thus, most studies correct for allometric effects by calculating encephalization quotients or residuals of log-log regressions of brain size versus body size. The main problem with statistically correcting for body size is that it assumes that there is no systematic relationship between cognitive abilities and body size. However, this can be problematic if larger species are typically smarter, as recently suggested by Deaner et al. (2007).

7.5.2. COSTS OF PARENTAL CARE

The role that parental care plays in the transition to possessing larger brains and more elaborated learning capacities remains poorly understood. A priori, parental care seems to be necessary for large-brained species, given the relatively undeveloped and vulnerable conditions of their typically altricial offspring. Moreover, parental care is essential if the youngsters are to develop the neural architecture and contingent skills necessary to survive in less risky conditions. Lefebvre et al. (2006) argue that the larger brain in altricial birds relative to precocial ones is due to extended parental care, which releases the offspring from the need for a functional forebrain. At the same time, providing parental care might require increased cognitive demands, favoring the evolution of larger brains. Gittleman (1994) found some evidence for a link between brain size and parental care: across carnivore species, more parental investment accompanies larger brains in females.

However, life history models suggest that, in long-lived species, parents should not invest in parental care if this entails a risk for their survival, a prediction confirmed in empirical studies (Roff 2002). The cognitive-buffer hypothesis offers three possible ways to resolve this apparent paradox. First, a large brain could serve to reduce the variance in the quantity and quality of the food acquired by the parents in the face of changes, reducing the costs of parental care while increasing offspring survival. Second, species with large brains may be better able to assess and choose where to raise their offspring, a decision that can affect offspring survival (T. Martin 2004). In some large-brained species of birds, such as parrots, woodpeckers, and raptors, parents tend to nest in secure places such as holes and cliffs, perhaps also contributing to a reduction in the cost of parental investment. These ideas remain to be tested. Third, the need for extra parental care in large-brained animals could be compensated for if both members of the pair provide such care (Pitnick et al. 2006). Current evidence in birds suggests that the demands of offspring have a major influence on whether a species exhibits biparental or uniparental care (reviewed in Székely et al. 2007). The high prevalence of pair-bonded monogamy and cooperative breeding among large-brained animals (Emery et al. 2007; Shultz and Dunbar 2007) is consistent with the view that the coordinated efforts of parents and/or family members are key in the evolution of large brains (but see Dunbar and Shultz 2007a), although more direct evidence is desirable.

7.5.3. THE DELAYED-BENEFITS HYPOTHESIS

To further develop the cognitive-buffer hypothesis, not only do we need more evidence for the existence of an association between brain size and life span, but we also need to address the intriguing possibility that a longer life itself contributes to selecting for larger brains. This idea, known as the delayed-benefits hypothesis (Deaner et al. 2003), is based on two observations. First, animals with longer lives are more likely to be exposed to changes during their lifetimes and hence gain greater benefits from information acquisition and flexible behaviors. Eliassen and colleagues (2007) considered a model that examined how foragers may benefit from utilizing a simple learning rule to update estimates of temporal changes in resource levels. The model shows that as life expectancy increases, learners invest more in information acquisition and exhibit better foraging performance as resource levels change through time.

The second observation behind the delayed-benefits hypothesis is the fact that long-lived animals have more time to learn important behaviors from their parents or other individuals, better preparing them to survive in nature at adulthood (Lefebvre et al. 2006). Large-brained species show remarkable improvements in behavioral performance with experience and age, which may be attributed to the acquisition of information and behavioral skills rather than physical development (Wunderle 1991). The improvement in these skills parallels a decrease in the probability of dying, suggesting that these learned behaviors are critical for survival (Wunderle 1991). If adult survival critically depends on learning some advanced skill, such as a complex foraging technique, having a large brain can still be adaptively advantageous even if the resource is stable and abundant.

The possibility that a large brain both selects for and is selected by a long life span would provide, if confirmed, additional evidence that a large brain is part of the life history strategy of some species. The implications for the evolution of large brains would be important. It would imply an accelerated brain increase in lineages characterized by long lives and high cognitive lifestyles, such as that observed in some lineages (e.g., primates, cetaceans, crows).

7.5.4. AUTOCATALYTIC PROCESSES

Implicit in previous sections is the recognition that the cognitive-buffer hypothesis would benefit from a better understanding of the nature of the selective forces favoring larger brains. Large-scale changes in the environment have likely been a powerful force in the evolution of enlarged brains, yet few studies have tried to identify the nature of these changes. In one of these few studies, Ash and Gallup (2007) suggested that early humans might have developed larger brains as they adapted to colder climates. Specifically, they found that, as the distance from the equator increased, north or south, so did brain size. The authors suggested that a key environmental trigger to the evolution of larger brains was the need to devise ways to keep warm and manage the fluctuations in food availability that resulted from cold weather.

There is nevertheless the intriguing possibility that evolution of larger brains would have been driven not only by the properties of the environment that an ancestral species finds itself in but also by the species ability to modify its own ecological niche. In their behavioral drive hypothesis, Wyles et al. (1983) suggested that, by expanding the brain's powers, the evolution of large brains would enable animals to exploit niches not readily available to smaller-brained animals. The invasion of these novel niches would in turn expose individuals to novel selective pressures, which would promote further increases in brain size. Wyles et al. (1983) suggested that this has led to an "autocatalytic process" that has accelerated the evolution of larger brains in some lineages, such as that of humans. A similar argument is the basis of two important hypotheses of brain size evolution: the extractive foraging hypothesis (Parker and Gibson 1977) and the technical intelligence hypothesis (Passingham 1975), both of which argue that the need for sophisticated technical skills to exploit resources or avoid predators would have been a major force selecting larger brains. The existence of an autocatalytic process in brain evolution has never been explicitly studied, but recent findings suggest that large-brained lineages have experienced a more extended evolutionary diversification (Nicolakakis et al. 2003; Sol 2003; Sol et al. 2005b; Sol and Price 2008).

7.6. Summary

Most life history thinking on brain evolution has been concerned with constrained evolutionary responses associated with the allocation of limited time and resources. While developmental constraints are likely crucial for adequately explaining the brain size–life history association, the possession of a large brain as an adult also requires some sort of benefit that counterbalances the costs of a longer developmental period. The cognitive-buffer hypothesis is the most general explanation for the benefits of the evolution and development of large brains, proposing that a major advantage of a large brain is to produce behavioral responses that protect the animal from the vagaries of the environment.

Mounting evidence supports the view that a large brain is representative of a more general adaptive strategy that buffers some long-lived species against vagaries in the environment and reduces mortality and lengthens life span. The buffer function of the brain has the potential to generate "autocatalytic" and positive-feedback processes that, although still not well understood, could accelerate brain evolution. Thus, the cognitive-buffer hypothesis offers a general explanation about why brain size has increased so quickly and so many times in independent lineages, including our own.

ACKNOWLEDGMENTS

I thank Reuven Dukas and John Ratcliffe for the invitation to contribute to this book, John Ratcliffe for exhaustive revision of early versions of the chapter, Robert Deaner and Robin Dunbar for sending papers of their work, Xavier Ferrer for lending me the oystercatcher picture, and Louis Lefebvre, Simon Reader, Trevor Price, Gray Stirling, Tamas Székely, Julie Morand-Ferron, Richard Duncan, Phill Cassey, Tim Blackburn, and Andras Liker for fruitful discussions over the past years. This work was supported by a Ramón y Cajal fellowship and a Proyecto de Investigación (CGL2007-66257) from the Ministerio de Educación y Ciencia (Spain) and was partially written during a stay at the University of Newcastle (New South Wales, Australia) financed by DURSI (Generalitat de Catalunya).

PART III

Decision Making

MATE CHOICE AND PREDATOR-PREY INTERACTIONS

8 Cognitive Mate Choice

MICHAEL J. RYAN, KARIN L. AKRE &
MARK KIRKPATRICK

8.1. Introduction

One of the most important decisions an animal makes is choosing an appropriate mate, and females often do this by assessing male sexual displays. Sometimes the females discriminate between conspecific and heterospecific males, and sometimes they choose from among a pool of conspecifics. Although these two processes result in species recognition and sexual selection, they are essentially the same behavioral process (M. Ryan and Rand 1993; Phelps et al. 2006). Here we primarily address mate choice among conspecifics, but the basic concepts extend to mate choice among species. We will also consider the canonical case of displaying males and choosing females, but the same principles apply when there is male choice of females and when both sexes choose simultaneously.

Mate choice can generate sexual selection and it genetically isolates species from one another; thus, mate choice is an intricate part of several fields in evolution and behavior. Mate choice can also be viewed as a problem in animal communication (M. Ryan and Rand 1993); senders transmit signals that have a probabilistic influence on the behavior of the receiver. Here we more generally consider mate choice in the context of interactions between signalers and receivers. Mate choice is a function of the receiver but is inextricably linked to the signal. Thus, our discussion of mate choice addresses issues relevant to both sender and receiver.

Mate choice is a cognitive process. Receivers perceive signals in the environment through the lens of their sensory processing machinery and then evaluate this information using a set of decision-making rules before acting upon it. Each point along the path of a mate choice is governed by a receiver's cognitive abilities and constraints. We follow this course of events as it occurs during a mate choice, beginning with detection and perception, and then moving to evaluating information and acting on it. Detection and perception are the initial steps in managing the information that could influence a mate choice decision. An individual's sensory system filters what information is available in the environment, and how that sensory information is then perceived sets off a chain of cognitive processing that makes up a female's evaluation of the information. We will consider intervening variables in this deceptively simple linear chain of reactions. We do not thoroughly review each of these components, but instead we consider topics that seem relevant, have not received sufficient attention, and might be ripe for integration into more mainstream studies of mate choice.

We will illustrate some of these issues with examples from the acoustic domain and others from the visual domain, using whichever data sets are most compelling. It should be obvious, however, that these principles are general to all modalities of communication and could influence mate choice based on acoustic, visual, olfactory, electrical, or tactile cues.

8.2. Detection and perception

Let us begin by considering mate choice in its simplest situation, a two-way interaction between a sender and a receiver. In his study that launched information theory, Claude Shannon (1948, 379) stated: "the fundamental problem of communication is that of reproducing at one point either exactly or approximately a message selected at another point." In mate choice, as in all communication, information transmitted between the sender and receiver must traverse a noisy channel where information can be lost (fig. 8.1). A fundamental problem for all communication is to enhance information transfer in a noisy world. This problem is solved by adaptations of both the sender and the receiver, and both must be considered to understand the interaction. We will illustrate these issues with examples in the acoustic domain, as did Shannon, as they are more numerous and detailed. More recently, analogous studies in the visual domain have become available (e.g., Seehausen et al. 1997; Persons et al. 1999; Endler et al. 2005; Cummings et al. 2006; G. Rosenthal 2007).

A female must perceive and detect a male's sexual display against an environmental background: noise in a forest, colors on a reef, and odors in the air. Male display traits should be conspicuous to the female in the relevant environment. The degree to which a signal is conspicuous is a product of the signal's contrast against the environment and the biases in the female's sensory system. We also expect females to have sensory and cognitive tools to detect and perceive these signals in that same environment. These general expectations apply to every mate choice: how do senders make their signals conspicuous and how do receivers increase their ability to perceive and detect them?

In a variety of animals, most notably many insects, frogs, and birds, mate choice decisions are based on acoustic signals. In most cases this occurs in a

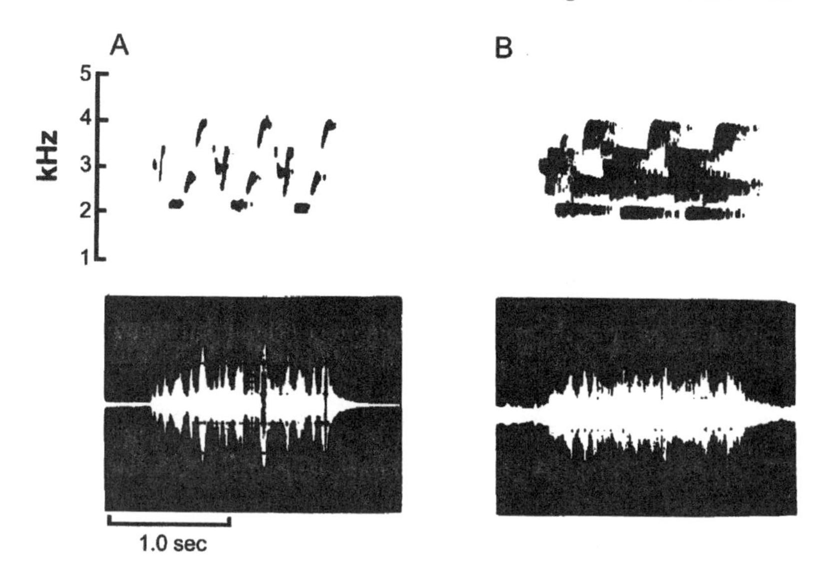

FIGURE 8.1. The song of a Carolina wren (top, sonograms; bottom, oscillograms) recorded (A) 10 m and (B) 50 m from the singing birds. Redrawn from D. Richards and Wiley 1980.

world where sound is degraded and attenuated by the habitat, and in which signals compete with abiotic noise and sound generated by both heterospecifics and conspecifics. It is a formidable challenge for senders to transmit signals that are salient to receivers over socially significant distances and for receivers to be able to parse these signals from the noise. The perception and detection of signals that are critical to mate choice do not occur unimpeded.

8.2.1. ENHANCING SIGNAL CONTRAST

Adaptations to enhance acoustic signal transmission are well known (Marler 1955; Morton 1975; Wiley and Richards 1978; Slabbekoorn and Peet 2003; Brumm and Slabbekoorn 2005). Habitats vary in how they affect signals, which results in some general patterns of signal evolution. For example, birds calling on the forest floor use lower-frequency calls than birds calling in the canopy; birds in open fields are more likely to use amplitude-modulated signals, whereas forest birds tend to use tonal signals (Morton 1975; see also Hunter and Krebs 1979). Birds can also use frequencies that avoid noise from insects (M. Ryan and Brenowitz 1985; Wiley 1991) or anthropogenic sources (Slabbekoorn and Peet 2003). In many taxa background noise influences a facultative shift in signal amplitude, known as the Lombard Effect (Lombard 1911; e.g., fish: Zelick et al. 1999; birds: Brumm and Todt 2002; Brumm 2004; Cynx et al. 1998; monkeys: Sinnot et al. 1975; whales: Scheifele et al. 2005;

humans: Tonkinson 1994; Winkworth and Davis 1997). Noise might also inhibit acoustic signaling in birds (Lengagne and Slater 2002), and some frogs have the ability to call in small gaps of silence between bursts of noise (Zelick and Narins 1985).

Interspecific competition can cause a diel shift in calling activity in some insects (Greenfield 1988, 2005), and intraspecific competition can cause some birds (e.g., Ficken et al. 1985) and frogs (e.g., Schwartz 1987) to interleave call notes with those of neighbors. Thus, there is a considerable amount of data allowing some understanding of how acoustic signals evolve to enhance detection and perception by the receiver. All of these general principles should apply specifically to sexual signals, and indeed, much of the empirical support is from studies of these types of acoustic signals. As mentioned above, there are analogous studies in the visual domain (reviewed in G. Rosenthal 2007).

8.2.2. ENHANCING SIGNAL DETECTION

Evolution of adaptations to noisy environments is not restricted to signals. There is also a suite of receiver adaptations that enhance detection and perception of acoustic signals. This begins with the transduction of the sexual display into a neural code that can be evaluated by the receiver during a mate choice. The simplest mechanism to enhance signal detection is when the tuning of the peripheral auditory systems matches the frequencies that characterize the species' mating signal, as is common in insects and frogs (Gerhardt and Huber 2002; Brumm and Slabbekoorn 2005). The more tuned the system, the more noise it rejects. The tuning can also be influenced by the environment. Witte et al. (2005) showed that the "noise reject" features of a cricket frog's peripheral auditory system are enhanced in populations in which signal detection is more difficult due to habitat acoustics.

Even with the most precisely tuned filters, noise will overlap signals. Noise can prevent a signal from being detected and perceived by a receiver due to signal masking (reviewed in Brumm and Slabbekoorn 2005); thus, mate choice can be obscured under noisy conditions. A more interesting phenomenon is when noise changes the rules of choice, as it does in female hourglass frogs (*Hyla ebracatta*). Wollerman and Wiley (2002) showed that in the absence of noise females preferred lower-frequency mating calls; at moderate noise levels they did not discriminate between call frequencies; and at higher noise levels they preferred the modal frequency of the population. Thus, it appears that, as the task of call discrimination becomes more difficult, the females switch from evaluating the quality of conspecific males to the more conservative task of ensuring choice of the correct species.

Noise need not be an insurmountable barrier to mate choice. If a signal is loud enough, it is released from masking. How loud it needs to be, in birds at least, depends on its frequency: higher frequencies need to be louder to be emancipated from masking. But exceptions to this general rule suggest that there are adaptations for perception and detection of social signals. In the parrot *Aratinga canicularis*, for example, release from masking for frequencies in social calls occurs at a lower amplitude than it does in other birds (T. Wright et al. 2003). Although these are not sexual signals in parrots, this finding identifies a fruitful area of investigation for cognitive adaptations for mate choice.

Another form of noise is the acoustic signals from other conspecifics. Sexual signaling in the acoustic domain often occurs in social aggregations or choruses; thus, animals must be able to parse signals of specific individuals from a complex acoustic background. In reference to human psychoacoustics, Bregman (1990) speaks of this general problem as auditory scene analysis, and Cherry (1953) refers to the specific challenge of assigning an auditory stream to a single individual as the "cocktail party effect." Although social aggregations of signaling animals share some of these same challenges with humans at cocktail parties, only recently has the concept of auditory scene been applied to animal choruses (reviewed in Hulse 2002; Bee and Micheyl 2008).

There have been some detailed studies of mechanisms that allow animals to assign acoustic signals to specific individuals (i.e., auditory streaming), especially in birds and monkeys (Bee and Klump 2004, 2005; Fishman et al. 2004). Studies show, for example, that emperor penguins can identify calls of offspring when mixed with other penguin calls (Aubin and Jouventin 1998), and finches can parse the songs of conspecifics mixed with heterospecific songs (Benney and Braaten 2000). In addition, Schwartz and Gerhardt (1989) suggested that the spatial origin of the calls of gray treefrogs (*Hyla versicolor*) can facilitate auditory streaming. On the other hand, H. Farris et al. (2002, 2005) showed that túngara frogs (*Physalaemus pustulosus*) are not able to correctly assign components of the same call to individuals using only spatial cues, and Bishop showed that in painted reed frogs (*Hyperolius marmoratus*) preferences for call parameters that emerge in two-choice tests are not exhibited in four-choice tests (e.g., Bishop et al. 1995). These last two studies suggest that social communication imposes serious constraints on a receiver; we cannot assume that adaptations always arise to allow receivers to successfully parse signals. On the other hand, there could be context-dependent use of information, such that as density of signalers, complexity of the chorus, and distance to the source change so does the relevancy of different types of information. Although there have been numerous studies of mate choice in acoustically chorusing species (Andersson 1994; Greenfield 2005), there has been little analysis of how this occurs in complex auditory scenes.

8.2.3. CATEGORICAL PERCEPTION

Once the sexual display is encoded by the receiver's sensory system, the neural code needs to be interpreted. In most cases it appears that stimuli can vary continuously and are usually perceived as such: louder-softer, brighter-darker, hotter-colder. There are, however, well-known exceptions in humans in which continuous variation in stimuli, especially in color, speech, and faces, is perceived categorically (Harnad 1987). Categorical perception can be identified by two components, labeling and discrimination. In the first, continuously variable stimuli on the same side of a border are labeled as being in the same category. In the second, the ability of receivers to discriminate between stimuli on different sides of the border, in different categories, is strong, while the ability to discriminate between stimuli of the same magnitude of difference on the same side of the border, within a category, is weak.

In a few cases, animals have been shown to exhibit categorical perception of human phonemes (e.g., Kuhl 1981; Kluender et al. 1987). There are several cases in which animals are known to show categorization of their own signals: syllables in songbirds (D. Nelson and Marler 1989), ultrasonic vocalizations in rat pups (Ehret and Haack 1981), and mating calls and bat echolocation calls in crickets (Wyttenbach et al. 1996).

We do not know if categorical perception of signals used in mate choice is common. Preference functions for simple signal parameters, such as pulse number (Ritchie 1996), call duration (Gerhardt et al. 2000), and tail length (Basolo 1990), or for suites of signal traits (M. Ryan et al. 2003) do not suggest categories of preferences. But data from studies of evoked calling in bullfrogs (Capranica 1966) and trill duration in red-winged blackbirds (Beletsky et al. 1980; reviewed in Ehret 1987) suggest that there are categories (i.e., that there is labeling), but discrimination within and between categories was not tested in these studies. A recent study of túngara frogs, however, demonstrates both criteria of categorical perception: labeling (fig. 8.2a) and discrimination (fig. 8.2; Baugh et al. 2008). Thus, categorical perception of variation in mating signals can occur. Evidence might be rare because few experiments have been designed specifically to test that hypothesis. The use of categorical perception of mating signals may be important in simplifying the processing of signals to make mating decisions. Assigning signals to categorical groups where each group triggers a different response, rather than determining a response to a signal that lies on a continuous scale, could be an effective way to reduce the time required to make a mating decision.

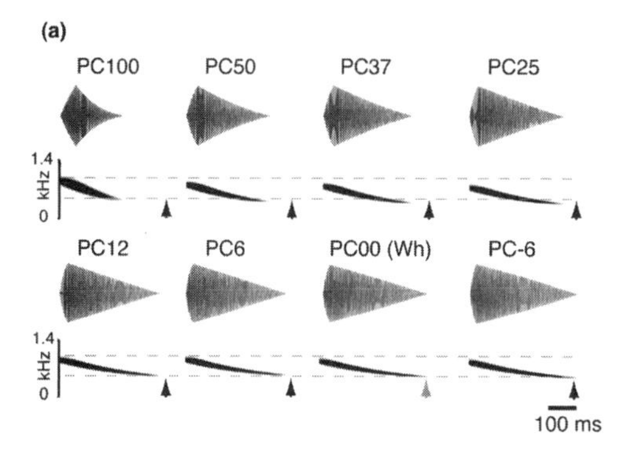

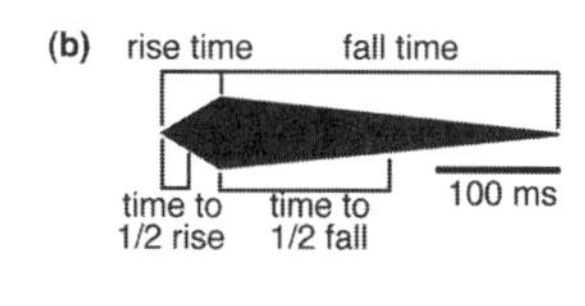

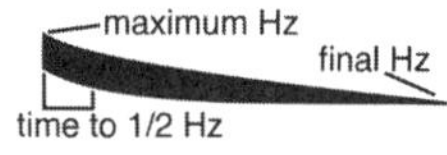

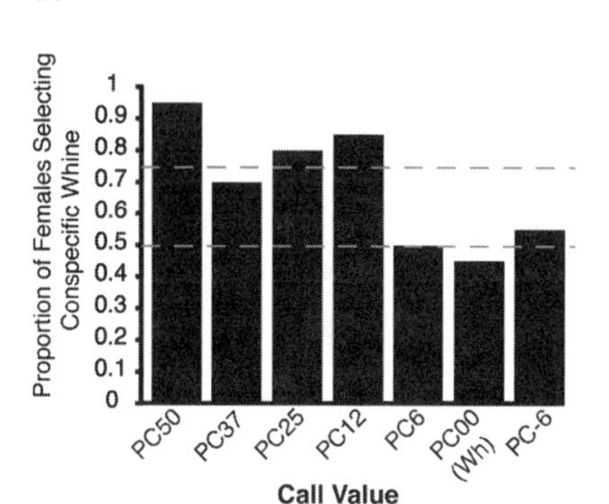

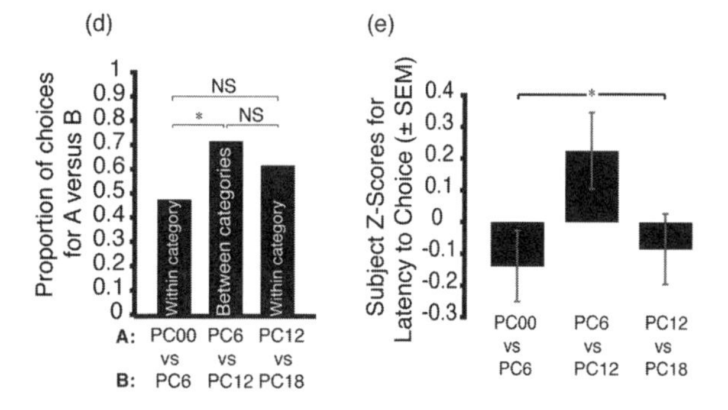

FIGURE 8.2. Categorical perception of mating signals in the túngara frog, *Physalaemus pustulosus*. a. Oscillograms and spectrograms of the seven synthetic stimuli used in the labeling component of the study. The heterospecific call, *Physalaemus coloradorum* (PC100), is presented for comparison. Dashed lines indicate the beginning and ending frequencies of the conspecific call, *P. pustulosus* (PC00), and arrowheads indicate the duration of the PC00. b. A stylized oscillogram (upper) and a spectrogram (lower) of the synthetic túngara whine are shown along with the seven acoustic parameters used to construct stimuli in this study. c. The number of túngara frog females (*N* for each experiment is 20) that prefer the conspecific call to each call variant shows that females label the calls PC – 0.06 through PC 0.06 as conspecific (they do not prefer the conspecific call to these variants), and PC 0.12 through PC 0.50 as not conspecific (they prefer the conspecific call to each of these variants). The dashed lines indicate the null expectation (bottom line) and the critical value for a significant preference (top line). d. Females were tested with pairs of stimuli in which one stimulus (row A on abscissa) was more similar to the conspecific call than the alternative was (row B). Females do not show a preference between a pair of calls that differ by 6% within the category that is labeled as conspecific (whine vs. PC6), whereas they do show a preference between calls that differ by 6% between the categories that are labeled as conspecific and not conspecific (PC6 vs. PC12). The difference in the strength of preference between the within-category and the between-category is statistically significant. There is, however, discrimination between calls that differ by 6% within the not-conspecific category (PC12 vs. PC18), but the strength of this discrimination between categories was not statistically significant (NS = not significant). e. The *z*-scores of the latency to respond from the same phonotaxis tests shown in fig. 8.2a show that females take significantly more time to make a choice in the between-category comparison than in either of the within-category comparisons (Baugh et al. 2008).

We caution that, if categorical perception occurs, it need not be present in perception of all stimuli. For example, in humans there is categorical perception of /ba/-/pa/ phonemes that is based on continuous variation in voice-onset time, but there are numerous other stimulus sets that are not perceived categorically (Harnad 1987). If categorical perception of mating signals were common, it would require that we seriously reevaluate how we examine mate choice rules that are based on continuously varying preference functions (section 8.3).

This abbreviated review clarifies where we are in our understandings of sexual communication and mate choice in naturalistic scenes, at least in the acoustic domain. Signal degradation and attenuation, a variety of noises, and animals signaling simultaneously are the rule rather than the exception in the wild. There is substantial knowledge of how senders have overcome these challenges in the acoustic domain, and there are analogous, although fewer, studies showing similar results in the visual domain (reviewed in G. Rosenthal 2007). On the receiver end of the equation, numerous studies show how peripheral tuning can increase signal-to-noise ratios prior to analysis in the central nervous system. We know very little, however, about how signal detection and perception might be enhanced or compromised in complex auditory scenes. It seems that studies of this aspect of mate choice should take their lead from studies of auditory (e.g., Bregman 1990) and visual scene analysis in humans (Bar 2004), which have made considerably more progress than have analogous studies in animals. The available data suggest that perception of sexual signals is usually continuous rather than categorical, but there are some exceptions and there has been little effort to directly assess the use of categorical perception in mate choice.

8.3. Evaluation and decision

Once a mating signal is perceived and detected, females use the information to make mate choice decisions. Much of the research on this part of the mate choice process is aimed at deducing the rules of mate choice without concern for the internal mechanisms involved. But even treating preferences as a black box has hidden subtleties.

8.3.1. RATIONAL MATE CHOICE

Almost all work on mate choice implicitly assumes that females act rationally. "Rational" here is a technical term borrowed from microeconomics and political science, and rational mate choice follows automatically if choice follows a simple two-step process (Samuelson 1947; Tversky 1969; Tversky and

Simonson 1993; M. Kirkpatrick et al. 2006). In the first step, all the signals and cues from a male are transduced into a single preference score that is independent of the scores of other potential mates. In the second step, the female chooses from among the males in such a way that males with higher scores are chosen more often than those with lower scores.

This paradigm leads to the idea of a "preference function" that shows the relation between a stimulus value (say, the fundamental frequency of an acoustic signal) and a female's preference (Lande 1981). Researchers have estimated the preference function in several species (e.g., Basolo 1990; Diekamp and Gerhardt 1992; Ritchie 1996). The most common procedure is to measure how often a female chooses a focal stimulus when it is paired with a reference stimulus (either real or simulated). We then plot the probability that a focal stimulus is chosen over the reference as a function of the value of the focal stimulus. Figure 8.3 shows Ritchie's (1996) results from estimating the preference functions of *Ephippiger* cricket females based on the number of syllables in the male call.

This approach seems so conceptually straightforward that it may be hard to believe that there could be a serious pitfall. But consider a case where a female chooses stimulus A twice as often as the reference display, and stimulus B four times as often. Does that guarantee that, when choosing between A and B, she will choose B exactly twice as often? Does it even guarantee that she will choose B more often than A? While both predictions seem plausible, in fact they are not experimentally tested in most studies. It is even possible that mating preferences are intransitive, such that a female prefers stimulus A over B, B over C, but C over A. Intransitive choice has been found in animal foraging (Shafir 1994) and human economic decisions (Tversky 1969), so it seems plausible that it also occurs in mate choice.

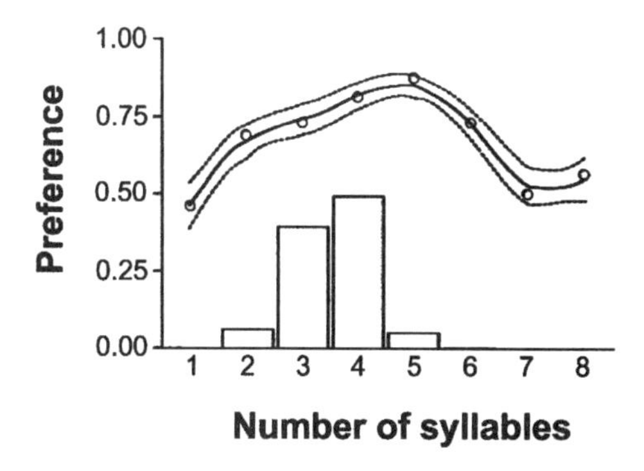

FIGURE 8.3. Preference function of *Ephippiger* cricket females based on the number of syllables in the male call. The reference stimulus was a call from a different population with a single syllable. The circles show the proportion of females preferring the test stimulus to the reference stimulus. The solid curve is fitted to the data with a cubic spline, and the dashed lines show ±1 standard errors from bootstrapped replicates. The bar graph shows the distribution of syllable number in the females' native population. Redrawn from Ritchie 1996.

A recent study of túngara frogs explicitly tested whether females use strict preferences or show evidence of intransitive choice (M. Kirkpatrick et al. 2006). We analyzed results from an earlier study in which females' choice was tested for all possible pairs of nine stimuli (M. Ryan and Rand 2003). Figure 8.4 shows how the data compare with the expectations based on strict preferences. If females did follow strict preferences, then the proportions of times they chose the focal male call (shown by the circles) would tend to fall on the curved surface. Statistical analysis shows that the deviations from that prediction are much greater than what sampling error alone would produce, and so we can say that females are not using strict preferences. Do they have intransitive preferences? A separate analysis gives an ambiguous answer: we cannot prove that these frogs have intransitive preferences, but we cannot reject the possibility either. In sum, there has been only one test to date of the near-universal assumption of strict mating preferences. The data reject that assumption, but unfortunately do not give a clear alternative paradigm that we can use for thinking about mate choice.

Understanding mate choice rules becomes yet more complex when there are more than two potential mates. If mate choice meets the test of rationality, then a female's preference for one male over another should not be reversed by the presence or absence of other males. Yet these kinds of reversals do in fact happen in foraging decisions. The quantity and quality of two sucrose awards can be matched to result in an equal preference by foraging hummingbirds. The presence of a third award of inferior value to the original two influences

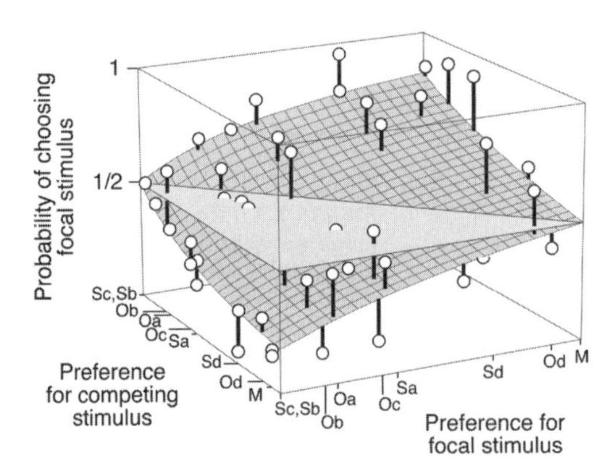

FIGURE 8.4. Fraction of times that females chose the focal male call (shown on the *x*-axis) for all possible pairs of nine call stimuli in the túngara frog. The curved surface shows the results expected if females follow the simple two-step rational preference rule described in the text. Relative preference scores for each stimulus (corresponding to positions on the horizontal axes) are assigned so that the data best fit the curve. The deviations are statistically significant, and so the simple preference rule is rejected. If females did follow strict preferences, then the proportions of times they chose the focal call (shown by the circles) would tend to fall on the curved surface. Redrawn from M. Kirkpatrick et al. 2006.

the preferences between the two superior awards (Bateson et al. 2002, 2003; Schuck-Paim et al. 2004). This principle of context-dependent choice is used to create "competitive decoys" in human marketing schemes. When there are two products equally preferred by consumers, placing a less-preferred object on the shelf can cause a preference for one of the original two products (M. Ryan et al. 2007).

We do not know of any demonstrations of irrational mate choice resulting from any of these mechanisms (M. Ryan et al. 2007). The possibility calls for investigation, however. If it occurs in foraging, it seems plausible that it occurs in mate choice. If it is a domain-specific phenomenon in animals (e.g., foraging but not mate choice), it would be important to know why.

8.3.2. THE SOCIAL CONTEXT

Mate choice often happens in an environment where there are several or many potential mates. When animals gather in groups to advertise and choose mates, receivers are faced with an enormous amount of potential information (Valone 2007). This situation poses challenges for both the sender and the receiver: the sender must compete with conspecific signals, and the receiver must parse individual signals from the conglomerate. In that sense, the social group is a source of noise, as we discussed earlier in section 8.2. But the group setting offers advantages as well as difficulties to mating systems. Females choosing a mate in a group environment are able to use information about both the general mating environment and individual potential mates in their mate choice decision.

Females can modify their receptivity to potential mates of different levels of attractiveness when given experience with variable mating conditions. Research in fruit flies (*Drosophila melanogaster*; Dukas 2005b), zebra finches (*Taeniopygia guttata*; Collins 1995), variable field crickets (*Gryllus lineaticeps*; Wagner et al. 2001), and painted reed frogs (*H. marmoratus*; Passmore and Telford 1981), among others, shows that, when females are exposed to groups of males that differ in the attractiveness of individuals, females modify their receptivity (fig. 8.5). In most of these cases, the females are always receptive to highly attractive mates, but experience influences how permissive they are to mating with less attractive males. One possible interpretation is that, in situations where highly attractive mates are uncommon, the cost of search time required to find highly attractive mates outweighs the benefits of passing up males of low attractiveness in order to mate with highly attractive individuals. This situation demonstrates one way that social context introduces more complexity to the dynamics of mate choice than is suggested by results of simple two-choice experiments.

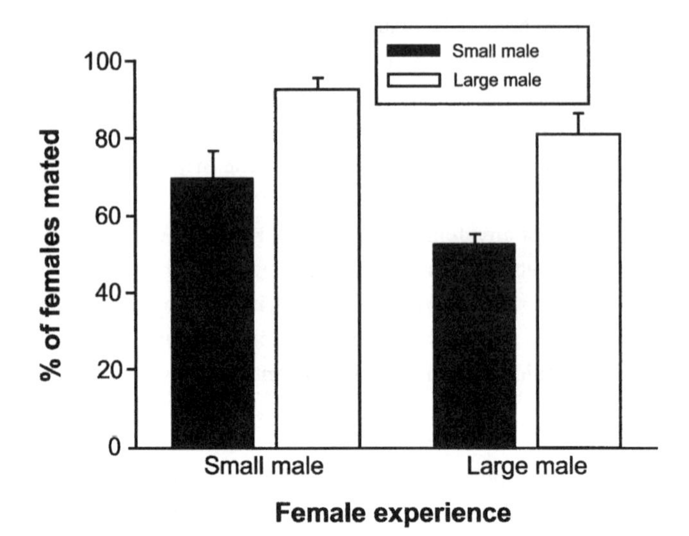

FIGURE 8.5. The mating behavior of female fruit flies depends upon previous experience with courtship by small (less attractive) or large (more attractive) males. Females that have been courted by small males mate more frequently with both small and large males. Redrawn from Dukas 2005b.

Mate assessment in a social group allows females to consider more information about potential mates than what can be gathered from a male's signal alone. A clear example of this use of social information is mate choice copying, which occurs when the choice of a mate is influenced by mating decisions of others. Male mating success on bird leks, where female choice reigns supreme, is sometimes too skewed to be explained by variation in male phenotypes alone. In sage grouse (*Aratinga canicularis*) leks, R. Gibson et al. (1991) suggested that skews emerge in part because females copy the mate choice of others (see also R. Gibson and Höglund 1992; D. White and Galef 2000). Strong support for this hypothesis came when Dugatkin (1992) showed experimentally that female guppies (*Poecilia reticulata*) show mate choice copying. In the standard experiment, versions of which have been repeated in numerous studies (e.g., Dugatkin and Godin 1992; Dugatkin 1992, 2007; Schlupp et al. 1994; Briggs et al. 1995; Grant and Green 1996; Schlupp and Ryan 1997; Witte and Ryan 1998, 2002; Witte and Noltemeier 2002), the focal female fish is in the center of a tank with only visual access to two potential mates, each one on the opposite side of the tank. The female's time spent associating with each male is used as a proxy for preference. The focal female is then allowed to see the less preferred male interact with a model female. The model female is removed, the mate choice experiment is repeated, and if mate copying occurs, the female increases the amount of time she now spends with what had been the less preferred male. Mate choice copying has been demonstrated in other taxa besides fish, mostly birds and humans (e.g., birds: Höglund et al. 1995; Galef and White 1998; D. White and Galef 1999; Ophir and Galef 2004; humans: B. Jones et al. 2007). Although there are few published studies of negative data, mate choice copying is not universal (e.g., Clutton-Brock and McComb 1993).

We can make some generalizations from studies of mate choice copying. It can be exhibited by both males and females (e.g., male and female mollies, *Poecilia latipinna*: Schlupp and Ryan 1997 and Witte and Noltemeier 2002 respectively). Females can copy mate choices of heterospecifics as well as conspecifics (Schlupp et al. 1994), and younger females are more likely to copy older females (Amlacher and Dugatkin 2005). The effects of mate copying are more pronounced when the difference between potential mates is greater (Dugatkin 1996) and when the perceived utility of the model female is higher (S. Hill and Ryan 2006). The effects of copying can be long lasting, persisting up to five weeks in one study (Witte and Noltemeier 2002). And, importantly, it has been shown in the field as well as in the lab (Witte and Ryan 2002). Some of the evolutionary dynamics of copying have been modeled, and it has been shown that copying can even spread through a population when there is mild direct selection against the copying allele (Servedio and Kirkpatrick 1996).

There is no question that mate choice copying occurs. Although further empirical research is necessary, especially to determine more clearly why females exhibit this behavior, there needs to be some method for including this phenomenon into a formal model of mate choice rules, such as a preference function.

8.3.3. LEARNING AND MEMORY

Mating in a social group is one way that females can increase the information they use in mate choice decisions. Increasing the amount of information available for use in decision making can improve one's ability to make the "best" decision, or the decision that maximizes fitness, so many females draw upon learning and memory to inform their evaluation of potential mates. Learning and memory are two cognitive tools that can increase the amount of information available to females that are making mate choice decisions. Learning allows females to use experience to inform their decisions, and memory allows females to store, retain, and recall information for use in decisions that are made after there is no trace of that information in the environment. Processing too much information in a decision, however, can make the decision process time-consuming and inefficient, such that at some point more information becomes costly rather than helpful (Bernays and Wcislo 1994; M. Sullivan 1994). Thus, learning and memory work together with the opposing forces of selective attention and forgetting to control the amount of information available to females for mate choice decisions.

The influence of learning and memory on natural decision making has been explored extensively with respect to foraging decisions (Guilford and Dawkins 1991; Speed 2000). This work has shown incredible variation in how animals learn and remember information related to food acquisition. Predators learn to avoid poisonous prey by recognizing conspicuous aposematic coloring (T. Roper and Redston 1987), while frog-eating bats use social learning to shape their response to potential prey (R. A. Page and Ryan 2006). Food-storing birds can remember specific cache sites for many months (Shettleworth 1990), and honeybees remember not only how to get to their food sources but also what kind of reward a flower produces and when the reward is produced (Boisvert et al. 2007). Learning and memory could have a dramatic influence on mate choice as well, but until recently research on mate choice has not focused on this question (Jennions and Petrie 1997; Bateson and Healy 2005). Mate choice decisions differ substantially from foraging decisions in three key ways that predict that learning and memory play different roles in these two domains.

(i) Most animals forage repeatedly, often for many hours of every day of their lives, but an animal's opportunity to mate is much less frequent, and some individuals mate only once in a lifetime. In these extreme semelparous cases, individuals have no opportunity to learn from actual mating experience. In these cases, individuals may learn information from social experiences other than mating itself, and short-term, or working, memory over the course of mate assessment may be more useful than long-term memory.

(ii) Many of the costs and benefits that result from a mate choice decision are more difficult to judge than those that result from a foraging decision, and so mating-related experiences may not be as easily learned. For individuals who mate multiple times, learning from previous mating experience may occur, but they can learn only about immediate costs and benefits rather than long-term costs and benefits such as the health of offspring resulting from a mate choice, which is apparent only long after the decision is made. In many species, males especially have the opportunity to learn from experience with multiple matings, because they mate or court more often than females do. Immediate costs and benefits of mate choice include factors such as time spent courting different kinds of females that vary in receptivity, as demonstrated in male fruit flies that learn from experience to modify their courtship behavior (Dukas 2004b, 2005a).

(iii) Many mate choice decisions are based on information received from communication signals. Communication related to foraging does occur but is less frequently critical to a foraging decision than to a mating decision. Signals used in mate choice are often ephemeral male advertisement signals—acoustic vocalizations or visual display behaviors that can influence a female's choice only if they persist in her decision process temporally beyond the moment of signal production to the moment of choice. This could occur through working memory. Thus, working memory of signals may be an important factor in mate choice decisions that is less relevant in foraging decisions.

Learning can influence both males (senders) and females (receivers) of a mating system. Males may learn what signals to produce, as has been demonstrated extensively in some birds and mammals (Janik and Slater 1997; Beecher and Brenowitz 2005; Searcy and Nowicki, chapter 5 in this volume), and they may learn which individuals to court, as mentioned above for fruit flies that learned about female receptivity (Dukas 2004b, 2005a). Females can apply learning to inform both preference and receptivity. The influence of early experience on female preference has been demonstrated in songbirds (reviewed in Riebel 2003) and also in invertebrates, as female wolf spiders' (*Schizocosa uetzi*) mate choice is dependent upon experience as a subadult (fig. 8.6; Hebets 2003). Not all learning about which mates are preferable occurs at an early age. As discussed above (section 8.3.2), once displaying males aggregate in groups, females can use short-term learning to inform their assessment of individuals that are potential mates.

The role of memory in mate choice has been less explored than that of learning. Each case of learning demonstrates some kind of memory, as the learned information must be retained in some form for animals to be able to use it. A study of memory, though, also considers when animals forget, shedding

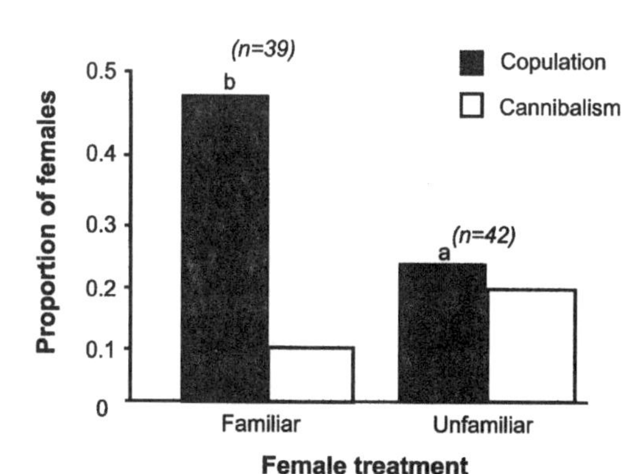

FIGURE 8.6. Mate choice is influenced by experience as a subadult in wolf spiders. Females mate more frequently with males of a familiar phenotype than with males of an unfamiliar phenotype, and they cannibalize males of an unfamiliar phenotype more frequently than males of a familiar phenotype. Redrawn from Hebets 2003.

the information they have stored. In a dynamic environment, memory of a temporary condition could become misleading once conditions have changed, so discarding memories at the appropriate time is adaptive, and an animal's performance is often enhanced by the ability to forget information that is no longer needed (West-Eberhard 2003). In social systems where individuals recognize each other and interact repeatedly, memories about a certain individual could influence the willingness to mate with that individual over a long period of time. For species that do not recognize individuals or encounter the same individuals repeatedly, remembering an evaluation of a potential mate over a long period would not be helpful. Retaining memories could even be costly due to mechanistic requirements of the physical process of memory (Dukas 1998b; Dukas, chapter 2 in this volume), such as the energy required for protein synthesis (Mery and Kawecki 2005), and out-of-date memories could clutter future mate choice decisions with irrelevant information that increases processing time (Bernays and Wcislo 1994; M. Sullivan 1994). These costs are especially relevant to working memory of an individual's signals because these signals rarely apply to future mate choice decisions and should be forgotten at some point.

Understanding the limits to working memory of mate choice–related information could help us understand mate choice strategies in general, because some mate choice strategies depend more than others upon working memory (Janetos 1980; Real 1990). For example, female gray treefrogs assess potential mates over a period of about 2 minutes, and an understanding of their memory capacity would help inform us about what occurs during this assessment period (Schwartz et al. 2004). During an assessment period, females could use a best-of-n strategy to compare males actively and then choose the most attractive one. This would require memory of the previous males that were assessed. If the memory capacity for this information is limited, it would impose a constraint on how long females could spend assessing each male or on how many males they could assess. As an alternative to simultaneous comparison of potential mates, a female could spend an assessment period attending sequentially to individual males until her attraction reaches a threshold sufficient to act upon, and then she would choose whichever male she is attending to. This sequential-search threshold model could operate without any working memory system, and in cases where memory is costly this threshold model may be favored. Figure 8.7 shows how mate search costs influence the expected fitness of the best-of-n versus the sequential-search strategies. Given the impact of mate search costs, females may even switch between strategies depending on what other factors demand usage of a limited memory capacity, such as when predators are present or when environmental conditions are especially treacherous.

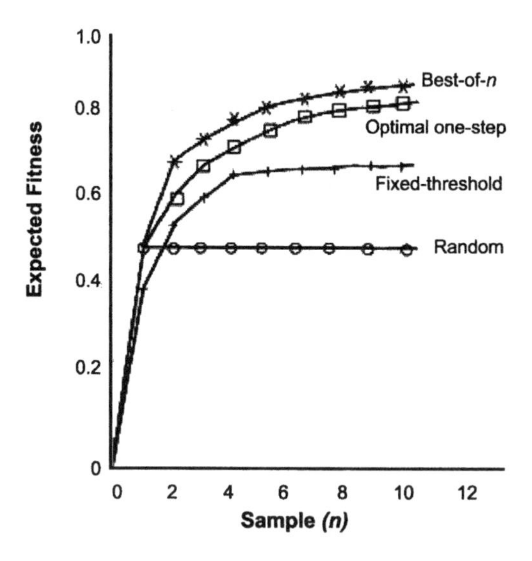

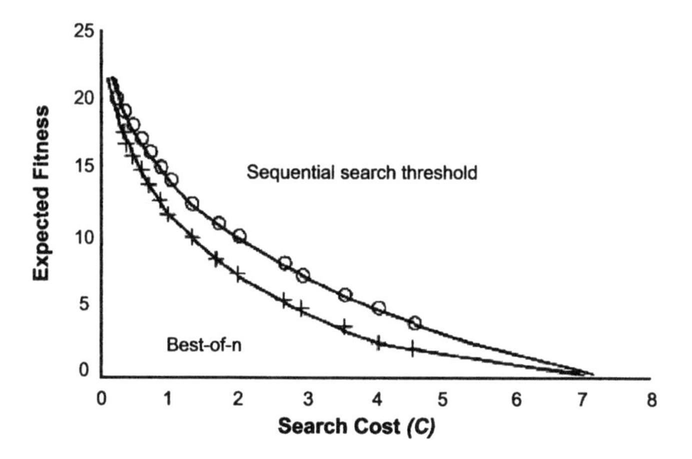

FIGURE 8.7. The costs of mate search influence the expected fitness of a mate choice strategy. Top: Initial models of mate choice strategy showed that a best-of-*n* comparison strategy would always be the strategy that resulted in the highest fitness for females, for any *n* number of males available to females. Redrawn from Janetos 1980. Bottom: But when the costs of searching for a mate are considered, models predict highest fitness for a sequential-search strategy, where the difference in expected fitness varies with the degree of cost entailed by mate search. Redrawn from Real 1990.

As mate choice research begins to address questions related to working memory, it will be helpful to consider when one would expect memory about potential mates to be beneficial to females. One relevant issue is whether an attractive signal indicates a potential mate that would increase a female's fitness. In cases where mating with the particular male whose signal is most attractive improves a female's fitness, she would benefit from the ability to remember the association between that male and his signal until she actually chooses her mate. In cases where females are attracted to a signal that does not relate to a male's quality—signals that reduce female search time, for example—there would be no selective force for her to remember which male makes an attractive signal, and signal memory would not be critical. Another relevant issue is whether signals are constant or variable through the course of a mate assessment, approach, and pairing. Females would benefit from remembering attractive signals that are patchy and variable through time, such as when attractive signals are rare and costly due to increased predation risk. Females would have no need to remember signals, though, when they are constant and uniform through time.

Exploring the influence of memory on mate choice could inform our understanding of sexual selection, as females' cognitive abilities and constraints pose yet another selective pressure on male advertisers. Given that certain stimulus configurations can be better remembered, this should guide the evolution of male display traits. Guilford and Dawkins (1991) broach this idea, arguing that how signals are remembered influences the evolution of warning coloration signals. The same process could occur in the evolution of mating signals, so research emphasizing how females' cognitive adaptations for learning and memory can bias display evolution would be valuable, akin to studies of preexisting sensory biases in auditory tuning and photopigment sensitivity.

8.4. Conclusions and future directions

In this review we have considered some of the mechanistic aspects of mate choice. Our main goal was to review topics addressing how the animal's cognitive biology contributes to the patterns of mate choice in the wild.

We know a lot about mate choice. It is clear that female choice generates sexual selection that is responsible for the evolution of many of the details of male mating signals. Simple two-choice experiments show that females exercise preferences and have allowed researchers to understand the details of what makes signals attractive to females.

For example, we know that males can increase the attractiveness of their signals by making them contrast more with the environment and by matching them to the sensory biases of females (M. Ryan 1998). On the other hand, females can evolve mechanisms to detect and perceive signals better at both the peripheral and the central nervous system. A popular approach to quantifying preferences has been the construction of preference functions.

But the principles of mate choice that have been established through all this research become less clear when we consider mate choice under conditions that approach the real world. Preferences exhibited in simple comparisons sometimes break down in more complicated ones, and we do not really know why. Preference functions need not be continuously variable but can be step functions, as when categorical perception takes place. Social groups can provide females with an enormous amount of information to consider in mate choice, as demonstrated by mate choice copying; thus, a male's utility is defined not only by his phenotype but by how others in the social group, besides the choosing female, react to his phenotype. To further complicate matters, the social group can also act as noise: male signals must compete to contrast against conspecifics, and females must parse signals from a complex social background.

As this review shows, a cognitive perspective on mate choice reveals some critical issues that have been virtually ignored. When female preferences change in social groups, is this adaptive plasticity or cognitive constraint? We have little idea as to the degree to which females' sensory and cognitive abilities are either adaptations to or constraints on mate choice. These abilities include auditory and visual scene analysis, auditory streaming, learning, and memory. When adaptations seem apparent, we must ask if these are domain specific or domain general. Have these adaptations arisen for mate choice specifically, or have they been co-opted from other domains such as foraging? Other major questions need to be addressed: Is female choice always rational? Do simple female preference functions really predict mate choice? Are preferences sometimes intransitive, and might choice be influenced by competitive decoys? Furthermore, we caution that there are other complicating factors in mate choice that we have not reviewed but have barely been probed, such as state-dependent (age, hormonal) changes in mate choice (e.g., K. Lynch et al. 2005), multimodal communication (e.g., Partan and Marler 1999), and the relationship between recognition and discrimination (Phelps et al. 2006.). Early studies of sexual selection and mate choice concentrated on demonstrating that they occurred and their potential adaptive significance. Given numerous insights into phenotypic plasticity, neural bases of signal decoding, and the cognitive biology of decision making, there are new frontiers in mate choice that will not only inform us more deeply about how it occurs but also give us a more realistic understanding of how and why it evolves.

9 Monogamous Brains and Alternative Tactics: Neuronal V1aR, Space Use, and Sexual Infidelity among Male Prairie Voles

STEVEN M. PHELPS & ALEXANDER G. OPHIR

9.1. Introduction

All of animal behavior can be regarded as a series of decisions—decisions about when to mate, forage, court, fight, or sleep, to name a few. The diverse decisions that underlie adaptive behavior often depend on the coordinated actions of many brain regions. Foraging and mate choice, for example, use sensory structures to recognize either food or prospective mates, reward structures to assign emotional valence to stimuli (Pfaus and Heeb 1997; Schultz 2006; Hoke et al. 2007), and motor regions to execute approach, avoidance, or the more specific actions required in each context. While each behavior relies on multiple mechanisms, each mechanism may also contribute to different decisions. Female water mites, for example, attend to vibrations on the surface of the water to detect their prey; males capitalize on female sensory design by using similar cues when courting (Proctor 1991). Although several authors have discussed the multiple uses of a sensory system (e.g., M. Kirkpatrick and Ryan 1991; Endler 1992; Christy 1995), the repeated use of other cognitive mechanisms has received less attention (see Sherry and Schacter 1987). For example, female guppies provided with orange foods develop incidental but significant preferences for orange males (Rodd et al. 2002), indicating that associations between color and value made when feeding are influencing seemingly unrelated mating decisions. To fully understand behavioral evolution, we must ask how distinct cognitive processes are coordinated to mediate a behavioral tactic; we must also ask how animals resolve conflicting demands made on common cognitive substrates. We explore the interactions among neural mechanisms and mating decisions in males of the socially monogamous prairie vole, *Microtus ochrogaster*, a species with alternative mating tactics. The results inform our understanding of both mating-system evolution and the complex interplay among the cognitive processes that govern animal behavior.

9.1.1. COGNITIVE ECOLOGY AND THE EVOLUTION OF MONOGAMY

A mating system emerges from the set of reproductive decisions made by individual males and females. The variables that shape individual tactics are subjects of a rich history of theoretical and empirical study that coincides with the very origins of behavioral ecology. Trivers (1972), for example, argued that sex differences in parental investment drive the elaboration of sexually dimorphic reproductive strategies; in this scenario, monogamy emerges when females and males are selected to exhibit similar levels of investment in young. Emlen and Oring (1977) emphasized the importance of ecological variables in shaping female distributions in space and time. In general, it is difficult to monopolize widely dispersed females, which favors monogamy; it is also difficult to defend females that breed synchronously, which favors a polygamous, nonterritorial, "scramble" competition for mates. Both treatments remain valid but have been refined by a series of theoretical studies that examine female polygamy (Kokko 1999; Ihara 2002; Wakano and Ihara 2005), mate guarding by males (Sandell and Liberg 1992; Kokko and Morrell 2005), and female mate choice (Kokko and Morrell 2005) in more detail. These studies demonstrate how mating patterns can emerge from complex interactions between decisions made by males and females—decisions influenced by a broad suite of social and environmental variables.

Given the multifactorial nature of mating decisions, they seem likely to rely on multiple cognitive processes. In vertebrates, for example, mate guarding emerges when males maintain a close proximity to mates and actively repel intruding males. One behavior requires selective social attachment; the other, selective aggression—two very different behavioral processes. The mechanisms of attachment are in many ways particularly interesting, because they utilize the neuronal circuits that underlie reward in many other contexts, including food, sex, and recreational drug use (Pfaus and Heeb 1997; Everitt and Robbins 2005; Schultz 2006; Hoke et al. 2007). Humans asked to view photographs of people with whom they are deeply in love show enhanced activation of a reward region called the ventral tegmental area (Aron et al. 2005; H. Fisher et al. 2005). This same region is activated by attractive faces in humans (Aron et al. 2005; H. Fisher et al. 2005) and by preferred mate signals in túngara frogs (Hoke et al. 2007). In prairie voles, pair-bonding and its attendant mate guarding are mediated by similar circuits (Young and Wang 2004). Not surprisingly, these reward mechanisms do not seem central to the neurobiology of aggression (reviewed in R. Nelson and Trainor 2007).

Attachment, attraction, and aggression, however, do not fully define mating tactics. Species differences in mating system, for example, are associated with differences in space use. Polygynous species are characterized by sex differences in home range size; in some taxa this is associated with sex and species differences in the size of the hippocampus, a brain region central to spatial navigation (Sherry et al. 1992; L. Jacobs 1996). Recent theoretical studies suggest that spatial navigation should prove central to mating tactics. The optimal level of mate guarding, for example, depends on how efficient males are when guarding mates (Kokko and Morrell 2005), on the number of females a nonterritorial male encounters when roaming (Sandell and Liberg 1992), and the number of females a territorial male can monopolize (Sandell and Liberg 1992). All three of these measures are influenced by a male's use of space, and all seem likely to draw on the mechanisms of spatial navigation. While a cohesive mating tactic requires coordinating multiple behaviors in response to diverse information, the contribution of multiple neural centers to individual and species differences in such tactics remains poorly understood.

9.1.2. A MODEL FOR MAMMALIAN MONOGAMY

Fewer than 3% of mammalian species are thought to be monogamous (Kleiman 1977; Komers and Brotherton 1997), and avian monogamy is much better studied in the field (e.g., H. Smith 1995; Temrin and Tullberg 1995; Owens and Bennett 1997; D. Westneat and Sherman 1997; Petrie and Kempenaers 1998; Griffith et al. 2002). Nevertheless, the extraordinary depth of work on the neurobiology of mammalian behavior is a useful resource for any exploration of neurobiology, cognition, and behavior. Nearly one-third of all mammals are Muroid rodents (Steppan et al. 2004), and this taxonomic diversity, coupled with a vast knowledge of their behavioral mechanisms, makes these species particularly promising models for the integrative study of a mating system. In this context, the prairie vole, *Microtus ochrogaster*, has emerged as a leading model for the study of monogamy.

Prairie voles are small Arvicoline rodents widely distributed in North America. Males and females form long-term associations characterized by shared nests, exclusive home ranges, and biparental care (Getz et al. 1981, 1993). Although most males pair-bond with females and exhibit a "resident" mating tactic, a significant minority forgo pair-bonding and adopt a "wandering" phenotype, characterized by large home ranges that overlap those of multiple males and females (Getz et al. 1981, 1993; Solomon and Jacquot 2002).

In the lab, male pair-bonding has been studied primarily in the context of the selective attachment formed after mating. Intracerebral injections of the neuropeptide vasopressin can promote such attachment, even in the absence of mating (Winslow et al. 1993). Similarly, vasopressin antagonists block male pair-bonding but leave mating behavior unaltered (Winslow et al. 1993). In the central nervous system, the main target of vasopressin is the V1a receptor (V1aR). Among voles, species differences in the distribution of V1aR contribute to variation in mating system (Insel et al. 1994). Indeed, genetic manipulations that increase the abundance of V1aR in particular brain regions can cause males of a polygamous species to exhibit the selective affiliation characteristic of monogamous species (Young et al. 1999; Lim et al. 2004). Among the brain regions that have been studied, V1aR expression in the ventral pallidum seems to be particularly important for male attachment. Similarly, the medial amygdala regulates paternal care, and the lateral septum contributes to both pairing and paternal care (Young and Wang 2004). There are many other brain regions that express V1aR but whose contributions to behavior have not been investigated (Insel et al. 1994; Phelps and Young 2003). How does variation in neuronal V1aR abundance relate to the behavioral diversity evident in natural populations? Does the circuitry that underlies spatial navigation contribute to male tactics or their efficacy? We address these questions through field studies of V1aR function in prairie vole behavior.

9.2. Reproductive decisions, space use, and mating tactics

9.2.1. AFFILIATION, ATTACHMENT, AND MATING SUCCESS

In the wild, monogamous resident males defend small territories and exclude intruding conspecifics (e.g., Getz et al. 1993). The exact proportion of males that become residents varies somewhat from study to study, but residency is generally the more common tactic (Solomon and Jacquot 2002; Ophir et al. 2008b). In seminatural enclosures, for example, roughly 75% of males adopt a resident strategy, while the remaining 25% become wanderers (fig. 9.1; Ophir et al. 2008b). Although the enclosures exclude predators and limit movements of subjects, the overall pattern of space use resembles that of free-living animals (e.g., Getz et al. 1993). Why do animals adopt these distinct tactics? Are they evolutionarily equivalent, or is one tactic favored?

To assess the reproductive consequences of resident and wandering tactics, we placed sexually naive, adult male and female prairie voles into seminatural enclosures at natural densities. We then examined the paternity of embryos sired during the experiment. We found that residents had a significantly higher probability of fertilizing females than did wanderers (fig. 9.1; Ophir et al. 2008b). This probability of fertilization translated into a larger number of embryos sired by residents than by wanderers (fig. 9.1; Ophir et al. 2008b).

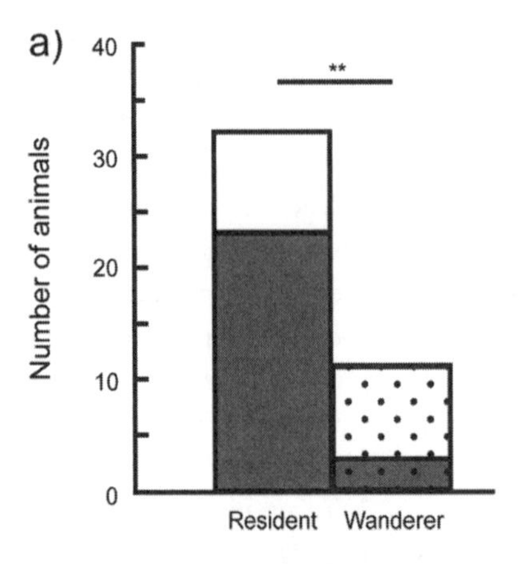

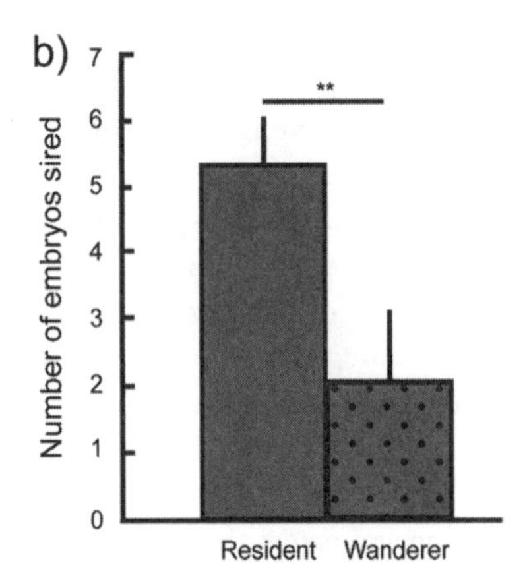

FIGURE 9.1. Mating success of residents and wanderers. a. Number of resident (solid) and wandering (stippled) males that mated successfully (gray) or did not mate successfully (white). Proportions were compared with a Fisher's exact text, $^{**}P = 0.01$. b. Mean (±SE) number of embryos from males who were residents or wanderers. Means and standard errors are based on enclosure means ($N_{enclosure} = 8$; *T*-test, $^{**}P < 0.01$). Modified from Ophir et al. 2008b.

Fertilization rates are important but incomplete components of an animal's fitness: differences in longevity, condition, or parental care could substantially alter lifetime reproductive success. With respect to longevity, our data do not address whether there are differences in predation. We found that wanderers were in no better condition than residents (Ophir et al. 2008b), suggesting that the energetic costs of territory defense do not reduce the benefits of pairing. Getz and McGuire (1993) report that wanderers live longer than residents, but these differences are smaller than those we report for mating success. Considering the added advantages residents gain by caring for their young (Wang and Novak 1992), we suspect that our data underestimate the value of residency.

On balance, it appears that residency is a favored tactic. If so, wanderers must be making the best of a bad situation. What prevents wanderers from becoming successful residents? One explanation is that there is a paucity of receptive females. Puberty and ovulation in female prairie voles are induced by adult male olfactory cues (Carter et al. 1980; Dluzen et al. 1981); if females were insufficiently exposed to male pheromones, the number of receptive females would become limiting. Similarly, females may prefer the males who ultimately become residents to those who become wanderers. Although residents and wanderers are not known to differ in most gross morphological attributes (Solomon and Jacquot 2002; Ophir et al. 2007), resident males do have a longer anogenital distance (AGD) than wanderers (fig. 9.2a). Long

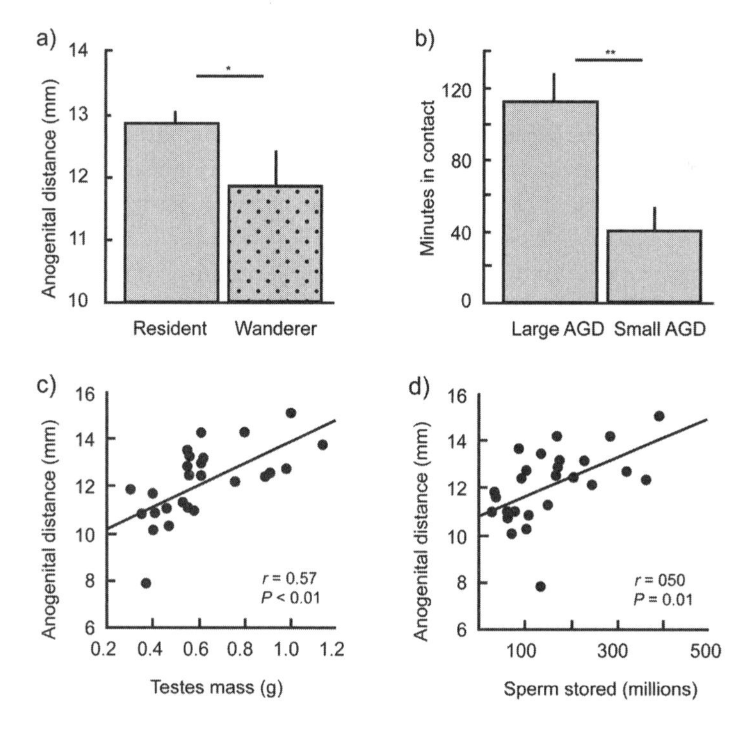

FIGURE 9.2. Anogenital distance (AGD), behavior, and fertility. a. Mean (±SE) AGD (mm) differed significantly between residents and wanderers (*$P < 0.05$). b. Mean (±SE) time (min) that females spent in side-by-side contact with males that had either longer or shorter AGD (**$P < 0.01$). AGD was significantly correlated with testis size (c) and sperm counts (d). Modified from Ophir and delBarco-Trillo 2007.

AGD is a marker for in utero masculinization and may serve as a proxy for other phenotypic differences detectable by females (Drickamer et al. 1995; Drickamer 1996; B. Ryan and Vandenbergh 2002). In the lab, prairie vole females prefer long-AGD males to short-AGD males (fig. 9.2b; Ophir and delBarco-Trillo 2007). Potential reasons for such preferences include a preference for more aggressive males, which may translate into better territorial defense or infanticide deterrence. AGD is also positively correlated with testis size and sperm count in this same sample of prairie voles (figs. 9.2c and 9.2d; Ophir and delBarco-Trillo 2007). Because prairie voles form long-term pair-bonds, increased sperm production may translate into higher female reproductive success over a lifetime. Although the availability of responsive females may be limiting, one additional possibility suggested by the AGD data is that residents are simply more competitive and thus exclude future wanderers from available females.

9.2.2. SPACE USE AND HETEROGENEOUS SELECTION

One key attribute of reproductive tactics is how animals use space. Mate guarding by resident males, for example, includes both maintaining a close proximity to a mate and aggressively expelling intruders. Wandering males, in contrast, adopt a scramble tactic characterized by large home ranges and interactions with multiple prospective mates. Although males adopt one of two alternative reproductive tactics, they vary in how strictly they conform to the behavior that characterizes each tactic. We examined our radio-tracking and paternity data to determine whether there were differences in space use that predicted patterns of paternity and reproductive success within and between the two tactics.

As expected, wandering males had larger home ranges and overlapped more conspecifics (fig. 9.3; Ophir et al. 2008c). When we examined which space use patterns predict mating success, we found that successful wanderers used space differently from either unsuccessful wanderers or successful residents. Wanderers who were able to sire young roamed more broadly than unsuccessful wanderers: they exhibited larger home ranges, overlapped the home ranges of more males, and exhibited a trend toward overlapping the home ranges of more females (figs. 9.3a and 9.3c; Ophir et al. 2008c). Presumably this pattern maximized the rate that wanderers encountered potential mates and thus increased the probability of successful mating (Sandell and Liberg 1992). In contrast, resident males effectively exclude intruding males from their territories; on average, each resident home range overlaps the home range of less than one male. Based on theoretical treatments, resident males should maximize guarding when females exhibit moderate levels of multiple mating (Kokko 1999; Kokko and Morrell 2005). This small, actively defended home range is certainly consistent with such a tactic.

One very interesting treatment of monogamy emphasizes the trade-off that monogamous males must make between intrapair and extrapair paternity (Kokko and Morrell 2005; also see H. Smith 1995). Males who opt for intrapair paternity focus more on mate guarding at the cost of potential extrapair matings. Males who roam do so at the expense of mate guarding. Comparing space use of males who engage in intrapair (IPF) versus extrapair fertilizations (EPF) confirms these expectations. We find that EPF males—including both successful wanderers and philandering residents—exhibit patterns of conspecific overlap that resemble those of wandering males (fig. 9.3; Ophir et al. 2008c). This difference persisted after correcting for overall differences between residents and wanderers in the degree of overlap (Ophir et al. 2008c). Thus, even resident males increase extrapair paternity by venturing more often into surrounding environments.

In general, successful males adopt one of two strategies: they focus efforts

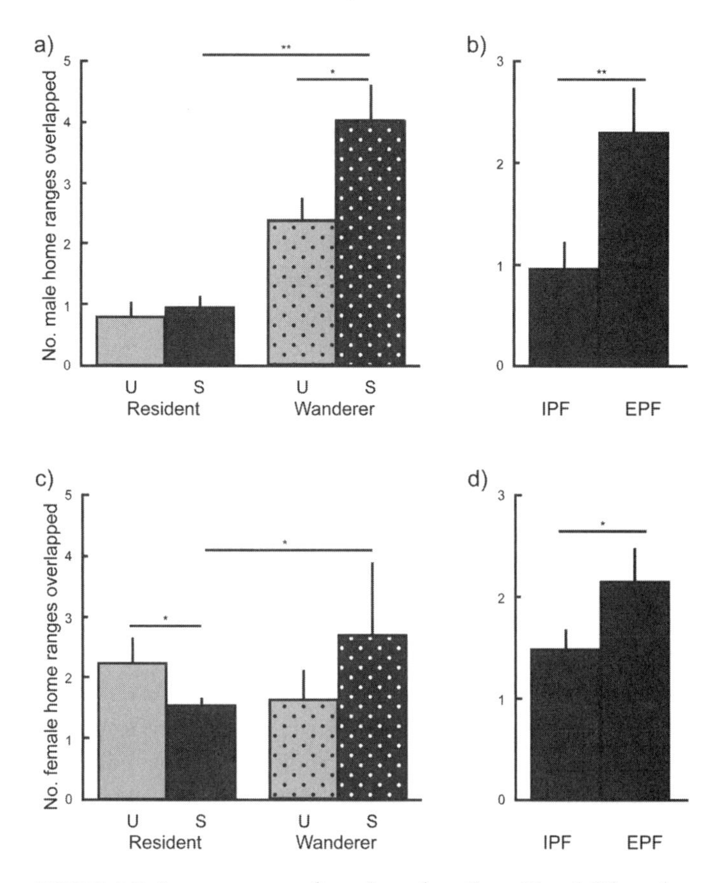

FIGURE 9.3. Space use among alternative male tactics. a. Mean (±SE) number of male home ranges overlapped by residents (solid) and wanderers (stippled) who were unsuccessful (U: light gray) or successful (S: dark gray) at siring young. An ANOVA revealed a significant effect of mating success ($P < 0.01$) and a mating success by reproductive tactic interaction (MS × RT, $P < 0.04$). b. The number of male home ranges overlapped by IPF and EPF males differed significantly (*T*-test, $^{**}P < 0.01$). c. The number of female home ranges overlapped by successful and unsuccessful residents and wanderers exhibited a significant MS × RT interaction (ANOVA, $P < 0.05$). d. The number of female home ranges overlapped by IPF and EPF males. IPF = resident males that mated successfully only with their partner; EPF = wanderers that mated successfully plus residents that sired offspring outside the pair. Post hoc *T*-tests are reported in each panel ($^{**}P < 0.01$, $^{*}P \leq 0.05$). Modified from Ophir et al. 2008c.

on mate guarding, or they maximize the number of females they encounter. This pattern suggests that selection operates against males with intermediate phenotypes. How do the brains of these animals reflect these evolutionary forces? In the next section, we examine natural variation in the neural expression of the vasopressin receptor, V1aR. We ask how natural diversity in brain phenotype contributes to the probability that a male will adopt a resident or a wanderer tactic and to the efficacy of the tactic a male adopts.

9.3. Neural substrates of alternative tactics

Across a broad range of taxa, arginine vasopressin (AVP) and its nonmammalian homologue, vasotocin, influence a diversity of social behaviors, including mating, territorial aggression, and social memory (Dantzer et al. 1988; Ferris and Delville 1994; Goodson and Bass 2001; H. Caldwell et al. 2008). In prairie voles, vasopressin antagonists block the pair-bonding that normally follows repeated mating (Winslow et al. 1993; Lim et al. 2004). Similarly, vasopressin alone is able to produce the specific social attachments characteristic of pair-bonding even in the absence of mating (Winslow et al. 1993). The effects of vasopressin are not limited to attachment, however. Injections of vasopressin also trigger the onset of intense, selective aggression directed at intruding males (Winslow et al. 1993). Vasopressin also contributes to male paternal care (Bamshad et al. 1994; Wang et al. 1994, 1998). Thus, the neuropeptide coordinates many attributes of the resident tactic.

The behavioral effects of vasopressin are generally mediated by the V1a receptor (V1aR), the predominant receptor in the central nervous system. Variation in V1aR expression profiles among *Microtus* species mirrors behavioral differences related to mating system. In addition to substantial differences between species, there are also profound differences among prairie voles (fig. 9.4; Phelps and Young 2003; Ophir et al. 2008c). While vasopressin synthesis and release vary dramatically with recent experience, V1aR expression seems to be stable throughout adulthood (e.g., Poulin and Pittman 1993; Wang et al. 1997). Given that vasopressin coordinates the transition to a resident tactic through its actions on V1aR, we investigated whether individual differences in adult V1aR explain variation in mating tactic or associated behaviors.

9.3.1. MECHANISMS OF MALE PAIR-BONDINGS

The effects of vasopressin on attachment have been directly linked to two neural structures. Vasopressin actions in the lateral septum influence social memory and aggression in several rodents (Dantzer et al. 1988; Everts et al. 1997; Bester-Meredith et al. 1999; Bielsky et al. 2005) and influence pair-bonding and paternal care in prairie voles (Wang et al. 1998; Y. Liu et al. 2001; Young and Wang 2004). More extensively studied, however, is the role of the ventral pallidum, a key node in the reward pathway (Cardinal et al. 2002; Everitt and Robbins 2005). Extensive vasopressin release during repeated mating seems to drive the formation of social preferences for a mate through its influence on

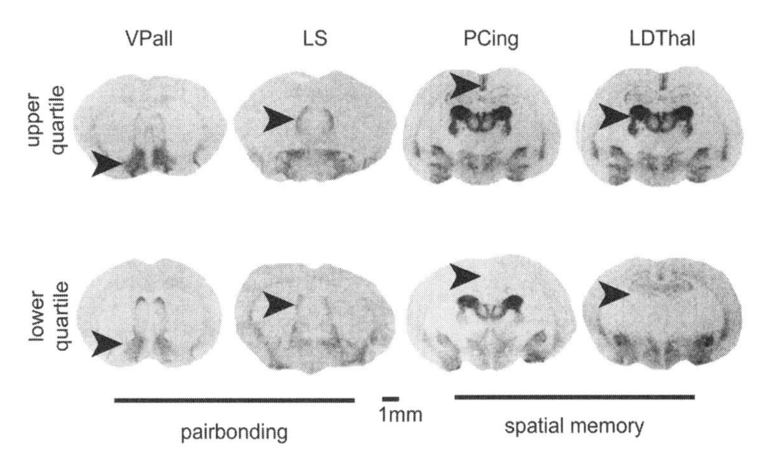

FIGURE 9.4. Natural variation in V1aR expression. Autoradiograms of brains in the upper and lower quartiles of ^{125}I-linear-AVP V1aR binding. Regions implicated in pair-bonding include the ventral pallidum (VPall) and lateral septum (LS). Regions implicated in spatial memory include the posterior cingulate/retrosplenial cortex (PCing) and laterodorsal thalamus (LDThal). Modified from Ophir et al. 2008c.

reward. This is supported by site-specific injections of hormone antagonists (Lim and Young 2004), by overexpression of V1aR receptors in the ventral pallidum (Pitkow et al. 2001; Lim et al. 2004), and by making transgenic mice that express V1aR under the control of the prairie vole V1aR regulatory sequence (Young et al. 1999). The ventral pallidum and lateral septum are part of a larger "pair-bonding" circuit, which links sensory information associated with a mate to the reward system (Young and Wang 2004). Given the extensive data on the importance of V1aR in monogamous behaviors, we first asked whether differences between wanderers and residents could be explained by natural variation in V1aR expression.

We predicted that the ventral pallidum and lateral septum would exhibit higher V1aR expression in resident males than in wandering males. This would suggest that wanderers were less responsive to vasopressin, and so less able to form pairs. To our surprise, we found no differences between residents and wanderers in the abundance of V1aR in either structure (fig. 9.5; Ophir et al. 2008c). These pair-bonding regions also failed to predict whether males exhibited sexual fidelity to their partners. Because we know that V1aR in these regions is necessary for pair-bond formation, the findings indicate that prairie vole males share a common propensity to form pair-bonds but that the use of these mechanisms is plastic. Presumably, this plasticity is attributable to variation in vasopressin release between tactics, though this has not been investigated. Because pairing substantially increases mating success, we suspect that

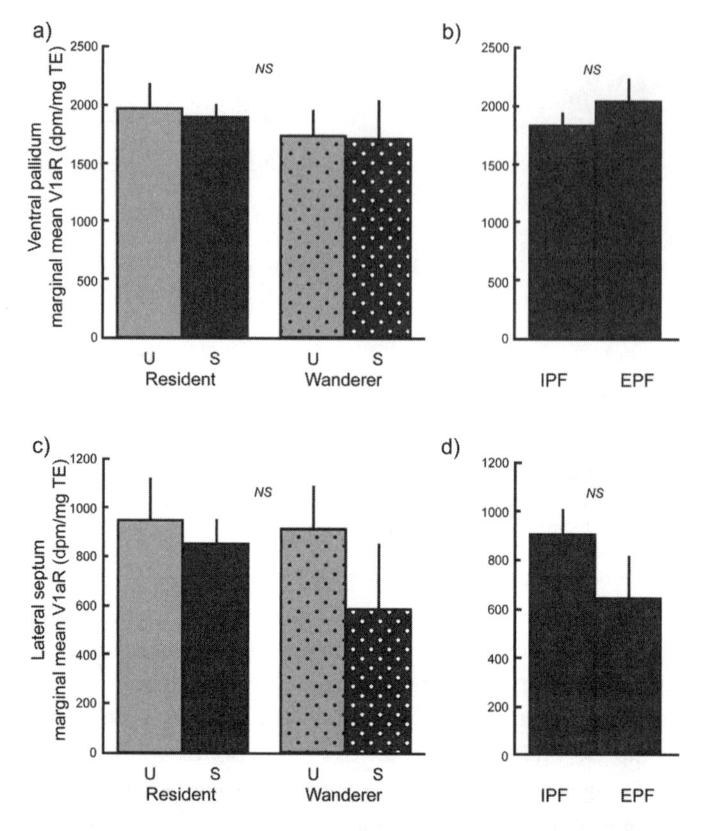

FIGURE 9.5. V1aR in pair-bonding regions does not predict tactic or fidelity. a. Mean (±SE) disintegrations per minute (dpm) in tissue equivalence (TE) of ^{125}I-labeled V1aR autoradiographic ligand binding in the ventral pallidum for successful (S) and unsuccessful (U) animals. Residents are depicted by solid bars and wanderers by stippled bars. S males fertilized at least one female. U males fertilized no females. b. Pallidal V1aR among IPF and EPF males. V1aR binding in the lateral septum among residents and wanderers (c) and among IPF and EPF males (d). There were no effects of residency, success, or sexual fidelity in either region ($P > 0.10$). Modified from Ophir et al. 2008c.

selection has cleared the standing variation in these structures. Thus, wanderers may be making the best of a bad situation, but their brains are ready to assume a favored strategy when the appropriate opportunity arises.

9.3.2. NEURAL SUBSTRATES OF SOCIOSPATIAL MEMORY

Vasopressin and V1aR are both integrally involved in pair-bond formation. However, pair-bonding is but one behavior modified by vasopressin. For example, a diverse literature documents vasopressin actions on memory consolidation and retention (DeWied 1971; Bohus et al. 1978; Egashira et al. 2004;

Hayes and Chambers 2005). Interestingly, two groups of structures vary dramatically in the abundance of V1aR between individual male prairie voles, both of which are implicated in memory (Phelps and Young 2003). These are the cingulate cortex and the dorsal thalamus.

Along the medial length of the cerebral cortex runs a fold called the cingulate cortex. Roughly midway between the rostral tip of the cortex and the caudal end is the posterior cingulate cortex, which then runs seamlessly into the more caudal retrosplenial cortex (Paxinos and Watson 2006). Both regions express V1aR in prairie voles, and both are highly variable in this expression (Insel et al. 1994; Phelps and Young 2003). The variation is present in both males and females, in mated and unmated animals, and in field caught and lab-reared animals (e.g., Phelps and Young 2003; Hammock and Young 2005; Ophir et al. 2008c). After examining well over 100 prairie vole brains, the two structures are always concordant—V1aR expression is either strong in both or weak in both (S. M. Phelps, personal observation). The posterior cingulate/retrosplenial cortex (PCing) has strong connections to the hippocampus, a brain region that maps the world in both space and time, thereby contributing to both spatial and "episodic" memory in humans and to spatial memory in many other taxa (Sherry et al. 1992; Cooper et al. 2001; Maguire 2001; Harker and Whishaw 2004). The PCing has been the subject of intensive recent interest for its involvement in both spatial and episodic memory in humans (Maguire 2001). In rats, lesions to PCing cause profound impairments in spatial memory and navigation (Harker and Whishaw 2004). In primates, the PCing sends projections to the laterodorsal thalamus (LDThal), which in turn projects to the hippocampus (van Groen and Wyss 2003; Shinkai et al. 2005). The LDThal has been the subject of less study, but given its neuroanatomical position and some limited behavioral data (van Groen et al. 2002), a role in spatial memory seems likely for this structure as well. Interestingly, the LDThal also exhibits profound variation in the expression of V1aR among prairie voles.

Given the natural diversity we observe in these structures, and the apparent disruptive selection we detected on space use patterns, we asked whether the abundance of V1aR in either structure was predictive of mating tactic, space use, or sexual fidelity among male prairie voles. We found that neither structure was responsible for becoming a resident or a wanderer (fig. 9.6; Ophir et al. 2008c). However, both were associated with the differences in space use that characterize successful wanderers and residents. Successful wanderers were characterized by larger home ranges and the overlap of more conspecifics; they also exhibited low V1aR expression in the PCing and LDThal. Residents were characterized by fewer conspecific overlaps and smaller home ranges; the PCing and LDThal of residents expressed higher levels of V1aR than did those of successful wanderers.

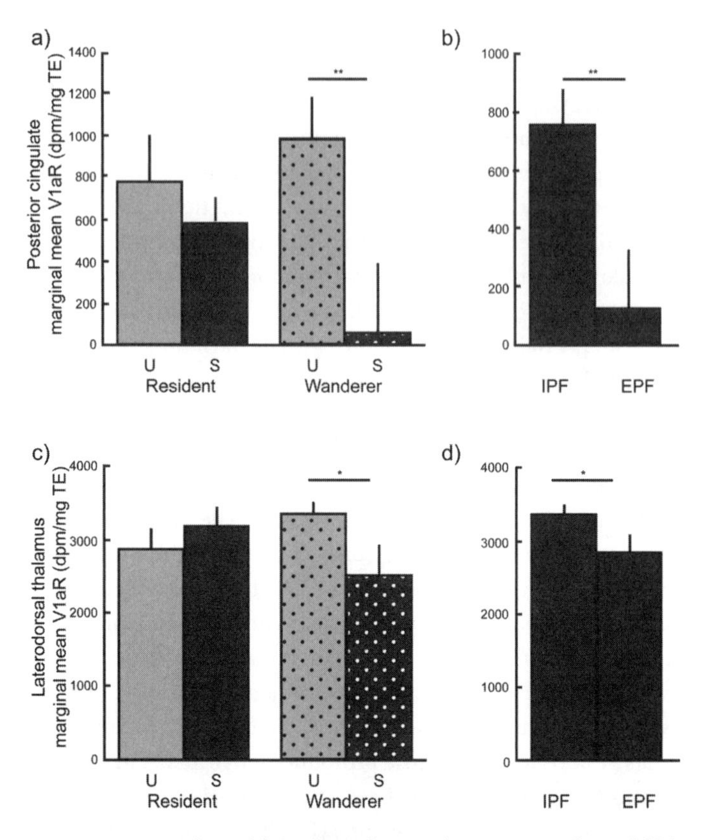

FIGURE 9.6. V1aR in spatial circuits predicts wanderer success and sexual fidelity. a. Mean (±SE) disintegrations per minute (dpm) in tissue equivalence (TE) of ^{125}I-labeled V1aR autoradiographic ligand binding in the PCing for successful (S) and unsuccessful (U) animals. Residents are depicted by solid bars and wanderers by stippled bars. MANOVA revealed a significant mating success by reproductive tactic interaction ($P < 0.05$) across both spatial memory structures. b. PCing V1aR in IPF and EPF males. V1aR binding in the LDThal among residents and wanderers (c) and among IPF and EPF males (d). Post hoc *T*-tests are reported in each panel (**$P < 0.01$, *$P < 0.05$). Modified from Ophir et al. 2008c.

Although V1aR in the pair-bonding circuit did not predict patterns of paternity, V1aR in the PCing was a particularly good predictor of male sexual fidelity. The prevalence of low V1aR expression among EPF males was attributable to both resident and wandering males, again mirroring patterns of conspecific home range overlap. This relationship between sexual fidelity and PCing V1aR expression also seems to hold across species (fig. 9.7). As discussed above, prairie voles are often socially monogamous but can exhibit alternative tactics. Their brains are characterized by high levels of PCing V1aR in most individuals and the persistence of low PCing V1aR at lower frequencies (Phelps and Young 2003). Pine voles, thought to be genetically monogamous, exhibit consistently high levels of PCing V1aR (Insel et al. 1994). The polygamous montane and meadow voles, in contrast, exhibit no PCing V1aR whatsoever (Insel et al. 1994). This striking convergence of space use, sexual fidelity, and V1aR expression in spatial memory circuits suggests a common link between the cognitive demands of spatial navigation and the opportunity for EPF.

Male prairie voles thus possess the substrates for exhibiting pair-bonding, a preferred tactic, reflected in their uniformly high levels of V1aR in the ventral pallidum. It seems, however, that the males exhibit plasticity in the deployment of this machinery. Pair-bonding requires repeated bouts of mating with a single individual, with each bout releasing AVP, which modulates reward responses and forms attachments. By linking AVP release to mating bouts, the males have a neural mechanism for pairing with responsive females when they can be assured of paternity. In contrast, V1aR expression in spatial circuits may mediate a trade-off between the spatial demands of effective mate guarding and those of maximizing encounters with multiple females. Given evidence that PCing expression is highly heritable (Hammock and Young 2005), this raises the possibility that the variability in this and other spatial circuits may persist in a sort of balanced polymorphism. We return to this prospect in our conclusion. We now move down a level of analysis to focus on genetic mechanisms that underlie V1aR expression, with the ultimate hope of identifying allelic variation that could contribute to neuronal polymorphism.

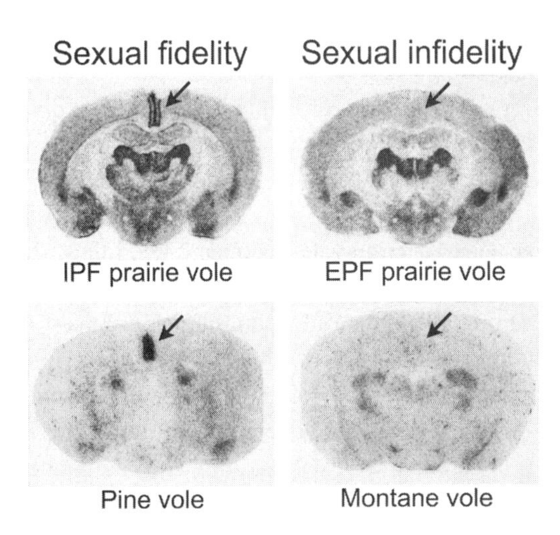

FIGURE 9.7. Individual and species differences in PCing. Sexually monogamous prairie voles and genetically monogamous pine voles both exhibit high levels of V1aR binding in the posterior cingulate/retrosplenial cortex. Sexually promiscuous prairie voles and polygamous montane voles both lack V1aR in the cingulate cortex. Pine vole and montane vole images from Insel et al. 1994.

9.4. Microsatellite polymorphisms and phenotypic diversity

The mapping of genotype to phenotype is central to an integrative understanding of evolution in any context. For cognitive ecology, this will necessarily require investigating the relationship between variation at the level of individual genes and their function in neural circuits. In this regard, prairie vole V1aR expression and male monogamy again provide a useful model. The emergence of the resident tactic in the ancestors of modern prairie voles clearly required evolutionary changes in the pattern of *avpr1a* gene expression rather than changes in coding sequence (Young et al. 1999). As we have already reviewed, prairie voles differ from polygamous congeners in their neuronal pattern of V1aR expression, and these differences are causally related to their capacity to form pair-bonds. In a seminal study, Young et al. (1999) generated a transgenic mouse that expressed the prairie vole *avpr1a* locus under the control of an upstream prairie vole noncoding sequence. The transgenic mouse more closely resembled the V1aR phenotype of prairie voles than it did a wild-type mouse. Critically, intracerebral injections of vasopressin caused the transgenic mice to form the specific social preferences characteristic of pair-bonding, while vasopressin injections into the wild-type mice had no effect. Within the ~1100 bases of the prairie vole regulatory sequence included in the transgene, the most conspicuous difference between prairie voles and promiscuous congeners lay in the expansion of a microsatellite repeat near the transcription start site. Further analyses revealed that this repeat was shared by the monogamous pine vole (Young et al. 1999).

A series of elegant studies demonstrated that in cultured cells, a common model for gene expression studies, the microsatellite could alter gene expression (Hammock and Young 2004, 2005). Hammock and Young (2005) found that male prairie voles with long and short microsatellite repeat lengths differed in their neuronal V1aR abundance. This led the researchers to suggest that the prairie vole microsatellite length might cause both interspecies differences in mating system and intraspecies variation in mating tactics. Whether the microsatellite caused interspecies variation was challenged by Fink et al. (2006), who demonstrated that a long microsatellite was a basal feature of the clade, and that the promiscuous meadow and montane voles shared a reduction in its length through descent. Although this made clear that having a long *avpr1a* microsatellite was not sufficient to predict monogamy, it did not address whether more subtle variation in length or sequence caused differences between Microtine species or among prairie voles (Young and Hammock 2007). We set out to examine the latter in natural settings by genotyping the animals in our preceding studies. Could long microsatellite alleles predict tactic? Or, more subtly, could they predict success within a tactic?

To simplify our analysis, we focused on males that had either two long alleles (both above median length) or two short alleles (both below median length). We found that long-allele males had generally higher V1aR abundance when averaged across all brain regions (Ophir et al. 2008a). We also found differences in two regions implicated in monogamous behavior, the ventral pallidum and medial amygdala (figs. 9.8a and 9.8c). The ventral pallidum is clearly causally related to pair-bonding (Lim et al. 2004). The medial amygdala conveys pheromonal information to the ventral pallidum and is important for paternal care (B. Kirkpatrick et al. 1994; Young and Wang 2004); thus, although the medial amygdala is not known to play a role in pair-bonding, a concordance between it and the ventral pallidum could coordinate residency.

Although the neuronal data were promising, our behavioral data revealed no significant differences between genotypes in any measure. We could not say that long-allele males behaved "more monogamously." Long-allele animals were no more likely to become residents, did not have smaller home ranges,

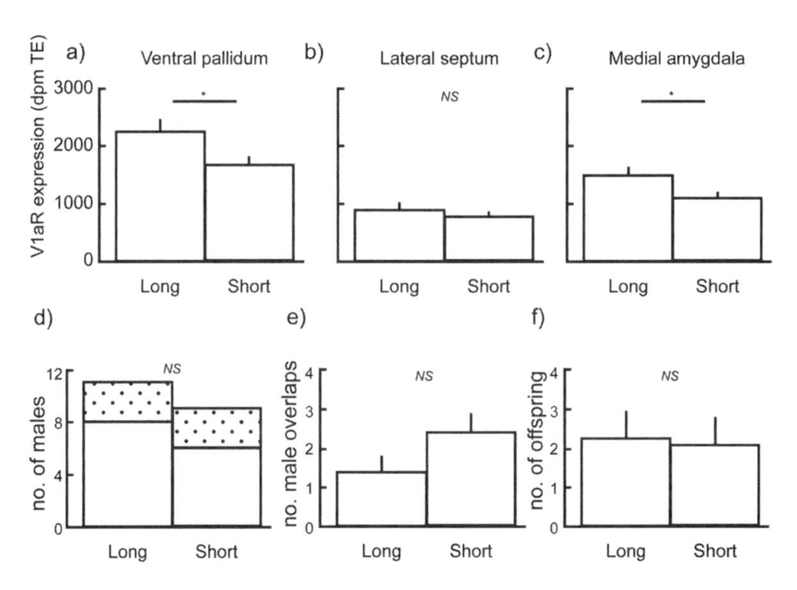

FIGURE 9.8. Microsatellite allele length influences V1aR abundance but not behavior or fitness. a–c. Long-allele males have higher V1aR in two nodes in the "pair-bonding circuit," the ventral pallidum, and the medial amygdala (*T* test, $^{*}P < 0.05$) but no differences in the lateral septum ($P > 0.10$). d–f. Despite differences in neural phenotype, we detect no effects on (d) the likelihood of becoming a resident (solid) or wanderer (stippled) (Fisher's exact, $P > 0.10$), (e) the number of male home ranges overlapped ($P > 0.10$), or (f) the number of offspring sired ($P > 0.10$). Data from Ophir et al. 2008a.

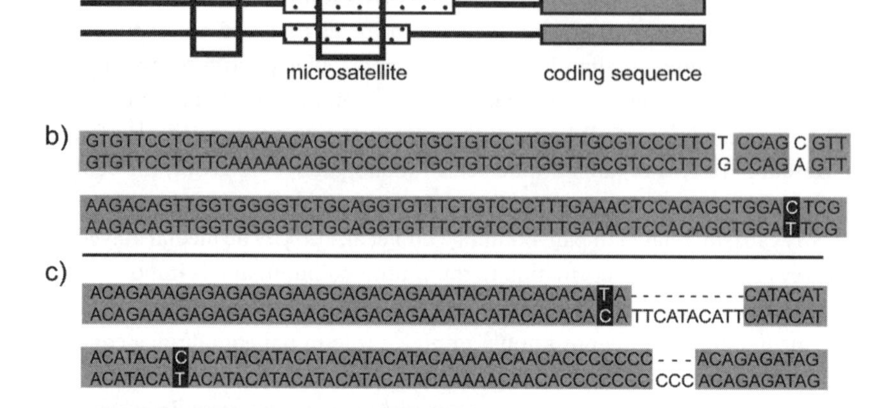

FIGURE 9.9. Nonhomologous origins of microsatellite length variation. a. Schematic of two *avpr1a* alleles with differing lengths of microsatellite repeats. Open boxes depict sites of sequences in panels b and c. b. Alignments of a long allele (top sequence) and short allele (bottom sequence), showing single nucleotide polymorphisms 5' of the *avpr1a* microsatellite. c. Alignments of the same two alleles within the complex *avpr1a* microsatellite. Although the long allele has an expanded GA repeat, it also has shorter CATA and polyC repeat lengths. Thus, different repeat motifs within the microsatellite can make unique contributions to length. Focusing on allele length misses potentially important sequence variation within the microsatellite and in neighboring sequences. Modified from Ophir et al. 2008a.

nor did they overlap the home ranges of fewer conspecifics (figs. 9.8d and 9.8e; Ophir et al. 2008a). The conclusion that length had no influence on natural behavioral variation was also supported by the lack of allele-length effects on PCing and LDThal variation, the only brain regions that were associated with behavioral differences (see Ophir et al. 2008c).

These data demonstrate that microsatellite variation influences *avpr1a* expression but that this does not translate into behavioral differences in natural settings. Thus, the extraordinary variation that persists at the microsatellite can conceivably be attributed to its lack of influence on fitness. Indeed, long-allele males were no more or less likely to sire young than were short-allele males (fig. 9.8f).

Finding that allele length did not predict behaviors in our study led us to reexamine the data from the transgenic study by Young et al. (1999). If the length of the microsatellite did not drive a prairie vole pattern of V1aR expression and social behavior in transgenic mice, something else within the sequence of the transgene must have. We suggest that sequence variation within or near the microsatellite, rather than length alone, is responsible for the prairie vole V1aR expression pattern and its influence on monogamy. The *avpr1a* microsatellite is a complex repetitive sequence, and there are many ways to produce equivalent lengths (fig. 9.9). Thus, within- and between-species differences in V1aR expression may yet map to sequence variation at this locus. If so, it will open up a rich array of experiments that can examine the persistence of genetic diversity and its contributions to the cognitive variation that underlies behaviors in natural environments.

9.5. Monogamy and cognitive ecology reconsidered

In a classic study, Bateman (1948) suggested that male fitness increases with number of mates, while female fitness does not. Emlen and Oring (1977) elaborated on this theme to suggest that females distribute themselves according to resources that influence their reproductive success, while males distribute themselves in a manner that maximizes the number of females they can monopolize. Thus, when females are widely dispersed in space, the effort required to monopolize females may preclude polygyny. Similarly, when female mating is synchronized with that of other females, males may be unable to monopolize multiple females simultaneously. Using game theory, Sandell and Liberg (1992) examined whether males should be territorial or nonterritorial as a function of female encounter rate, female defensibility, and the degree of resident advantage (the ability of a resident to dominate an intruder). The model demonstrated that the strategies reflected trade-offs in each of these domains, with increases in resident advantage favoring territoriality, and increases in female encounter rate favoring wandering.

Among socially monogamous animals, the common occurrence of female polygamy presents a conceptual challenge because it undermines male paternal investment and contradicts Bateman's view of female fitness as unaffected by the number of mates. One explanation emphasizes conflict over paternity, in which it benefits females to mate with extrapair males to gain either "good genes" or some other benefit, such as infanticide deterrence (Wolff and Macdonald 2004). Kokko and Morrell (2005) note that attractive males must make trade-offs between intrapair and extrapair paternity in order to maximize their fitness. The ideal male strategy emerges as a complex interaction between male attractiveness, female fidelity, and male mate-guarding efficacy. According to these models, males should mate-guard more when females occasionally attempt extrapair matings and when resident males are good at excluding intruders.

Many of the variables that should shape male tactics have correlates in the data we have reviewed. For example, the ultimate decision to mate-guard requires a male to be confident of paternity (Trivers 1972; Kokko 1999; Kokko and Morrell 2005) and to be effective at monopolizing the female (Sandell and Liberg 1992; Kokko and Morrell 2005). From a proximate perspective, males form pair-bonds and become territorial only after 24 hours of mating (Insel et al. 1995), and this is mediated by the prolonged release of vasopressin during this period (Winslow et al. 1993). Vasopressin acts on reward structures to promote proximity, and this in turn facilitates mate guarding. Thus, high levels of V1aR in the ventral pallidum enable males to become mate-guarding residents once a prolonged mating has ensured both paternity and the more general ability to monopolize the female during estrus.

Because pheromonal cues activate vasopressin cells that project to many parts of the forebrain (Murphy et al. 1997; de Vries and Miller 1998; Young and Wang 2004), it seems likely that vasopressin modulates brain regions in response to social encounters of multiple sorts. Indeed, vasopressin has been implicated in a diverse suite of social behaviors (e.g., Dantzer et al. 1988; Ferris and Delville 1994; Goodson and Bass 2001). If V1aR in the ventral pallidum increases responses to mating rewards, what is the function of V1aR in the PCing? We hypothesize that high PCing V1aR facilitates memory for the locations of territory intrusions, which in turn increases a resident's ability to exclude intruders. If this is the case, why would low PCing V1aR persist? One clue comes from the fact that every male that obtained an EPF lacked PCing V1aR altogether (fig. 9.10; Ophir et al. 2008c). Wandering and paired males who intrude onto a neighbor's territory are likely to encounter a local resident. In most species, including many rodents, residents are dominant over intruders (Maynard Smith and Parker 1976; Gauthreaux 1978; Wolff et al. 1983; Yoder et al. 1996). By recalling the spatial context of such social defeats, intruding males could avoid repeated encounters with these males, but they will also encounter fewer females. Thus, low PCing might be regarded as a means of "adaptive forgetting" by wandering males and may incidentally promote extrapair paternity among resident males as well.

To examine this more closely, we calculated the proportion of matings obtained by males with high or low PCing V1aR expression. ("High" V1aR was defined as above 400 dpm/mg TE; "low" expression was defined as below 400 dpm/mg TE and is at or near background levels.) We found that half of wanderers with low PCing were able to mate, but none of the high-PCing wanderers were successful (fig. 9.10). Resident males with high PCing mated about as often as did residents with low PCing. This seems to suggest that high PCing is not advantageous for resident males. However, a closer look reveals that low-PCing males are nearly three times as likely to be cuckolded. Moreover, roughly one-third of the low-PCing resident matings were EPFs. Thus,

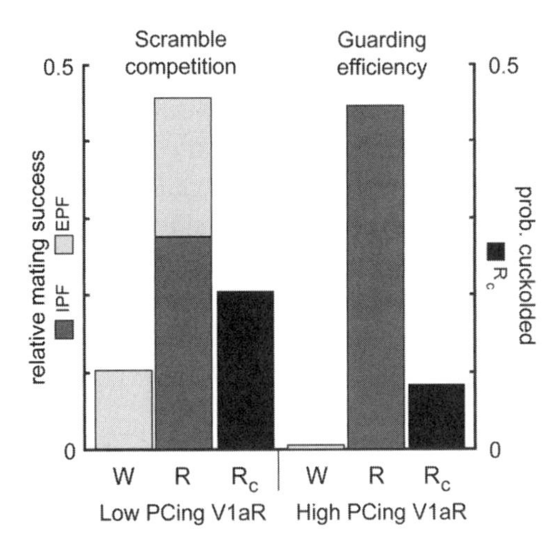

FIGURE 9.10. A balanced polymorphism in PCing V1aR abundance? Relative mating success was defined as the proportion of total fertilizations obtained by males of each class. W = wanderers; R = residents; dark gray bars correspond to success obtained through IPFs, and light gray bars to EPFs. On the same scale we have plotted the probability of being cuckolded by males of either brain phenotype (R_c, black bars). Half of the low-PCing wanderers were successful, while none of the high-PCing wanderers fertilized embryos. Similarly, all residents engaging in EPFs exhibited low PCing V1aR. While high-PCing males gained slightly fewer matings on average, they were much less likely to be cuckolded. Low PCing V1aR seems to be suited to scramble competition, and high V1aR to mate guarding.

it seems that high PCing V1aR maximizes mate-guarding efficiency, while low PCing facilitates a scramble tactic by increasing female encounter rates.

Interestingly, this interpretation is also supported by cross-species comparisons. Pine voles are widely dispersed (Fitzgerald and Madison 1983), presumably reducing potential female encounter rates and favoring mate guarding over wandering. This is reflected in both genetic monogamy (Marfori et al. 1997) and high levels of PCing V1aR (fig. 9.7; Insel et al. 1994). Montane voles have a very short breeding season triggered by new shoot growth, resulting in the synchronous breeding of many females (Negus and Berger 1977; Negus et al. 1977). Synchronously breeding females are difficult to defend, and this situation should favor a scramble tactic (Emlen and Oring 1977). Accordingly, the species is polygamous, is nonterritorial, and does not express V1aR in the PCing (Insel et al. 1994, 1995). The interspecific data, like the intraspecific data, suggest a trade-off between efficient mate guarding and female encounter rates mediated by cingulate V1aR.

Together these findings provide insight into both the evolution of mating systems and animal behavior more generally. The interaction between vasopressin and the ventral pallidum provides an interesting example of phenotypic plasticity in the neuroendocrine regulation of behavior. By making pair-bonding and residency contingent on repeated and prolonged mating, the mechanism ensures that a male is likely to be a successful resident before committing to the tactic. The association of cingulate V1aR variation with intrapair and extrapair paternity highlights how behavioral specializations can make conflicting demands of cognitive substrates. Lastly, the combination of reward and spatial memory systems reveals how diverse mechanisms are needed to execute a cohesive and successful tactic. Although there are many causal details that remain to be explored, our results provide a glimpse of how integrative approaches may yield a more complete understanding of animal behavior. As new methods permit the manipulation of V1aR and other genes in natural environments, such studies promise to clarify both the mechanisms of natural behavior and the origins of behavioral diversity.

ACKNOWLEDGMENTS

We would like to thank Dr. P. Campbell and K. Hanna for *avpr1a* microsatellite analyses, and Dr. A. B. Sorin for her help with paternity assignment. Dr. Jerry Wolff, who passed away while this volume was in press, was an integral part of this work. Among his innumerable contributions, he served as collaborator, mentor, and friend. He is much missed.

10 Assessing Risk: Embryos, Information, and Escape Hatching

KAREN M. WARKENTIN & MICHAEL S. CALDWELL

10.1. Introduction

Animals are killed by predators, pathogens, and abiotic environmental conditions, all of which impose selection for traits that improve survival. These defenses are often costly or conflict with other fitness-enhancing functions and confer a benefit only in the context of specific threats. Thus, inducible defenses have evolved, allowing animals to balance the costs and benefits of defense as risks vary. Inducible defenses include changes in morphology, biochemistry, and life history, as well as behavior (Tollrian and Harvell 1999). Effective deployment of such defenses depends on information about risks, that is, the probabilities of mortality from particular sources at a given time.

Prey respond to several types of information about predation risk. For instance, their activity patterns are shaped by environmental conditions, such as light level and proximity to refuge, as well as by aspects of their own condition that affect vulnerability to predation (Vasquez 1994; Kahlert 2003; Stankowitch and Blumstein 2005). These sorts of factors appear relatively easy to assess, being either long-lasting, cyclical, or properties of the prey itself. Animals also attend to signals specifically associated with predators. Some prey use stereotyped alarm calls to convey information about predators and pathogens (e.g., C. Evans et al. 1993a; Rosengaus et al. 1999; Manser, chapter 12 in this volume). Prey also eavesdrop on the intraspecific signals of their predators (B. Wilson and Dill 2002; Remage-Healey et al. 2006; K. Schmidt 2006; Ratcliffe, chapter 11 in this volume). In both cases, selection on signalers for efficacy in information transfer tends to make the signal more distinct from background noise and the task of the receiver easier (J. Bradbury and Vehrencamp 1998). Many prey, however, face a more challenging problem.

Selection on predators for crypsis should make them less conspicuous to their prey, and many prey gain no benefit from providing risk information to other individuals. Thus, prey also rely on a variety of nonstereotyped cues associated with the presence or activity of their predators. There is a substantial literature documenting responses of prey to predator cues and identifying the sensory modalities involved (e.g., visual: Stankowitch and Blumstein 2005; chemical: Kats and Dill 1998; Wisenden 2000; vibrational: Bacher et al. 1996; Castellanos and Barbosa 2006). Not surprisingly, defensive responses are sometimes elicited by stimuli that are not specific to predators. For instance, hatchling snakes freeze in response to substrate vibrations from a rock falling (Burger 1998), calling frogs fall silent in response to human footfalls (E. Lewis and Narins 1985), and molting geese flee to water in response to passing aircraft (Kahlert 2006). Similarly, snails develop defensive morphology in response to chemicals from both molluscivorous and nonmolluscivorous sunfish (Lagerhans and DeWitt 2002). Understanding mechanisms of risk assessment requires knowledge both of what cues defensive responses to real threats and of the ways animals avoid unnecessary defensive responses. We know relatively little about the specific mechanisms that animals use to assess risk based on nonstereotyped cues.

10.2. Cognitive strategies to assess risk using nonstereotyped cues

For discrete defenses, which are either deployed or not, the basic problem can be defined as follows: an animal has some prior information, either learned or innate, about its environment and the associations of cues with environmental conditions. It perceives a cue or stimulus with certain properties and must make a decision to either ignore the cue or defend itself. The defense increases its chance of surviving if indeed the source of risk (e.g., predator) is present, but it also carries a cost, reducing fitness in some other way.

Detection theory, originally developed in the context of stereotyped communication signals, offers a conceptual framework for decisions about defense (Macmillan and Creelman 2005). Animals experience stimuli from a variety of benign sources that, in terms of detection theory, we can consider part of the background. They also experience cues from predators or other sources of risk. They can correctly classify the source of a stimulus, responding defensively to risk cues (a "hit" in detection theory) and ignoring benign-source stimuli. They can also make two different errors. If they show a defensive response to a benign-source stimulus, it is a "false alarm." If they fail to respond defensively to a risk cue, it is a "miss," or missed cue. These two types of errors can also be defined in statistical terms. Since the majority of stimuli an animal experiences are not associated with danger, we consider the null hypothesis to be that the stimulus is from a benign source. The alternative hypothesis, then, is that the stimulus is caused by a predator or other threat. In this framework, false alarms correspond to type I errors and missed cues to type II errors.

For any property that an animal could assess, such as the frequency, amplitude, or duration of a vibration, or the shape, color, or looming rate of a visual stimulus, there is likely to be overlap between the distributions of that property in risk cues and in benign-source stimuli. Such overlap occurs even with communication signals selected to be conspicuous and is more likely with selection for crypsis (e.g., Fleishman 1992; Wilcox et al. 1996). If there is overlap, any criterion that an animal could use as a decision rule will result in some missed cues and/or false alarms (fig. 10.1a). Based on both the consequences of each type of mistake and its likelihood, as informed by prior information, the criterion can be adjusted to optimize the decision rule (J. Bradbury and Vehrencamp 1998). For instance, if animals that miss predator cues always get eaten, such errors can be avoided by relaxing the criterion so that a broader range of stimuli elicit defensive behavior. However, this also causes an increase in false alarms. The amount of overlap in the stimulus distributions determines the severity of this trade-off; if risk cues are less distinct from the background, then any given reduction in missed cues will result in more false alarms.

The evolution of risk assessment mechanisms is, therefore, shaped by the consequences of both missed cues and false alarms. Researchers modeling prey behavior have debated whether prey should overestimate risk, erring on the side of caution (reviewed in Bednekoff 2007); either over- or underestimation can be favored, depending on the cost of each error. For instance, some brief reversible defensive behaviors appear relatively low cost, and for these a relaxed criterion may be an effective solution. Thus, we might expect animals to freeze, deploy a defensive posture, or momentarily stop calling as a rapid response to a broad range of stimuli, including some that are not true indicators of danger (e.g., frogs: E. Lewis and Narins 1985; crickets: Gnatzy and Kämper 1990; katydids: Faure and Hoy 2000). At the other extreme are defenses that are costly and irreversible, such as permanent changes in morphology that compromise fecundity (e.g., Lively 1986a, 1986b) or shifts in the timing of life stage transitions that expose vulnerable animals to a new suite of predators (Warkentin 1995, 1999a). Strong selection against both missing cues to danger and responding to false alarms should favor improvements in cue discrimination, that is, mechanisms to reduce the overlap with benign-source stimuli by using more, or better, information.

Animals can get more information in parallel or serially (figs. 10.1b and 10.1c; Macmillan and Creelman 2005). One way to gather more information is to simultaneously attend to a greater number of cue properties. If the properties vary independently, then each property added will reduce the overlap between the joint distributions, reducing the unavoidable errors (fig. 10.1b).

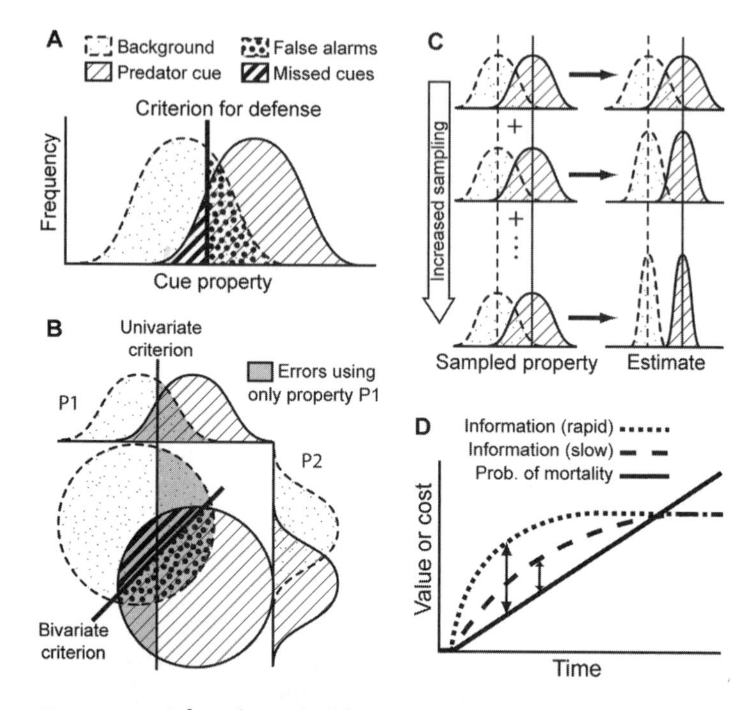

FIGURE 10.1. Information use in risk assessment. A. According to detection theory, when the distributions of cue properties overlap between benign background stimuli (dashed curve) and predator or other risk cues (solid curve), any criterion used to express a defense will result in missed cues and/or false alarms. For a given amount of information, shifting the criterion reduces the incidence of one type of error and increases that of the other. Reducing total errors requires more information to achieve better separation of predator cues from background. B. More information can be acquired in parallel if multiple cue properties vary independently. Errors using one cue property (P1) are shown in gray. Concurrent use of a second property (P2) gives better separation of the joint distributions. C. More information can be acquired serially. Increased sampling of a cue property reduces overlap between the distributions of the estimated property, reducing unavoidable errors. D. Graphical model of the trade-offs in information sampling, redrawn from Warkentin et al. 2007. Information has value in reducing the incidence of potentially fatal errors in decisions about defense. This value accrues as a decelerating function of the number of samples, either rapidly (dotted line) or slowly (dashed line) over time, depending on the duration of each sample. The cost of information accrues as probability of mortality (solid line), a function of the time animals are exposed to risk while sampling cues. When information accrues more slowly, animals must either base decisions on less information, accept a higher information acquisition cost, or both. Here maximizing the net benefit of information use (i.e., value minus cost, arrows) requires changes in both.

However, integrating information from multiple cue properties likely requires more complex neural processing than attending to a single property. Another way to get more information is through increased sampling. Rather than basing their decision on a single, brief sample, animals could use some estimate constructed from longer or repeated sampling. As sampling increases, the variance around an estimate will decrease, resulting in better separation of partially overlapping distributions and reducing the unavoidable errors (fig. 10.1c). However, any cost of sampling will increase concurrently.

Sampling costs may be particularly important when the source of a cue is also a source of mortality, since increasing the lag time before defense increases the likelihood of death. Animals should therefore balance this increased probability of mortality against the value of information, or fitness increment gained by avoiding decision errors. In assessing predation risk, the value and cost of information likely accrue at different rates (fig. 10.1d). In general, serial samples of a cue provide diminishing amounts of information, and thus diminishing improvements in decision making (J. Bradbury and Vehrencamp 1998). In contrast, probability of mortality may more often accrue as a linear, or even increasing, function of time spent sampling, although this can vary with the foraging strategy of the predator. The trade-off between the cost and value of information will determine the optimal amount of information on which a decision should be based (J. Bradbury and Vehrencamp 1998). Typically, this will be less than perfect information. In particular, we expect animals to use less information when the cost of gathering information is higher and when the consequences of decision errors are less severe (Koops and Abrahams 1998; Dall et al. 2005).

10.3. Adaptive responses of embryos in heterogeneous environments

Most research on plastic responses to risk focuses on postembryonic developmental stages. Embryos are, nonetheless, subject to many of the same conditions that have shaped risk sensitivity in later life stages and could similarly benefit from inducible defenses. Both eggs and newly hatched young often suffer high mortality from predators, pathogens, and/or abiotic factors such as desiccation and hypoxia (Rumrill 1990; Houde 2002). Importantly, many of these threats are stage specific because hatching both frees animals from the physiological, spatial, and locomotor constraints of the egg and deprives them of the physical protection it offers. Thus, accelerating hatching can allow embryos to escape from dangers specific to the egg stage, while delaying hatching postpones their exposure to larval-stage threats (Sih and Moore 1993; Warkentin 1995; K. Martin 1999).

The basic requirements for the evolution of adaptive plasticity are environmental heterogeneity, a trade-off such that different phenotypes are most fit in different environments, a cue that provides information about the selective environment, and sufficient genetic variation (Pigliucci 2001; DeWitt and Scheiner 2004). In general, embryos within eggs, as well as many newly hatched larvae, neither feed nor reproduce. The rate and quality of their development are important, but survival is the variable most immediately affecting their fitness. Most sources of embryo mortality are, at the level of the egg or egg clutch, heterogeneous. For instance, even if egg predators or pathogens are abundant in a habitat, they are unlikely to attack every egg (Williamson and Bull 1994; Kiesecker and Blaustein 1997; Chalcraft and Andrews 1999). Likewise, eggs may be drowned or die of desiccation, but these risks fluctuate with the weather (Pyburn 1970; Banks and Beebee 1988; Losos et al. 2003). Clearly, different life history phenotypes—hatching stages or timings—are favored under different environmental conditions as mortality risks vary. Can embryos perceive cues indicative of risk, assessing this environmental variation? And can they produce appropriately different hatching phenotypes in response?

Although there must be developmental constraints on the ability of embryos to acquire and process information and to alter hatching timing, hatching plasticity is well documented in a variety of taxa in response to a diverse array of threats and opportunities (table 10.1). Embryos alter hatching in several ways in response to environmental cues. Some species hatch only when environmental conditions have become suitable for larval life (K. Martin 1999). Some species time hatching in response to cues that indicate periods of reduced larval risk (S. Morgan 1995; I. Bradbury et al. 2004). Some delay hatching until they are larger or more developed if cues indicate increased posthatching risk (Sih and Moore 1993; Ireland et al. 2007). Others hatch prematurely in response to cues indicating egg-stage risk (Wedekind and Müller 2005; Gomez-Mestre et al. 2008). Some embryos slow development and delay hatching in response to conspecific cues associated with larval food limitation (Voronezhskaya et al. 2004). Others wait to hatch until a parental signal indicates larval food is available (Clare 1997). In some turtles and many precocial birds, embryos synchronize hatching timing with their clutch-mates across a range of developmental stages, so that hatchlings leave the nest together (R. Spencer et al. 2001; Brua 2002). In a spider with parental care, plasticity in hatching timing appears to be solely under maternal control (Li 2002; Li and Jackson 2005). In contrast, some crab and lobster

TABLE 10.1. Examples of apparently adaptive embryonic regulation of hatching timing in response to environmental conditions

Environmental conditions (specific cues)	Responses	Taxa	Sources
		Physical conditions directly affecting development and survival	
Flooding of terrestrial eggs (hypoxia)	Induced hatching (end delay)	Fishes: *Fundulus heteroclitus*, *F. confluentes*, *Adinia xenica*, *Galaxias maculatus*; salamanders: *Ambystoma opacum*, *A. cingulatum*, *A. gracile*; Frog: *Pseudophryne bibroni*	Reviewed in K. Martin 1999; see also DiMichele and Taylor 1980; Petranka et al. 1982
		Pig-nosed turtle, *Carettochelys inscuplta*	Webb et al. 1986; Doody et al. 2001
		Dragonfly, *Potamarcha congener*	P. Miller 1992
Flooding of terrestrial eggs (wave action)	Induced hatching (end delay)	California grunion, *Leuresthes tenuis*	Griem and Martin 2000
Repeated flooding of high intertidal eggs	Induced hatching (early, typical, or after delay)	Snail, *Melampus bidentatus*	Russell-Hunter et al. 1972
Flooding of terrestrial eggs (hypoxia)	Induced premature hatching	Phyllomedusine treefrogs: *Agalychnis callidryas*, *A. moreletii*, *A. annae*, *A. saltator*, *A. spurrelli*, *Pachymedusa dacnicolor*, *Cruziohyla calcarifer*	Warkentin 2002; Gomez-Mestre et al. 2008
Flooding of terrestrial eggs	Induced premature hatching	Lizard, *Anolis sagrei*	Losos et al. 2003
Aquatic hypoxia	Induced premature hatching	Quacking frog, *Crinia georgiana*	Seymour et al. 2000
Aquatic oxygen level (hypoxia/hyperoxia)	Premature/delayed hatching	Rainbow trout, *Salmo gairdneri*	Latham and Just 1989
Air surface exposure of aquatic eggs	Induced premature hatching	Salmonid fishes: *Salvelinus alpinus*, *Salmo trutta*, *Coregonus* sp.	Wedekind and Müller 2005
		Physical conditions associated with risk of predation or resource availability	
Flooding + bacterial growth (hypoxia)/high larval density	Induced/inhibited hatching	Mosquitoes, *Aedes* spp.	Gjullin et al. 1941; Livdahl and Edgerly 1987 and references therein
Diel cycle (darkness)	Synchronous hatching at low-risk time	Rainbow smelt, *Osmerus mordax*	I. Bradbury et al. 2004
Diel (light-dark), tidal, and tidal amplitude cycles (entrained rhythms)	Synchronous hatching timed to reduce risk of predation	Many crabs, spiny lobster (hatching involves embryo-mother interactions, embryo role varies)	Reviewed in S. Morgan 1995; see also Christy 2003; Ziegler and Forward 2007

TABLE 10.1. *(continued)*

Environmental conditions (specific cues)	Responses	Taxa	Sources
		Physical conditions associated with risk of predation or resource availability	
Diel cycle (darkness, light, endogenous rhythm)	Induced and inhibited hatching	Monogenean parasites of fishes (hatching timing varies with host activity pattern)	Gannicott and Tinsley 1997 and references therein
Seasonality (photoperiod, temperature)	Induced hatching of resting eggs	Zooplankton: cladocerans, rotifers, copepods	Reviewed in Gyllström and Hansson 2004
		Larval food resources	
Vertebrate host (heat)	Induced hatching (end delay)	Botflies, Cuterebridae	Catts 1982; Cogley and Cogley 1989
Phytoplankton bloom (maternal chemical)	Induced hatching (end delay)	Barnacle, *Semibalanus balanoides*	Clare et al. 1985; Clare 1997
Starved conspecifics (chemical cue)	Slowed development, delayed hatching	Snails: *Helisoma trivolvis, Lymnaea stagnalis*	Voronezhskaya et al. 2004
Larval host (chemical cues)	Induced hatching (of developed eggs)	Monogenean parasites of fishes	Gannicott and Tinsley 1997 and references therein
		Pathogen infection of eggs	
Fungus (Dothideales)	Induced premature hatching	Red-eyed treefrog, *Agalychnis callidryas*	Warkentin et al. 2001
Fungus (*Fusarium* and *Gliocladium*)	Induced premature hatching	Rock lizard, *Lacerta monticola*	Moreira and Barata 2005
Oomycetes (*Saprolegnia* and *Achlya*)	Induced premature hatching	Amphibians: *Bufo americanus, Rana sylvatica, Ambystoma maculatum*	Gomez-Mestre et al. 2006; Touchon et al. 2006
Bacteria (*Pseudomonas*)	Induced premature hatching	Whitefish, *Coregonus* sp.	Wedekind 2002

		Egg predator attack or predation risk	
Snake attack (vibration)	Induced premature hatching	Phyllomedusine treefrogs: 5 species of *Agalychnis* (above), *Pachymedusa dacnicolor*	Warkentin 1995; Gomez-Mestre et al. 2008
Wasp attack	Induced premature hatching	Red-eyed treefrog, *Agalychnis callidryas*	Warkentin 2000
Fly larvae	Induced premature hatching	Reed frog, *Hyperolius cinamomeoventris*	Vonesh 2000
Frogs, fly larvae	Induced premature hatching	Reed frog, *Hyperolius spinigularis*	Vonesh 2005
Leeches (chemical)	Induced premature hatching	Frogs: *Rana arvalis, R. cascadae, R. clamitans, R. temporaria, Hyla regilla*	Chivers et al. 2001; Laurila et al. 2002; Capellán and Nicieza 2007; Ireland et al. 2007
Crayfish (chemical)	Induced premature hatching	Leopard frog, *Rana sphenocephala*	J. Johnson et al. 2003; Saenz et al. 2003
Crayfish (chemical)	Induced premature hatching	Fathead minnow, *Pimephales promelas*	Kusch and Chivers 2004
Simulated predation (injured egg chemical)	Induced premature hatching	American toad, *Bufo americanus*	Touchon et al. 2006
		Larval predation risk	
Flatworm, sunfish (chemical)	Delayed hatching	Streamside salamander, *Ambystoma barbouri*	Sih and Moore 1993; R. Moore et al. 1996
Dragonfly nymph (chemical)	Delayed hatching	Green frog, *Rana clamitans*	Ireland et al. 2007
Stickleback fish (chemical)	Delayed hatching	Common frog, *Rana temporaria*	Laurila et al. 2002
Salamander (chemical)	Inhibited hatching of resting eggs	Crustaceans: *Arctodiaptomus similus, Ceriodapnia quadrangula, Cyzicus sp.*	Blaustein 1997; Blaustein and Spencer 2001
Turbellarian (chemical)	Inhibited hatching of resting eggs	Fairy shrimp, *Branchiopodopsis wolfi*	De Roeck et al. 2005
		Conspecifics	
Clutch-mates hatching (acoustic cues)	Delay and/or accelerate to synchronize hatching	Precocial birds	Reviewed in Brua 2002
Clutch-mates hatching	Accelerate hatching to synchronize	Turtle, *Emydura macquarii*	R. Spencer et al. 2001

embryos interact with their egg-carrying mothers to play a role in the timing and synchrony of their own hatching, or "larval release," which is adjusted in response to environmental cycles affecting predation risk (S. Morgan 1995; Ziegler and Forward 2007).

Embryo hatching responses are cued by a variety of stimuli, including waterborne chemical cues, direct physical disturbance or vibration, hypoxia, and acoustic cues (table 10.1). In many cases the responsive embryos are already well developed, and the response is primarily behavioral (Warkentin 1995; R. Brown and Iskandar 2000; Brua 2002). These late-stage embryos are likely capable of considerable neural processing of information. In other cases, embryos respond to risk at very immature developmental stages. For instance, in response to infection with pathogenic water mold, American toad embryos hatch at the tail bud stage, shortly after closure of the neural tube and before the development of muscular response (Gomez-Mestre et al. 2006; Touchon et al. 2006). The potential mechanisms for such early responses are limited by the poorly developed condition of the nervous system.

As well as altering hatching timing, embryos show other apparently adaptive responses to varying conditions inside the egg. For instance, some cooling bird embryos vocalize to solicit parental care, improving thermoregulation (reviewed in Brua 2002). Snail embryos, *Helisoma trivolvis,* move to the better-oxygenated side of their spacious eggs and increase their rate of rotation under hypoxia (Kuang et al. 2002). As early as the neural tube stage of development, red-eyed treefrog embryos use ciliary rotation to position their developing heads in the air-exposed region of their eggs, where oxygen is highest; well-developed embryos position their external gills in the same location (Rogge and Warkentin 2008).

Embryos from a diverse array of taxa respond to environmental conditions that affect their survival or development, in ways that appear adaptive. This is what we would expect from a consideration of selective pressures, and in some ways embryos are perhaps not so different from later life stages. In other ways they are very different. Their sensory systems are not fully developed and are changing rapidly, affecting the information they can gather. Their prior opportunities for learning are very limited. Their response options are also more limited than those of later stages, and the implications are different; hatching is irreversible and precipitates substantially greater ecological and physiological changes than do most defensive behaviors.

While the rapidly changing capabilities of developing embryos may complicate the study of their responses to risk, these early developmental stages offer certain advantages. Because reproduction and even feeding will not become relevant until later, their fitness components often simplify to survival, and potentially the rate or quality of development. This means that the trade-offs that shape plastic responses often occur in a common currency (death in one context vs. death in another), facilitating their analysis. Embryos thus offer excellent and underutilized opportunities to study how animals use information in a naturally simplified, yet ecologically important, context.

10.4. Hatching decisions: Information use by red-eyed treefrog embryos

10.4.1. THE ECOLOGICAL CONTEXT OF ESCAPE HATCHING

Red-eyed treefrogs, *Agalychnis callidryas,* are currently the best-studied case of embryo responses to multiple sources of egg-stage mortality. These frogs live in Neotropical, low to mid-elevation, wet forests from the Yucatán through Panama (Duellman 2001). They attach their gelatinous egg masses to vegetation overhanging ponds and swamps, and the tadpoles fall into the water upon hatching. The eggs are preyed on by arboreal and aerial predators, including several species of snakes and wasps. Egg predation rates can be high: during field monitoring, over half the clutches at one pond were eaten by snakes, and about half the clutches at another pond were attacked by wasps (Warkentin 1995, 2000). Egg clutches can also become infected with pathogenic fungus or infested with egg-eating fly larvae (Villa 1979, 1980; Warkentin et al. 2001). Terrestrial incubation is obligate. Even in air, oxygen levels within eggs can be very low (Warkentin et al. 2005), and eggs that are submerged underwater prior to hatching competence drown (Pyburn 1970).

Embryos escape from all of these egg-stage threats by hatching prematurely, as much as 30% before the peak of spontaneous hatching. At our field site in Panama, embryos are capable of hatching at 4 days but typically hatch at 6–7 days if undisturbed (Warkentin 2000). Hatching-competent embryos do not drown; all hatch shortly after submergence (Warkentin 2002). Similarly, hatching-competent embryos are rarely if ever killed by fungus; they hatch prematurely as the fungus begins to grow over their egg capsule (Warkentin et al. 2001). Through most of the plastic hatching period escape success in snake attacks is high, averaging 80%; however, at the onset of hatching competence escape success is lower (60%) and more variable (Gomez-Mestre and Warkentin 2007). Similarly, escape success in wasp attacks increases from 38% at the onset of hatching competence to 85% a day later (Warkentin et al. 2006a).

Agalychnis callidryas embryos hatch by performing specific movements that rapidly rupture the egg capsule and propel them from it, typically in under a second (Warkentin et al. 2007). Unlike some other amphibians (e.g., Carroll and Hedrick 1974), *A. callidryas* egg capsules do not lose integrity as embryos mature, and neither growth nor routine movements rupture them. Embryos exit the egg only when they perform the hatching behavior, and the timing of hatching is therefore a behavioral decision.

Both within *A. callidryas* egg clutches and among clutches laid at the same time at a breeding site, development is remarkably synchronous. Thus, animals hatching at different ages enter the water at different sizes and stages of development (Warkentin 1999b, 2007). These differences among hatchlings affect their vulnerability to aquatic predators; older hatchlings are better able to evade fish, shrimp (Warkentin 1995, 1999a), odonates, belostomatids (J. Vonesh and Warkentin, unpublished data), and backswimmers (Warkentin, unpublished data). Red-eyed treefrogs breed throughout the rainy season (usually May to November in Gamboa, Panama) in a wide variety of habitats, including temporary and permanent ponds. Tadpoles entering a recently filled pond may enter an essentially predator-free environment, but because predators rapidly colonize ponds, most hatchlings face aquatic predators (J. Vonesh and Warkentin, unpublished data).

Thus, *A. callidryas* experience a risk trade-off at hatching. Hatching early allows embryos to escape from multiple sources of egg mortality, whereas delaying hatching allows tadpoles to avoid aquatic predators until they are developmentally more competent to evade them. Hatching plasticity is adaptive in this context, and appropriate hatching timing depends on information about risk. It is unlikely that arboreal embryos have access to current information about the aquatic environment. Nonetheless, their evolutionary history with aquatic predators (as well as with egg-stage threats and other stimuli) has presumably shaped their hatching behavior. This history provides their prior information about risk, in the Bayesian decision theory sense (J. Bradbury and Vehrencamp 1998; Dall et al. 2005). Individual embryos also have access to immediate, local information about egg-stage risk that allows behavioral modulation of hatching stage.

10.4.2. HOW DO EMBRYOS DECIDE WHEN TO HATCH?

Red-eyed treefrog embryos must use multiple types of information in their hatching decision because they respond to different sources of risk, such as flooding and snake attacks, that provide different types of cues. Hypoxic stress appears to be both the cause of death of flooded embryos too young to hatch and the stimulus that cues hatching when older eggs are flooded; the hatching response of eggs in hypoxic gas mixtures is very similar to that under flooding (Warkentin 2002). The egg masses are attached to vegetation and eggs adhere closely to each other, leaving only a fraction of each egg's surface exposed to air. This generates strong oxygen gradients within individual eggs, so embryos can experience local hypoxia if they move their gills away from the air-exposed surface (Warkentin et al. 2005). After both experimental repositioning and spontaneous movements that place gills in poorly oxygenated regions, embryos move themselves back, within seconds, to place their gills near the air (Rogge and Warkentin 2008). Embryos that hatch under hypoxic stress move frequently within the egg in the last few minutes before hatching, suggesting that they extensively sample the respiratory environment within the egg before they decide to leave it (Rogge and Warkentin, unpublished data). These position changes are distinct from the movements that cause hatching.

Snake-induced hatching occurs over a different time frame and depends on different information than flooding-induced hatching. Experimentally flooded embryos began hatching in ≈2 minutes, 55% hatched by 5 minutes, and 99% by 25 minutes (Warkentin 2002). In videotaped snake attacks, embryos began hatching in 16 ± 3 seconds, and those that escaped (78%) had hatched in 4.8 ± 0.8 minutes (Warkentin et al. 2007). It is unlikely that snakes alter gas exchange of eggs, except perhaps for those already in their mouth. Predator-induced premature hatching appears, rather, to be a response to the physical disturbance of egg clutches. Embryos are not induced to hatch when snakes are nearby, even within a few centimeters and looking at the clutch, but only after physical contact, as the snake attacks the clutch and moves eggs. It is also possible to induce hatching with artificial mechanical disturbance (Warkentin 1995, 2005; Gomez-Mestre et al. 2008). Nonetheless, some intense but benign physical disturbances, such as torrential tropical rainstorms, do not induce premature escape hatching.

10.4.3. DO EMBRYOS USE VIBRATIONAL CUES IN PREDATOR ATTACKS?

The physical disturbance of egg clutches in predator attacks is complex: eggs are moved in three dimensions as well as deformed by compression and stretching. Thus, embryos within eggs might perceive motion, pressure, and/or tactile cues. Many animals use vibrations as a source of information, both in predator-prey interactions and in intraspecific communication (P. Hill 2008), and vibrations are more experimentally tractable than other elements of the physical disturbance. We therefore recorded vibrations from egg clutches during snake attacks and rainstorms and played them back to embryos to determine (*i*) if vibrations alone are sufficient to elicit rapid, premature hatching and (*ii*), if so, if the hatching response of embryos to vibrations is general, or if it depends on the specific properties of the vibrations.

We used miniature accelerometers embedded in egg clutches to record vibrations and an electrodynamic shaker connected to an interface of blunt

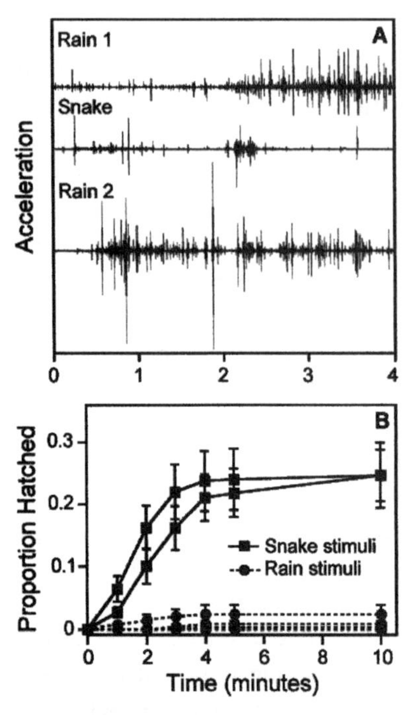

FIGURE 10.2. A. Vibrations recorded from *Agalychnis callidryas* egg clutches. Middle: Complete recording of an attack by a parrot snake, *Leptophis ahaetulla*, in which 76% of the eggs hatched and the rest were eaten. Top and bottom: Periods of rain from two different storms, recorded from different clutches. These vibrations were used as stimuli in playback experiments. B. Hatching response of *A. callidryas* egg clutches to playbacks of recorded vibrations from A and to a second set of duration- and amplitude-matched parrot snake and rain recordings. More embryos hatched in response to snake playbacks than in response to rain. Data are means ±SE. Redrawn from Warkentin 2005.

metal tines inserted among the eggs to play vibrations (Warkentin 2005). Neither the recording nor the playback system achieves the fidelity possible with sound recordings and playback in air; nonetheless, they allow us to test hypotheses about embryo behavior. Our first playback experiment used two exemplars of snake vibrations, each paired with two rain exemplars recorded from different clutches and matched to the snake recordings for duration and root-mean-square acceleration (fig. 10.2a). We played the vibrations to 5-day-old egg clutches in Gamboa, Panama, which are fully hatching competent but would not normally hatch for another day or two if undisturbed.

Snake vibrations elicited substantially more hatching than rain, and all exemplars of rain vibrations elicited almost no hatching (fig. 10.2b; Warkentin 2005). This demonstrates that embryos can hatch in response to vibrations alone, without other cues that are present in real predator attacks, and that they discriminate between vibrational patterns in a manner appropriate to the original sources of vibration. The absolute level of hatching in response to these snake vibration playbacks was lower than average escape success in real attacks, including in the clutches used in recordings. Nonvibrational cues may also play a role in the response to snakes, or the weaker response to playbacks may reflect infidelities of the playback system, particularly attenuation of low frequencies by the amplifier used in those playbacks (Warkentin 2005). Whether or not other cues also contribute to escape hatching in snake attacks, vibrations are clearly an important source of information for red-eyed treefrog embryos deciding when to hatch.

10.4.4. WHAT PROPERTIES DISTINGUISH BENIGN AND DANGEROUS-SOURCE VIBRATIONS?

The most frequent source of intense, benign vibrations in *A. callidryas* egg clutches is rainstorms. Based on rainfall data collected by the Panama Canal Authority Meteorological and Hydrological Branch at our least rainy field site (Gamboa, Panama) during the rainy seasons of 1993–2003, the likelihood of eggs being rained on at least once during the 2-day period when they are hatching competent but still premature averaged 79% (±1% SE across years). During the same developmental period at our study pond with the highest incidence of snake predation (in Corcovado National Park, Costa Rica), ≈20% of clutches were attacked by snakes (Warkentin 1995; Gomez-Mestre and Warkentin 2007). Considering the elevated predation risk for premature hatchlings in the water, it may be as or more important for *A. callidryas* to avoid hatching early in response to rain as it is to hatch in snake attacks.

The vibrational cues available to embryos in any disturbance result from the combination of the forcing pattern (i.e., snake feeding movements or raindrop impacts) and the biomechanical characteristics of egg clutches. Our recordings reveal that clutch vibrations caused by snakes and rain differ statistically in several properties (Warkentin 2005), and there are also differences among vibrations from other sources, such as wind, routine movements of embryos, and attacks by egg-eating wasps (M. Caldwell et al. 2009). Nonetheless, there is substantial overlap between benign-source and dangerous-source vibrations for every cue property we have examined. This overlap derives in part from the large variation (lack of stereotypy) in forcing patterns by each source of disturbance and in part from the shared effects of clutch biomechanics on vibrations excited by any force. It also creates an information-processing challenge for embryos.

Our recordings of rain and snake vibrations suggest that amplitude is uninformative in solving this problem. We found almost complete overlap in the range of amplitudes of rain and snake disturbances, and no consistent trend between them (Warkentin 2005). Excluding attempts by snakes to eat our accelerometers, the highest peak accelerations we recorded from egg clutches were in rainstorms; thus, harder vibrations do not indicate greater risk.

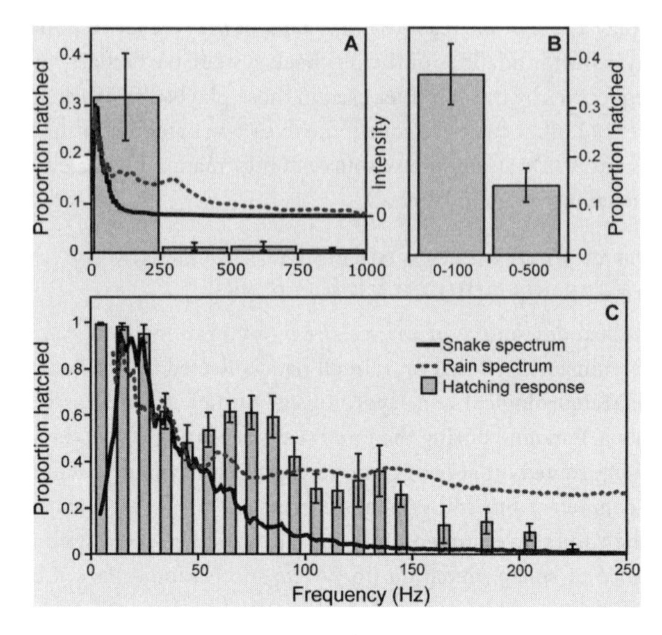

FIGURE 10.3. A. Hatching response of *Agalychnis callidryas* egg clutches to vibrational stimuli consisting of bursts of band-limited white noise in different frequency ranges (gray bars, left axis). The average frequency spectra of vibrations recorded from egg clutches during parrot snake attacks (solid line) and rainstorms (dashed line) are superimposed, with the zero point offset for clarity (right axis). Embryos hatch in response to low-frequency vibrations, which are strongly represented in snake and rain vibrations, and not in response to vibrations over 250 Hz, which do not occur in snake attacks. B. Hatching response to a low-frequency (0–100 Hz noise) vibrational stimulus and to a broader-band (0–500 Hz noise) stimulus, intensity matched for the overlapping range. Adding energy at higher frequencies decreased the hatching response. C. Frequency response curve of escape hatching, with superimposed snake and rain vibration spectra, for the 0–250 Hz range. Playback stimuli were 10 Hz wide bands of noise, at different frequencies, matched for peak acceleration using rerecordings. Data are means ±SE. Redrawn from M. Caldwell et al. 2009; M. Caldwell et al., submitted.

Vibration frequency is more informative. On average, clutch vibrations from individual raindrops have higher dominant frequencies than vibrations caused by snakes, and spectra from storms typically contain energy at higher frequencies that are not represented in snake spectra (figs. 10.3a and 10.3c). Nonetheless, many individual raindrops produce vibrations with dominant frequencies indistinguishable from those of snakes, and overall storm and snake spectra overlap substantially; both have strong low-frequency peaks and most of their energy is in a similar range (Warkentin 2005). Importantly, there is no frequency that could unambiguously implicate a predator as the source of a vibration.

The time domain also provides embryos with informative cue properties. On average, vibrations from rain are shorter in duration and separated by shorter intervals than vibrations caused by snakes (Warkentin 2005). In addition, vibrations of extremely short duration are more likely to occur in rainstorms than in snake attacks, and the longest interdisturbance intervals characteristic of snake attacks, occurring while snakes swallow eggs they have removed from clutches, are absent in hard rain. Nonetheless, the large amount of overlap in these temporal properties means that many individual vibrations, as well as some longer periods of intermittent disturbance, cannot be unambiguously classified using either property.

10.4.5. WHAT CUE PROPERTIES DO EMBRYOS USE IN VIBRATION-CUED HATCHING?

We are using vibration playback experiments to elucidate how *A. callidryas* embryos use vibrations to decide when to hatch. Specifically, we have tested several different cue properties for their effect on escape-hatching behavior, assessed how embryos combine information from two cue properties, and examined their sampling strategy. Our research is ongoing, and the picture emerging is of a multifaceted, complex risk assessment strategy. Such a strategy is in keeping with the irreversibility of hatching, high level of both egg and larval predation, and strong risk trade-off across life stages (i.e., with the high cost of both missed cues and false alarms). This complex strategy is also surprising for animals at such an early stage of development.

We have compared hatching responses to simple synthetic stimuli, in which we independently manipulated different vibrational properties. We have also used vibrations recorded from snake attacks and rainstorms that we played both in their original form and as edited versions, altering one or more properties. Based on these experiments, at least four different cue properties affect the proportion of embryos that hatch in response to playbacks, including two frequency properties and two temporal properties (Warkentin 2005; Warkentin et al. 2006b; M. Caldwell et al. 2009; M. Caldwell et al., submitted).

10.4.5.1. *Frequency response*

We used bursts of synthetic white noise that were band-pass filtered to create stimuli in different frequency ranges, and we used these to construct frequency response curves for the escape-hatching response (M. Caldwell et al. 2009). All stimuli had the same temporal pattern, known to elicit hatching, and were matched for peak amplitude based on rerecordings. With 250 Hz wide bands, embryos showed a strong hatching response to the lowest-frequency stimulus (0–250 Hz), and essentially no response to higher frequencies (up to 1000 Hz; fig. 10.3a), consistent with the absence of those higher frequencies in predator attacks (Warkentin 2005). Similarly, 10 Hz wide bands of noise at very low frequencies that are strongly represented in snake attacks elicited the most hatching, and the hatching response declined through higher frequencies that contain progressively less energy in snake attacks (fig. 10.3c). The strong hatching response to vibration frequencies that predominate in snake attacks enables escape from snakes; however, these frequencies are also strongly represented in rain vibrations (figs. 10.3a and 10.3c; Warkentin 2005). If this were the only information embryos used, they would also hatch prematurely, and unnecessarily, in storms.

The frequency spectrum of rain is broader than that of snake attacks, extending to higher frequencies with substantial energy above 100 Hz (figs. 10.3a and 10.3c). To test if these higher frequencies might reduce the hatching response to lower frequencies presented concurrently, we played embryos stimuli constructed from 0–100 Hz noise and 0–500 Hz noise. Stimuli were intensity matched for the overlapping range, so the broader band had substantially more energy. We found that adding high-frequency energy to the low-frequency base stimulus substantially decreased the hatching response (fig. 10.3b; M. Caldwell et al., submitted). Similarly, we were able to increase the hatching response to playbacks of recorded rain vibrations by low-pass filtering them to remove high frequencies (M. Caldwell et al., submitted). The inhibitory effect of high frequencies would reduce the incidence of false alarms and is consistent with selection against early hatching by aquatic predators. However, embryos show a strong hatching response to snake attacks during rainstorms (Warkentin, personal observation). It is unclear how they achieve this or exactly how they integrate information from high- and low-frequency vibrations.

10.4.5.2. *Response to temporal properties*

We assessed how hatching varied with two simple temporal properties that differ between snake and rain vibrations: the duration of vibration events and the interval, or spacing, between them. We used amplitude-matched stimuli made from bursts of 0–100 Hz synthetic white noise and independently varied these two properties (Warkentin et al. 2006b). First, we asked simply if more vibration elicits more hatching. In fact, except at extreme values, the duty cycle (vibration duration/cycle length) was unrelated to the hatching response. The proportion of embryos that hatched varied substantially across stimuli with the same duty cycle, while other stimuli differing in duty cycle elicited similar levels of hatching. Stimuli with very short duty cycles elicited little or no hatching, but so did those with very long duty cycles. Thus, more vibration per se was not perceived as more dangerous, which allowed us to examine the effects of duration and interval across stimuli differing in total energy (Warkentin et al. 2006b). We found that both vibration duration and interval affect the hatching response, and they function not as redundant cues but as two necessary elements of a composite cue. Figure 10.4 shows the hatching response as a function of these two properties, revealing a single peak of high hatching, surrounded by a parameter space in which the hatching response declines to zero. If the value of either property is sufficiently distant from the peak, then embryos do not hatch, even if the other property is in the range that elicits highest hatching. Moreover, there is an interaction effect: at longer durations, the interval that elicits most hatching is also longer. This interaction could reduce unnecessary early hatching in hard rain, when vibrations from individual drops combine to produce longer vibrations, but the spacing between vibrations shrinks. Playbacks of snake and rain recordings with altered and original temporal patterns also corroborate that embryos attend to the duration of and intervals between vibrations (Warkentin 2005).

The hatching response of red-eyed treefrogs to vibrations is thus very specific. Each vibrational property that affects hatching adds information and narrows the range of stimuli that elicit the defense, reducing the chance of false alarms. To date, we have identified no redundancy, as occurs in communication when the same information is conveyed in multiple ways, reducing the chance

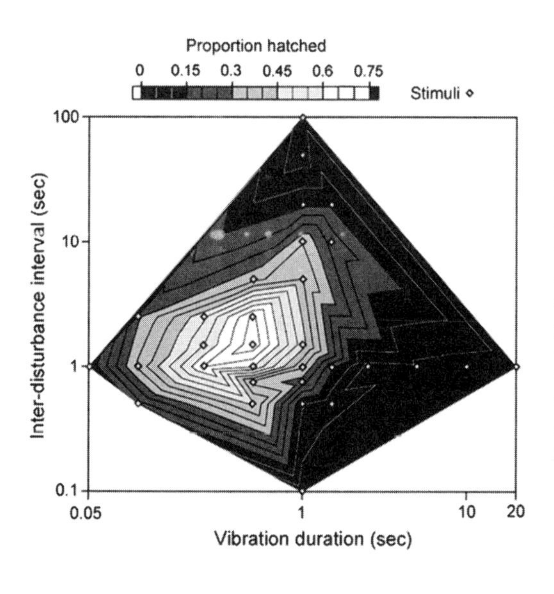

FIGURE 10.4. Contour plot of the hatching response of *Agalychnis callidryas* embryos to playbacks varying in vibration duration and the interval, or spacing, between periods of vibration. The 31 stimuli were bursts of amplitude-matched 0–100 Hz noise in rhythmic patterns. A small range of durations and intervals elicited high hatching (lighter shades) and if either property was out of this range there was little or no hatching. These properties therefore function as nonredundant elements of a composite cue. Redrawn from Warkentin et al. 2006b.

of missed cues (Partan and Marler 2005). Given the substantial overlap in vibrational properties and the high probabilities that embryos will experience rainstorms and snake attacks, the use of multiple cue properties may be essential to effectively discriminate predator attacks from benign disturbances.

10.4.6. HOW MUCH INFORMATION DO EMBRYOS USE IN VIBRATION-CUED HATCHING?

The mean latency for the first embryo to hatch in snake attacks, as measured from videotapes, is 16 seconds, and many embryos wait longer, so that the last embryo is not gone from the clutch until on average 4.8 minutes after the snake began its attack (Warkentin et al. 2007). This delay is not due to the process of hatching, which is very rapid (< 1 second) once embryos begin hatching movements. Rather, it suggests that *A. callidryas* are not hatching as rapidly as possible once a disturbance starts. Instead, they appear to sample a period of vibration before deciding to hatch. As with use of multiple cue properties, an extended sampling period is consistent with the difficulty of the discrimination task and the high cost of both decision errors.

Nonetheless, staying in an egg clutch to gather information during a snake attack is dangerous. At least for temporal properties (duration and interval) information accrues as a decelerating function of cycles of the pattern (fig. 10.1d). In contrast, the probability of being eaten likely accrues over time, as snakes consume eggs. Thus, gathering an equivalent amount of information entails longer exposures to risk, and greater probability of mortality, for vibrational patterns with longer cycles. The net benefit from information use is the reduction in potentially lethal missed cues and false alarms, minus the increased probability of predation that accrues in gathering the information (Warkentin et al. 2007). To maximize net benefit embryos should adjust their information use as the cost of sampling changes.

To assess if embryos use a flexible sampling strategy, basing their hatching decision on less information when that information is more costly, we examined the time course of hatching in response to stimuli that differed in cycle length, or the rate at which information accrues (Warkentin et al. 2007). Stimuli were matched for total energy and duty cycle and chosen to elicit a similar overall level of hatching, but they differed across an order of magnitude in cycle length (1.1–11 seconds). As cycle length increased, the lag time before hatching also increased and the number of vibration cycles before hatching decreased (fig. 10.5). This flexible sampling strategy is consistent with balancing a trade-off between the value of information in reducing decision errors and its potential cost in terms of predation risk. Embryos appear to partition their costs, accepting both a somewhat-higher total cost of sampling

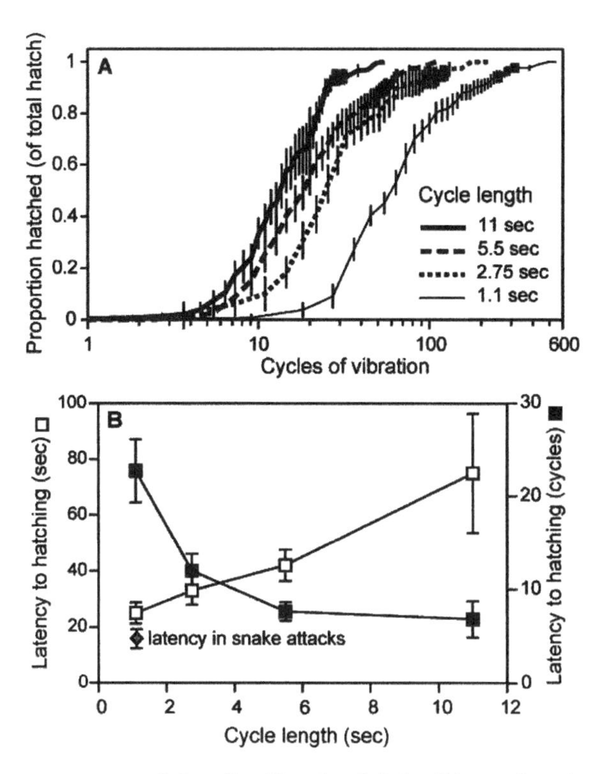

FIGURE 10.5. Timing of hatching of *Agalychnis callidryas* embryos in response to vibration playbacks varying in cycle length. Temporal pattern information accrues more rapidly with shorter cycles. Stimuli were matched for duty cycle and elicited similar overall levels of hatching over a 5-minute playback period. A. Cumulative hatching curve, as a function of cycles of vibration. B. Latency to first hatching, a measure of the minimum sampling period, plotted as time (open squares) and as cycles of vibration (black squares). The latency to first hatching in snake attacks, measured from videotapes, is included for comparison (gray diamond). When cycles were longer, thus entailing longer exposures to risk if the vibration source were a snake, embryos sampled fewer cycles before hatching. They also sampled a longer period of time, thus accepting greater exposure to risk to gather information before their decision. Data are means ±SE. Redrawn from Warkentin et al. 2007.

and greater uncertainty in their decisions when each increment of information is more costly.

This demonstrates an additional level of behavioral flexibility in *A. callidryas* embryos. Not only do these embryos use information on egg-stage risk to adjust their hatching timing, but they also adjust this information use based on the cost of sampling.

10.4.7. SUMMARY OF INFORMATION USE IN RED-EYED TREEFROG HATCHING TIMING

Hatching is both a momentary behavior for *A. callidryas* and an irreversible life history transition. Appropriate hatching timing can be critical for the survival of these animals, poised between opposing sets of egg- and larval-stage threats. Hatching is also well informed, affected by cues in multiple sensory modalities. In two very different risk contexts—predator attack and flooding—embryos appear to sample a relatively large amount of information for their decision.

The response of *A. callidryas* to vibrational cues demonstrates surprising complexity, not previously associated with behavioral decisions in such early developmental stages. Their response specificity and pattern of information use are consistent with the high cost of mistakes and with selection to minimize both missed cues and false alarms. By attending to multiple nonredundant cue properties, *A. callidryas* embryos effectively reduce the overlap between the benign background of their vibrational environment and the cues that indicate imminent predation. In addition, their flexible sampling strategy permits them to base their hatching decision on a substantial amount of information, balancing the value of better-informed decisions against the immediate risk associated with sampling.

10.5. Conclusions and future directions

10.5.1. COGNITIVE ECOLOGY OF EMBRYOS

Learning, memory, and information use are sometimes considered beyond the capabilities of embryonic life stages. However, there is substantial evidence that embryos utilize information gathered from the environment surrounding their eggs. Considering their ecological circumstances, this should not be surprising. Hatching is a major life history transition that changes both a developing animal's physical environment and its biotic interactions. Moreover, early life stages typically experience high mortality, and thus potentially strong selection. In this context, the appropriate timing of hatching can be critical. Embryos from a diverse array of taxa alter hatching timing in response to cues associated with conditions affecting their survival and development. In some cases, parents are involved in gathering information and transferring it to embryos (Crisp and Spencer 1958; Saigusa 2002), but the embryos nonetheless respond to the parental signal. In many other cases, embryos themselves gather and respond to ecologically relevant information. Embryos use environmental cues not only to inform hatching but also to regulate their behavior and development within the egg (e.g., Kuang et al. 2002; Voronezhskaya et al. 2004). Furthermore, there is evidence that some embryos learn and retain information relevant for later life (e.g., parental vocalizations; Brua 2002).

The fact that embryos use information makes them appropriate subjects for studies in cognitive ecology. The relative simplicity of their lives, without feeding or reproduction, narrows the range of ecologically relevant factors to consider in such studies. Moreover, the large numbers of embryos produced in many taxa and their high natural mortality rates facilitate experimental studies of risk. Insights gained from embryos can inform our general understanding of how animals use information in an ecological context. Moreover, because most animals begin life as eggs, and most eggs face variable environments, understanding how embryos use information is essential for a complete cognitive ecology.

10.5.2. INFORMATION AND INDUCIBLE DEFENSES

The effective use of inducible defenses depends on information about risk. Often this information takes the form of nonstereotyped cues. Discriminating such cues from overlapping background noise is a general, and still poorly understood, problem in information processing. Thus, well-designed studies of animal mechanisms of risk assessment offer opportunities for novel, broadly important insights. We expect risk assessment solutions to vary with the nature of defenses. Low-cost, reversible defenses are forgiving of false alarms and could be based on little, or poor-quality, information. In contrast, when defenses involve costly irreversible decisions, selection should favor improved information use.

Bioacoustics has contributed substantially to our understanding of animal communication (J. Bradbury and Vehrencamp 1998). In part, this is because sound is an important signaling modality, but the experimental power of sound recording, analysis, manipulation, and playback methods have also been critical. These methods have been productively applied to study prey eavesdropping on predator acoustic signals (Hoy et al. 1998; Ratcliffe, chapter 11 in this volume). In contrast, little work has addressed prey use of incidental sounds made by predators, perhaps because hunting predators are often effective at remaining acoustically inconspicuous. Substrate vibrations, although less obvious to human observers, may be comparable to sound in their importance as an information channel for animals (P. Hill 2008). Many animals, vertebrates and invertebrates alike, use stereotyped vibrational signals to communicate. They also use the nonstereotyped vibrations inevitably produced when other animals move as cues in both foraging and defense

(P. Hill 2008). Vibration-cued defenses and foraging offer excellent, underexploited opportunities to apply the power of recording and playback methods to study nonstereotyped forms of information.

Information, detection, and decision theory have played key roles in studies of animal communication. They also provide a theoretical framework in which to approach information use in predator-prey interactions and other contexts of risk assessment. Students of predator-prey interactions have considered how prey should behave under the typical context of imperfect information about risk (reviewed in Bednekoff 2007). These models are not, however, integrated with explicit consideration of the mechanisms by which, and the background environmental context within which, prey acquire that information. Together these will determine both the quality of information about risk and how much that information costs to acquire. In particular, to elucidate decision mechanisms in inducible defense, understanding why animals make the mistakes they do may be as informative as understanding what cues their correct responses. Linking empirical studies of risk assessment mechanisms to the theoretical framework of information, detection, and decision theory both provides testable hypotheses for empirical work and new contexts in which to assess the generality of these bodies of theory.

ACKNOWLEDGMENTS

We thank Mike Ryan, Stan Rand, Steve Phelps, and members of the Gamboa Frog Seminar group for discussion of these ideas in various stages of development, and John Ratcliffe for comments on the manuscript. Our collaborators on studies described in this chapter were Cameron Currie, Steve Rehner, Ivan Gomez-Mestre, Alison D'Amato, Timothy Siok, Jessica Rogge, and especially vibrations engineer J. Gregory McDaniel. We also thank all of the other students and interns who helped with egg testing. Our research has been supported by the National Science Foundation, Smithsonian Tropical Research Institute, Natural Science and Engineering Research Council of Canada, and Boston University.

11 Predator-Prey Interaction in an Auditory World

JOHN M. RATCLIFFE

11.1. Of bats and moths and coevolution

Echolocating bats and moths with bat-detecting ears are often cast as predator and prey in an iconic example of a coevolutionary arms race (e.g., Ratcliffe and Fullard 2005; Windmill et al. 2006). Yet whether any bat species possesses traits evolved specifically to circumvent the auditory defenses of moths is unclear (G. Jones and Rydell 2003). What can be said is that bats exploited the unrealized foraging niche of nocturnally active flying insects through two key innovations: powered flight and echolocation (fig. 11.1; Fenton et al. 1995; N. Simmons and Geisler 1998; G. Jones and Teeling 2006). On the other hand, while many flying insects possess ears able to detect the echolocation calls of bats (Hoy 1992; L. Miller and Surlykke 2001), for almost all eared moths, the sole purpose of their ears is to detect bat echolocation calls and initiate evasive flight and other anti-bat defensive behaviors (Fullard 1988, 1998; see Conner 1999 for review of secondary social functions of ears for a small number of species). By setting the analogy to an arms race aside, the interactions of bats and moths may be appreciated under different lights, where the unique properties of bat echolocation and the neural simplicity and singular purpose of the ears of moths take center stage. On the outside, bat-moth interactions last only a few seconds and are often over in just hundreds of milliseconds (Kalko 1995; Reddy and Fenton 2003). Thus, for both parties, auditory information must be processed and decided upon within a fraction of a second. For a coevolutionary perspective and a more thorough account of the neuroethology of auditory-evoked anti-bat flight in moths, see Fullard 1998. In this chapter, I will first introduce predator and prey and then discuss the ways bats detect, discriminate, and catch insects and how moths avoid being eaten by bats.

11.1.1. THE PREDATORS: ECHOLOCATING BATS

Bats are the most geographically widespread mammals and second only to rodents in sheer species number (≈1100 species; K. Jones et al. 2002). They

FIGURE 11.1. A little brown bat, *Myotis lucifugus*, emitting echolocation calls while flying through an abandoned mine in Ontario, Canada. Photograph by M. B. Fenton.

are long-lived, relatively slow to sexually mature, and typically bear a single pup once or twice a year (Kunz and Fenton 2003). Bats, insects, birds, and pterosaurs are the only animals to have evolved powered flight. Of these, only bats are known to echolocate their prey. Echolocation is the determination of spatial arrangements and characteristics of nearby objects using the echoes of one's own vocalizations, and it allows bats to orient and forage in darkness (D. Griffin 1958; Schnitzler et al. 2003). Of the 950 or so extant species of bats using laryngeal echolocation, more than 800 eat animals and almost all of these are expected to take insects. The dominant means by which these bats take prey are aerial hawking (the capture of airborne prey on the wing) and substrate gleaning (the capture of prey from surfaces). For most predatory bats, these two strategies demand different echolocation call designs, emission rates, and flight behaviors. Until recently, most research has focused on the sensory ecology of aerial-hawking bats; the use and integration of echoic and prey-generated sounds for a successful gleaning attack are less well understood. Echolocation is an active sensory system and the individual bat is both the sender and intended receiver of its own signals. Echolocation signals are thus susceptible to interception and some would-be prey can use this auditory information to evade capture (L. Miller and Surlykke 2001; ter Hofstede et al. 2008a).

11.1.2. THE PREY: EARED MOTHS

It is estimated that roughly 92,000 of the 200,000 species of the world's extant moths and butterflies possess ears (fig. 11.2; Spangler 1988; Fullard 1998). That these ears evolved solely as a result of selective pressures from bats is supported by several lines of evidence: (*i*) moths' simple ears face rearward and (*ii*) are sensitive only to high-frequency, high-intensity sounds, and for the vast majority of species ears (*iii*) apparently serve no other function than to evoke evasive flight and other defensive behaviors (Fullard 1988). As a result of energy redistribution through spherical spreading (i.e., each successive doubling of distance traveled reduces the signal's sound pressure level by 6 dB along its acoustic axis), an eared moth will detect an aerial-hawking bat's echolocation calls at greater distances than will the bat detect the moth from the returning signals (echoes) (Roeder 1966; Surlykke 1988). For most eared moths, the sounds of distant, searching bats elicit negative phonotaxis; those of proximate and attacking bats elicit erratic flight, power dives, and flight cessation (Roeder 1967). These auditory-evoked evasive flight behaviors confer considerable benefits to moths. Palatable moths able to effect acoustically mediated evasive flight maneuvers in the face of bats are more than 40% more likely to survive an encounter than are moths not so able (Roeder and Treat 1962; Roeder 1967; Dunning et al. 1992; Acharya and Fenton 1992, 1999). Evasive flight is, however, energetically expensive (Srygley and Chai 1990; Marden and Chai 1991) and will reduce the time available for searching for

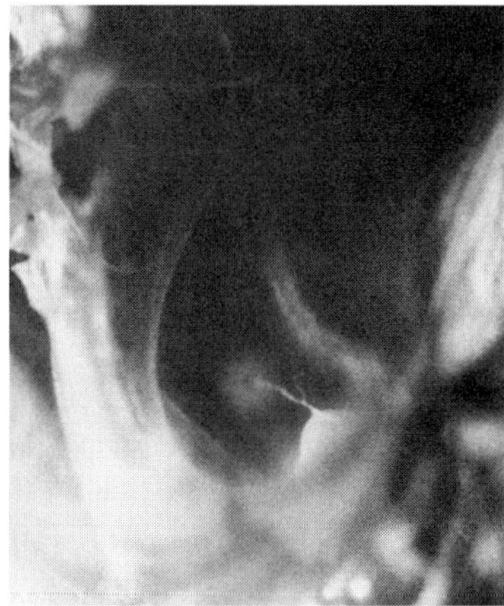

FIGURE 11.2. A noctuid moth (the herald, *Scoliopteryx libatrix*) and one of its tympanal ears. This Holarctic species overwinters as an adult and will do so in caves used by bats as hibernacula (Roeder and Fenton 1973). Photograph of moth by M. B. Fenton; photograph of ear by M. McCully.

pheromone plumes and egg-laying sites (Acharya and McNeil 1998; Fullard and Napoleone 2001). Some tiger moths (Arctiidae) that advertise their toxicity through acoustic warning signals effective against hawking bats (section 11.3.1) are not so easily deterred from flight (Ratcliffe et al. 2008).

11.1.3. EVOLUTIONARY AND ECOLOGICAL BACKDROP

Echolocating bats and moths have coexisted for more than 50 million years and are geographically cosmopolitan: species from both groups are found on every continent except Antarctica and, in the case of bats, some isolated oceanic islands. Moths precede bats in the paleontological record, and ears have apparently evolved independently several times in this group of insects (Hoy 1998; Fullard 1998; Yack et al. 1999). Laryngeal echolocation in bats has likely evolved only once (Teeling et al. 2002, 2005), and today most bats can be classified as being either low- or high-duty-cycle species (Fenton 1995). While in transit and in search of prey, low-duty-cycle species, which use frequency-modulated, downward-sweeping calls of short duration (≈1–20 ms) and long interpulse interval (≈50–200 ms; duty cycle: 5%–15%), calculate the delay between call and echo to estimate a target's location and, as they approach an object of interest, use spectral cues encoded in the echoes to resolve target characteristics (J. Simmons and Stein 1980). All else being equal, high-duty-cycle species use longer calls of more constant frequency and shorter interpulse intervals (duty cycle: >40%; Fenton 1995). Horseshoe bats (Rhinolophidae) and other high-duty-cycle bats calculate the frequency difference between a call and its echoes to estimate a target's relative velocity (Doppler-shift measurement), while amplitude fluctuations in these echoes (acoustic glints) are used to determine the wing beat frequency of would-be prey (Schnitzler and Henson 1980; Bell and Fenton 1984). Low-duty-cycle bats are found almost everywhere bats exist; with the exception of at least one species from the Neotropics, *Pteronotus parnellii* (Mormoopidae), high-duty-cycle bat species are found only in the Old World and are classified as belonging to the closely related families Hipposideridae and Rhinolophidae (G. Jones and Teeling 2006).

The range of call frequencies used by bats in a given community describes the echolocation assemblage (Fullard 1982). For example, in northeastern North America, aerially hawking bats emit echolocation calls with peak frequencies of between 20 and 55 kHz. The greater the number of bat species present, the greater the range of the echolocation assemblage tends to be (Fullard 1998). The upper bound can extend as high as 212 kHz in those parts of Africa where *Cleotis percivalis* (Hipposideridae) is found (Fenton and Bell 1981); the lower bound can extend as low as ≈11 kHz along the western coast of the continental United States and southern British Columbia, much of Europe and Asia, southern Australia, and southern Africa, where the bats *Euderma maculatum* (Vespertilionidae), *Tadarida australis, T. teniotus*, and *Otomops martienseni* (Molossidae) respectively are found (Rydell and Arlettaz 1994; Fullard and Dawson 1997; Fenton et al. 2004; Fullard et al. 2005). *E. maculatum* and *O. martienseni* are thought to be rare and may have little impact on the survivorship of moths. Some of the best evidence that insectivorous bats are the selective pressure maintaining ears in moths is that the overall range of auditory sensitivity for the nocturnally active species of a given moth community well reflects the echolocation assemblage of the insectivorous bat species living alongside them (Fullard 1998). Community-level analyses demonstrate that bats using frequencies at the limits of their respective assemblages consume more moths than those using frequencies nearer the middle of the range (Schoeman and Jacobs 2003; Pavey et al. 2006).

Auditory-evoked defenses aside, moths are an excellent food resource for bats: they are relatively large, slow-flying, soft-bodied insects (Fullard and Napoleone 2001). They are also short-lived (most surviving less than 1 week as adults) and, as many species do not feed as adults, essentially vehicles for reproduction. As prey, moths are therefore an interesting group in which to study potential trade-offs between finding mates and avoiding capture on a very tight energy budget and time schedule (e.g., Skals et al. 2005).

While bats and moths are found almost everywhere, the lion's share of published research on bat-moth interaction has used species native to temperate regions of the Northern Hemisphere (G. Jones and Rydell 2003), and my chapter relies heavily on data collected from wild-caught animals at Queen's University Biology Station in southeastern Ontario, Canada (44°3′ N, 79°15′W). Consequently, most of the bat species referred to are low-duty-cycle echolocators from the family Vespertilionidae (globally distributed family; all members primarily insectivorous). Due to their abundance and diversity in southeastern Ontario, and because of the neural simplicity of their one- or two-sensory-cell ears, the moths discussed here are arctiids, noctuids, and notodontids, families subsumed by the superfamily Noctuoidea.

11.2. Sensory ecology and behavioral flexibility of predatory bats

The progenitors of echolocating bats were likely nocturnal and insectivorous, constrained to short flights from perches and taking prey based on passive-acoustic cues and possibly visual information (N. Simmons and Geisler 1998; N. Simmons et al. 2008). The fossil record and phylogenetically based ancestral state reconstruction suggest that all extant bats evolved from a lineage of aerial-hawking species whose wing designs would have made them poorly suited to gleaning (N. Simmons and Geisler 1998; Safi et al. 2005). Today, many high- and low-duty-cycle species effectively use both strategies (Bell 1982; Bell and Fenton 1984; Schumm et al. 1991; Faure and Barclay 1994; Ratcliffe and Dawson 2003; Siemers and Ivanova 2004; Ratcliffe et al. 2006). Some bats also take prey from the surface of water, an uncommon strategy demanding flight and echolocation behaviors somewhat similar to those of aerial hawking (Siemers et al. 2001). In general, moths do not spend a lot of time over water (Yack 1988), and so this foraging strategy will not be considered here. Some bats use a sit-and-wait strategy and take insects that fly past their perch. Perch hunting is also akin to hawking, the difference being that the search phase (section 11.2.1) occurs while the bat scans its immediate environment from a fixed position rather than on the wing. Perch hunting may precede hawking in the evolutionary history of bats (N. Simmons and Geisler 1998) and demand exceptional spatial memory for relocating previously profitable foraging sites (Ratcliffe 2009). However, modern bats should not be defined by their expected or even an observed foraging strategy. Many bats appear to be behaviorally flexible (section 11.2.3), and gleaning behaviors have reevolved independently in multiple lineages of predatory bats (Norberg and Rayner 1987; N. Simmons and Geisler 1998).

11.2.1. AERIAL HAWKING: THE EVOLUTIONARY IMPETUS FOR EARS IN MOTHS

Aerial hawking is a complex sensoribehavioral process in which the relative positions of predator and prey change dynamically up to the point of capture or escape (Ghose et al. 2006). While the results of laboratory studies suggest that bats should be able to use echoic information to discriminate preferred from unprofitable insect prey in the wild, field studies show that experimentally naive bats attack small stones and leaves tossed into their predicted flight path (Acharya and Fenton 1992; Barclay and Brigham 1994). Given that the time elapsed from insect detection to capture in the wild is often less than 1 second (Schnitzler and Kalko 2001), what bats do in the lab reveals ultimate information-processing capabilities rather than how they should be expected to perform in nature (Barclay and Brigham 1994).

Bats that are specialized for aerial hawking in open spaces typically have long, narrow wings and high wing loading, resulting in aerodynamically efficient and thus energetically inexpensive flight characterized by high speed and agility and well suited for intercepting insects in obstacle-free space (Norberg and Rayner 1987; Fenton 1990). Examples include *E. maculatum*, *T. australis*, *T. teniotus*, and *O. martienseni* (section 11.1.3), all of which are relatively large bats (≈20–40 grams) (Norberg and Rayner 1987). Correspondingly, open-space aerial-hawking species tend to produce long-duration, narrowband echolocation calls much less than 50 kHz in maximum frequency (Fenton 1990; Neuweiler 1989). Calls of this design produce strong echoes and maximize target detection distance because total energy is concentrated in a narrow frequency band (J. Simmons and Stein 1980; Neuweiler 1990) and because lower frequencies are less susceptible to the effects of atmospheric attenuation than higher frequencies (D. Griffin 1971; Lawrence and Simmons 1982). Further, the calls produced by bats in open space are often of very high intensity (>120 dB peSPL; Holderied et al. 2005; Surlykke and Kalko 2008). The combination of fast flight and long-range echolocation should make these predators particularly deadly to insects aloft in unobstructed spaces (Fenton and Fullard 1979).

Bats that hawk prey in closed habitats (e.g., within forest or at forest edge) face a more challenging task because echoes from emitted calls will often be returning not only from potential prey but also from other nearby objects (i.e., clutter) (Neuweiler 1989). These species tend to have shorter, broader wings and lower wing loading than do open-space aerial-hawking species, allowing for slower, more maneuverable flight (Norberg and Rayner 1987; Norberg and Fenton 1988). Clutter-tolerant species are able to hawk prey in open space, albeit probably less efficiently, while closed habitats appear to be unavailable to open-space aerial hawkers (Fenton 1990; Brigham et al. 1997). Wing beat and call emission rates are usually higher in clutter-tolerant species (Kalko 1995; Holeried and von Helversen 2003), and echoic information is thereby updated more frequently at the bats' ears. In low-duty-cycle bats (e.g., *Pipistrellus* spp.), the calls produced in clutter are shorter in duration (reducing target- and clutter-echo overlap) and often higher in peak frequency and greater in bandwidth (improving target localization and spectral resolution of prey/background matrix) (Neuweiler 1990; Schnitzler and Kalko 2001). The short-eared trident bat, *Cleotis percivalis* (section 11.1.3), and other high-duty-cycle species are clutter resistant because they can, additionally, separate call and echo using differences in frequency (section 11.1.3). Regardless of habitat, many high-duty-cycle bats typically use the same carrier frequency in their mostly constant frequency calls and do not exhibit the marked changes in frequency structure characteristic of low-duty-cycle species.

High- and low-duty-cycle echolocating bats hawking in either open or cluttered habitat stereotypically show evidence of at least three behavioral phases during attack sequences: (*i*) search, (*ii*) approach, and (*iii*) terminal (fig. 11.3; D. Griffin et al. 1960; J. Simmons et al. 1979), although it is difficult to tell when one phase ends and another begins from a given attack sequence re-

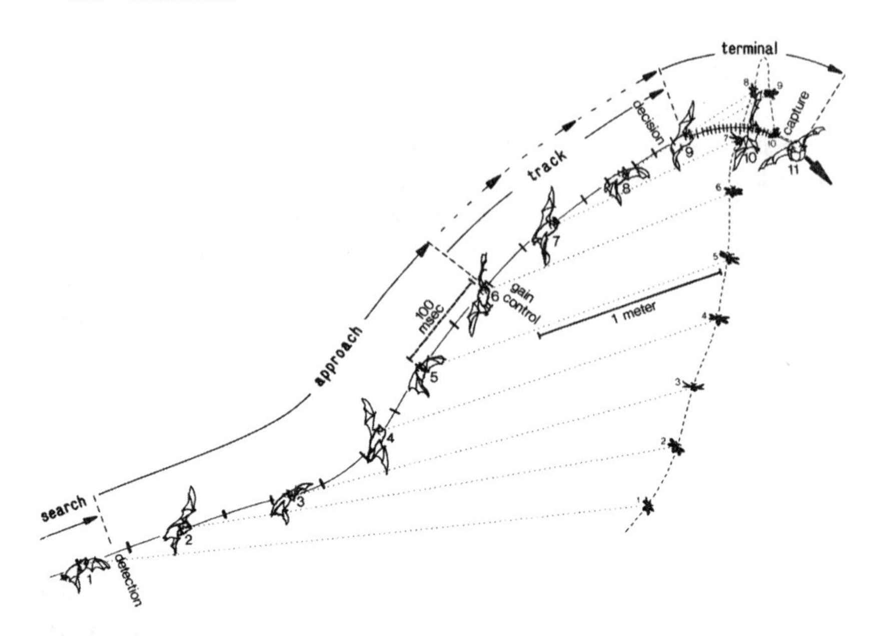

FIGURE 11.3. An aerial-hawking sequence displaying the phases of attack (images separated by 100 ms; distances between bat and insect indicated by dotted lines); the approach phase has been further divided into "approach" and "tracking." It is during this latter subphase that the dogbane tiger moth, *Cycnia tenera*, most reliably emits its defensive clicks (section 11.3.1). From Kick and Simmons 1984 with the permission of the Society for Neuroscience.

corded in the field (Surlykke and Moss 2000; D. Jacobs et al. 2008). Prior to the detection of prey (search phase), many bats produce one call per wing beat and so minimize the energetic cost of emitting echolocation calls (Speakman and Racey 1991). After prey detection (approach phase), bats first localize and then plot a course to intercept the insect and typically produce broader-bandwidth, shorter-duration calls every 10–50 ms (D. Griffin et al. 1960; Kick and Simmons 1984). During the terminal phase of attack (also called the buzz), the bat emits calls at rates surpassing 150 calls/second (J. Simmons et al. 1979). These buzz phase calls mark the end of a successful or unsuccessful attack (Surlykke et al. 2003). The insensitivity of moths' ears even to the frequencies to which they are best tuned and the antipredator behaviors these ears evoke suggest that aerial-hawking bats have been the primary selective pressure for the evolution and maintenance of these sensory structures. Intriguingly, the auditory afferents of noctuoid moths do not encode bats' buzz phase calls as expected to be necessary to evoke evasive flight (Fullard et al. 2003a; D. Jacobs et al. 2008). Lack of selective pressure to track these short, high-repetition-rate calls provides a plausible explanation: at this stage (≈150 ms from contact in the event of capture) the moth will be either out of harm's way or effectively already dead (Fullard et al. 2003a).

11.2.2. GLEANING: THE USE OF BATS' ECHOLOCATION CALLS FOR BATS AND MOTHS

In contrast to hawking, during substrate-gleaning attacks bats attempt to capture stationary or slow-moving prey from the ground, foliage, or tree trunks. Gleaning—more so than even clutter-resistant hawking—demands short, broad wings with rounded tips and low wing loading to allow bats to hover in front of or above prey just prior to capture (Norberg and Rayner 1987; Norberg and Fenton 1988). For ground gleaning, species that land on their prey must possess wings of this design to enable them to take off quickly to minimize risks imparted by terrestrial predators (G. Jones et al. 2003). Generally, gleaning also differs from aerial hawking with respect to call design (shorter, broader band, higher peak frequency) and repetition rate (usually lacking a buzz phase). Like hawking bats, however, gleaning bats from several families continue to emit echolocation calls up until ≈40 ms before capture (Ratcliffe et al. 2005); some continue to do so even on the ground (G. Jones et al. 2003; Ratcliffe et al. 2005).

In earlier literature, it had been proposed that bats use prey-generated sounds alone to detect and localize substrate-borne prey. Two premises were used in support of this argument. First, it was assumed that low-duty-cycle bats are unable to tolerate backward masking, a limitation preventing the discrimination of prey from background echoes (see Schnitzler and Kalko 2001 for review). This assumption now appears incorrect, as low-duty-cycle bats from at least three families have been observed to take silent, stationary prey on or close to surfaces all the while producing echolocation calls (S. Schmidt 1988; Siemers and Schnitzler 2004; E. K. V. Kalko, personal communication). Second, in complete darkness the gleaning bats *Megaderma lyra*, *Myotis myotis*, and *My. blythii* had been reported to stop echolocating and still go on to successively capture prey (Fiedler 1979; Arlettaz et al. 2001). This has also been refuted: under conditions similar to those found in nature, these three species have been shown to echolocate throughout gleaning attacks on noisy prey but to use call intensities much lower than those used by hawking bats (S. Schmidt et al. 2000; Ratcliffe et al. 2005; Russo et al. 2007). Bats long in captivity and hunting in familiar spaces may stop echolocating and instead rely on spatial memory (Ratcliffe et al. 2005); another possible explanation for the misunderstanding is that the audio-recording arrangements used in the original studies may have been insensitive to the low-intensity calls typically used by gleaning bats (S. Schmidt et al. 2000; Ratcliffe et al. 2005).

All of these studies may oversimplify the problem and a more correct assessment might be as follows: while (*i*) prey-generated sounds may be necessary for some species to detect and localize prey on or near substrate, (*ii*) the derived call designs used when gleaning should allow bats to restrict the location of prey because of the finer resolution of detail and spatial configuration potentially available from the echoes (Ratcliffe and Dawson 2003). For those bats unable to either detect or localize stationary silent prey, when potential prey moves against a static background, (*iii*) echoic information returning from each successive call will change the auditory scene describing the prey/background matrix and may thus allow for prey detection and/or localization (Ratcliffe et al. 2005). A promising study species for investigating auditory scene analysis (Bregman 1990; Moss and Surlykke 2001) through the use of multiple echoes returning from multiple calls emitted during gleaning attacks might be *Myotis nattereri*, a species found throughout much of Europe (see Siemers and Schnitzler 2000, 2004; Siemers 2001).

The calls that gleaning bats produce are less intense than those produced by aerial-hawking bats (Schnitzler and Kalko 2001) and over the course of an attack are relatively inaudible to the ears of very sensitive noctuid moths (Faure et al. 1990, 1993). The ears of moths are sensitive to only high-intensity sounds (fig. 11.4); if it were otherwise, evasive flight would be elicited by the high-intensity search phase calls of very distant bats. Reacting to this negligible

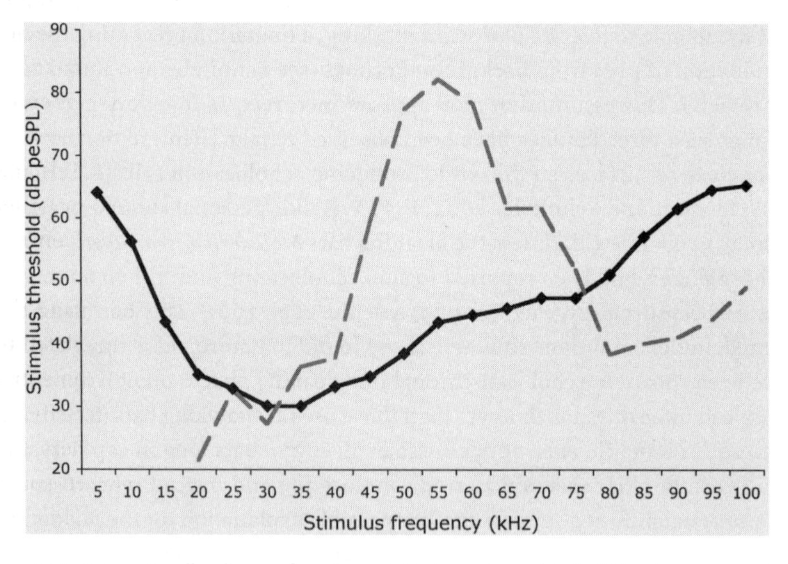

FIGURE 11.4. A1 cell audiogram from the noctuid *Caenurgina erechtea* (solid line) superimposed by power spectrum (dashed line) of an echolocation call from the most common sympatric bat, *Myotis lucifugus*. Redrawn from ter Hofstede et al. 2008b.

threat would have negative effects on a moth's reproductive success (section 11.4.2). Thus, even if hunting bats do in fact always emit echolocation calls, gleaning appears to be a strategy for which no tenable auditory-evoked defense currently exists in moths (Faure et al. 1993). That is, from the moths' perspective, the echolocation calls of gleaning bats are of no use at all. Some insects (e.g., katydids) have more sensitive ears and stop singing as bats approach (Faure and Hoy 2000; ter Hofstede and Fullard 2008), an effective defensive strategy against gleaning bats that have detected and are attempting to localize male insects based on the source of their song (ter Hofstede et al. 2008a).

11.2.3. BEHAVIORAL FLEXIBILITY, LEARNING, AND INNOVATION IN BATS

The little brown bat, *Myotis lucifugus,* and the northern long-eared bat, *M. septentrionalis,* are 5–8 grams, similar in ecomorphology, and found throughout much of Canada and the United States (figs. 11.1 and 11.5). *M. lucifugus* was assumed to take prey only on the wing, while *M. septentrionalis* was reportedly a specialized gleaner. In a large, screened, outdoor flight room we presented these bats with airborne and perched moths and found that both species were able to successfully hawk and glean prey (Ratcliffe and Dawson 2003). Similarly, Siemers and Ivanova (2004) have shown that the horseshoe bat, *Rhinolophus blasii,* is behaviorally flexible with respect to the use of these strategies. Comparing the echolocation call attack sequences of hawking *R. capensis* (D. Jacobs et al. 2008) with those of gleaning rhinolophid species (Neuweiler et al. 1987; Pavey and Burwell 2004; Siemers and Ivanova 2004) suggests that horseshoe bats do not need to change either call design or emission rate between aerial-hawking and gleaning strategies. Such behavioral flexibility is apparently especially common in myotids and rhinolophids and may account for these groups' unmatched species radiations among other primarily insectivorous groups of bats (Ratcliffe et al. 2006; B. M. Siemers, personal communication).

For a variety of vertebrates, species-specific behavioral flexibility and the ability to deal effectively with environmental change are intimately related to brain size (Sol, chapter 7 in this volume). We recently investigated the potential relationships between foraging behavior and relative brain volume, classifying each of 59 predatory species as one of the following (*i*) ground gleaners, (*ii*) behaviorally flexible, (*iii*) clutter-tolerant aerial hawkers, or (*iv*) open-space aerial hawkers. A review of the published literature suggests that, while all insectivorous species hawk some of their prey, more than 40% are also facultative gleaners (Ratcliffe et al. 2006; see also Kalka et al. 2008; Williams-Guillén et al. 2008). Species-level data and phylogenetically independent

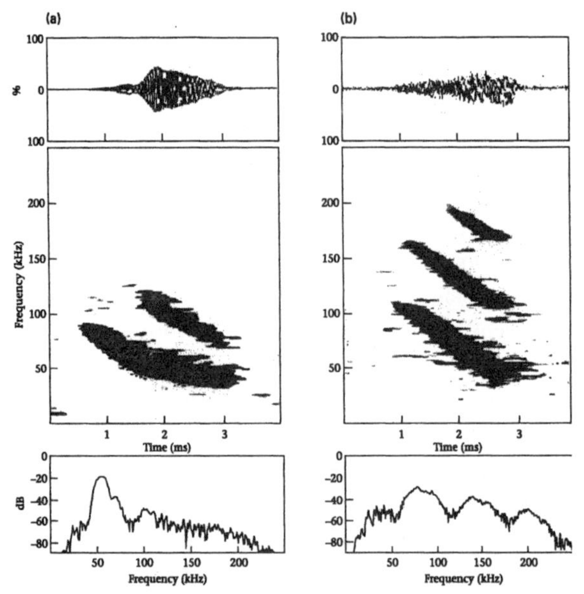

FIGURE 11.5. A northern long-eared bat, *Myotis septentrionalis*, eating a large noctuid moth and call designs used by (a) *M. lucifugus* and (b) *M. septentrionalis* during gleaning attacks. The call of *M. lucifugus* does not differ from those emitted while hawking; that of *M. septentrionalis* is representative of the derived call designs used by many specialized gleaning species. Photograph by M. B. Fenton; oscillograms (top), spectrograms (middle), and power spectra (bottom) redrawn from Ratcliffe and Dawson 2003.

contrasts showed that relative brain volume is significantly greater in behaviorally flexible species. This positive correlation between relative brain volume and behavioral flexibility suggests that larger brains, though costly, confer the ability to use multiple foraging strategies (Ratcliffe et al. 2006). Given the life history of bats and the overall success of this group of mammals (see section 11.1.1), we might expect bats to readily form novel associations and to exhibit flexibility in foraging behaviors not limited to simply gleaning or hawking (Sol and Federspiel et al., chapters 7 and 14 in this volume). R. A. Page and Ryan (2005) demonstrate that the fringe-lipped bat, *Trachops cirrhosus,* learns to associate the calls of frogs and toads with differential profitability and as quickly relearns new associations when consequences are reversed. Naive *T. cirrhosus* also learn about beneficial prey through eavesdropping on the calls of frogs and toads and the subsequent chewing sounds of experienced conspecifics (R. A. Page and Ryan 2006).

Eastern red bats, *Lasiurus borealis,* are migratory species found at different times over much of the Americas (Shump and Shump 1982). Streetlights are an anthropogenic trap for many moths, which apparently confuse them with the moon, and red bats often use these structures as favorite feeding sites. The bats may be exploiting not only an abundant resource but also moths' decreased cognitive capacity, a result of the moths' attraction to light (A. Svensson and Rydell 1998). Reddy and Fenton (2003) have shown that red bats are more sagacious still. The ears of moths found in southwestern Ontario are well tuned to the echolocation calls of red bats, and the aerial-hawking attacks of these bats elicit evasive flight behaviors (Acharya and Fenton 1992, 1999; Obrist and Wenstrup 1998). However, individual bats increase their foraging success by attacking the same moth a second and even a third time, catching the freefalling or diving—and thus more easily tracked—insect through pure persistence. Moreover, when more than one bat hunts at the same streetlight, bats exploit not only the auditory defenses of moths but each other. Should an eavesdropping bat be within earshot of the first (failed) attack, it will often beat the first bat to the now-vulnerable moth (Reddy and Fenton 2003). Prey switching in *T. cirrhosus* and exploitation in *L. borealis* illustrate novel strategies for environmental change; outside our hearing range and cloaked in darkness, other innovative solutions have surely gone unnoticed.

11.3. Neuroethology of auditory-evoked defenses in noctuoid moths

The bat-detecting ears of field crickets each have ≈70 auditory afferents (Hoy et al. 1985), a value orders of magnitude lower than that of vertebrates. By comparison, the ears of arctiid and noctuid moths contain only two auditory afferents (Roeder 1967). The more sensitive is called the A1 cell; the less sensitive, the A2 cell (fig. 11.6). The activity of both neurons can be recorded simultaneously from the auditory nerve in live preparations using standard electrophysiological techniques. Though difficult to assess in the wild,

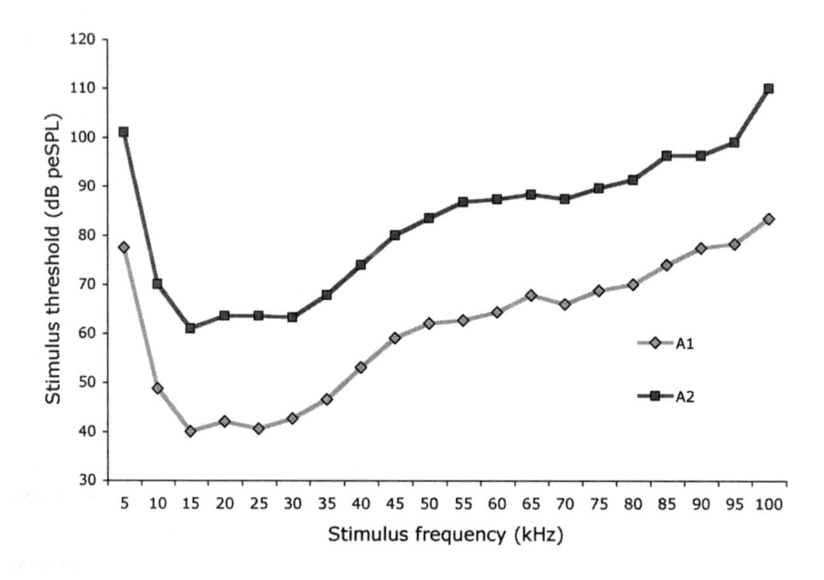

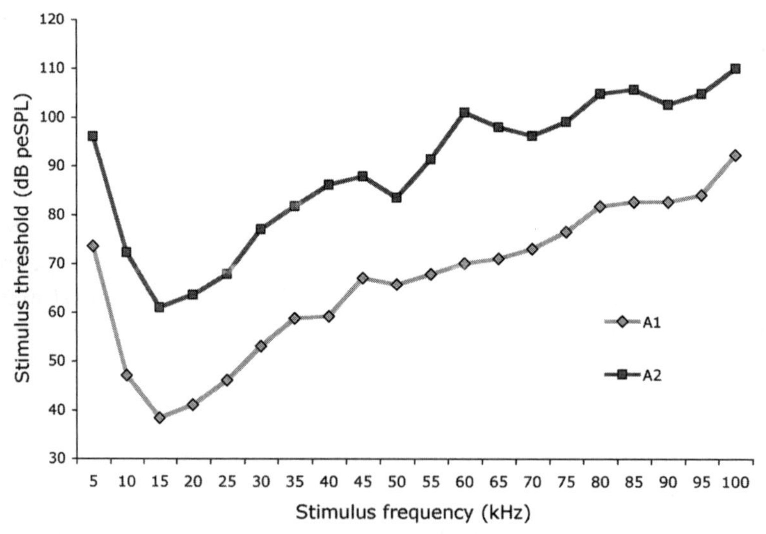

FIGURE 11.6. Average A1 and A2 cell audiograms from (a) 3 adventive and (b) 3 endemic moth species ($N = 5$ moths per species) found on the islands of Tahiti and Mo'orea. Data from Fullard et al. 2007b.

echolocation call-evoked A1 activity is expected to initiate flight away from the bat before the bat has detected the moth. Confronted with an attacking bat, the bat's emitted approach phase echolocation calls will elicit A2 activity. Most, perhaps all, noctuoid moths in this predicament often initiate erratic flight behaviors (Roeder 1967). Roeder (1974) proposed that the evasive behaviors of noctuoid moths are thus bimodal and that flying away is elicited by A1 activity and erratic flight, power dives, and flight cessation by A2 activity. Like noctuids and some arctiids, notodontids exhibit a bimodal response to calls of distant versus proximate bats; however, their ears contain only a single auditory afferent (Surlykke 1984). While further experiments are necessary to refute Roeder's hypothesis for noctuoids with both neurons (e.g., noctuids and arctiids), different levels of A1 activity may be sufficient for initiating both classes of defensive flight in these moths, too (Surlykke 1984; Fullard et al. 2003a).

11.3.1. DEFENSIVE SOUND PRODUCTION IN TIGER MOTHS

Many species of arctiid are unpalatable to bats as a result of toxins sequestered from host plants as a caterpillar or produced de novo (Weller et al. 1999; Nishida 2002; Hristov and Conner 2005a) and can produce defensive sounds (Fullard and Fenton 1977; Barber and Conner 2006). The North American dogbane tiger moth, *Cycnia tenera,* sequesters cardiac glycosides, the same class of toxins as those used by the monarch butterfly, *Danaus plexippus* (Cohen and Brower 1983), and, using modified metathoracic episterna called tymbals, produces ultrasonic clicks when handled and in response to the echolocation calls of bats (fig. 11.7; Blest et al. 1963; Fullard and Fenton 1977). As an adult, when in flight *C. tenera* repels (using sound) and is repellent (taste/toxicity) to the bats *Eptesicus fuscus, Lasiurus borealis,* and *Myotis septentrionalis* (fig. 11.8; Hristov and Conner 2005b; Ratcliffe and Fullard 2005; Barber and Conner 2007).

The dogbane tiger moth phonoresponds preferentially to pulse repetition rates of high-intensity, batlike sounds matching the approach phase of aerial-hawking attacks (Fullard 1984; Fullard et al. 2007a), a result confirmed using playbacks of aerial-hawking echolocation call attack sequences (Fullard et al. 1994, 2003a; Barber and Conner 2006). Using a behavioral paradigm (stimulus generalization, controlling for intensity and duty cycle), we have shown that *C. tenera* is able to assess the relative risk of hunting bats, distinguishing not only the calls of a searching bat from those of one attacking (Fullard et al. 2007a) but also those of a bat plotting a course for interception from those of one about to make contact, two stimulus patterns to which the moth is initially as responsive (Fullard and Ratcliffe, unpublished data).

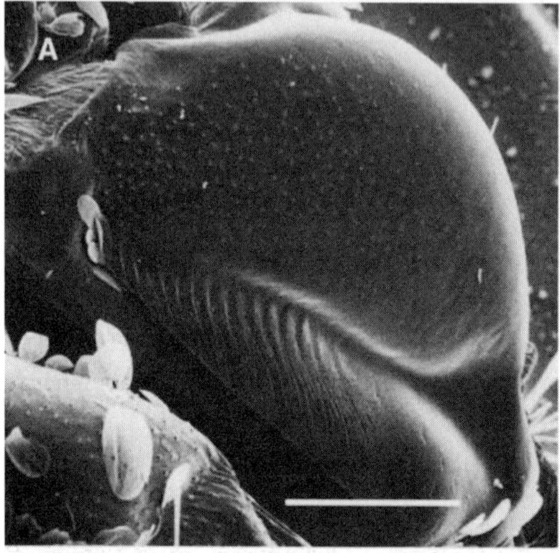

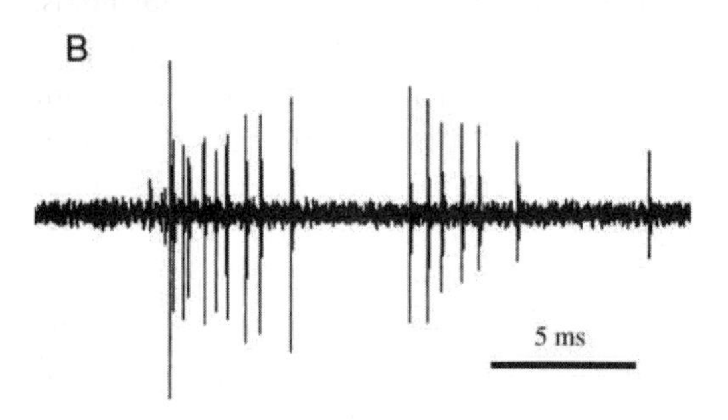

FIGURE 11.7. Top: Dogbane tiger moth, *Cycnia tenera*. Photograph by J. M. Ratcliffe. Bottom: Striated tymbal of *C. tenera* (A) and an oscillogram of a complete click modulation cycle (B). From Conner 1999 with the permission of the Company of Biologists.

Past behavioral studies have used intense, short-duration pulses (well reflecting the calls of some sympatric bats), and A2 activity has consistently been observed to precede sound production (Fullard 1984, 1992; Dawson and Fullard 1995; Fullard et al. 2003a). We have recently demonstrated that *C. tenera* decide whether or not to produce ultrasonic clicks on the basis of acoustically evoked superthreshold A1 activity alone, indicating that this single sensory neuron provides sufficient information to initiate this highly effective defense against aerial-hawking bats. These results allow us to reject the hypothesis that A2 activity is necessary for eliciting the dogbane tiger moth's phonoresponse under acoustic stimulation and support the hypothesis that A1 activity alone is sufficient for eliciting defensive sound production. But our results do not eliminate the possibility that the dogbane tiger moth's A2 cell plays a supplementary role in anti-bat defensive sound production. A2 spikes appear to be summated with A1 spikes by the interneuron/interneurons responsible for initiating and maintaining the phonoresponse (Ratcliffe et al. 2009). All told, Roeder's bimodal theory for the neural control of anti-bat flight behaviors and the role, if any, of the A2 cell remain enigmatic questions for neuroethologists interested in predator-prey interactions.

11.3.2. BAT–TIGER MOTH INTERACTIONS IN REAL TIME

Tiger moths have long been known to contain toxins (see Rothschild et al. 1970) and to produce sounds (see Blest et al. 1963). Around streetlights, red bats, *Lasiurus borealis,* and hoary bats, *L. cinereus,* have been observed to abort attacks on the tiger moths *Hypoprepia fucosa* and *Halysidota tessellaris* but to catch and drop individuals whose tymbals had been surgically ablated (Acharya and Fenton 1992; Dunning et al. 1992). However, until recently, the interaction of bats and tiger moths had never been considered under (*i*) controlled conditions while (*ii*) real bats and real tiger moths interacted (Hristov and Conner 2005b; Ratcliffe and Fullard 2005; Barber and Conner 2007). Dunning and Roeder (1965) found that, when played back the sounds of tiger moths, little brown bats, *Myotis lucifugus,* would abort attacks on mealworms projected approximately 1 meter above the speaker and proposed that tiger moth clicks act as acoustic aposematic signals (Dunning 1968). Fullard et al. (1979) interpreted these and other data differently, suggesting that clicks may interfere with bats' ability to process echoic information. Cross correlation suggests that clicks may be interpreted as echoes (Fullard et al. 1994), and the clicks of moths with complex tymbals (e.g., *C. tenera*) should often fall within the window of time during which neurophysiological studies suggest they will negatively affect the ranging abilities of aerial-hawking bats (see Ratcliffe and Fullard 2005 and references therein).

Sensoribehavioral evidence of jamming/interference from actual bat-moth interactions is, however, wanting (fig. 11.8). Ratcliffe and Fullard (2005) showed that the call emission patterns of wild-caught northern long-eared bats, *M. septentrionalis,* are significantly different (e.g., time between calls is greater) when attacking clicking dogbane tiger moths than when attacking those that have been muted. Similarly, Hristov and Conner (2005b)

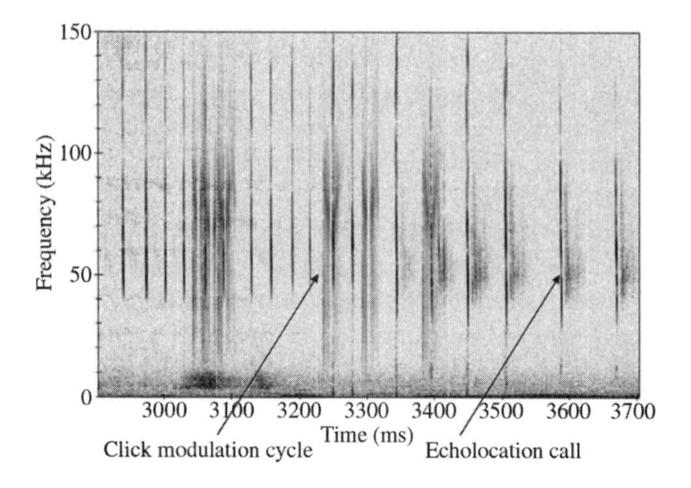

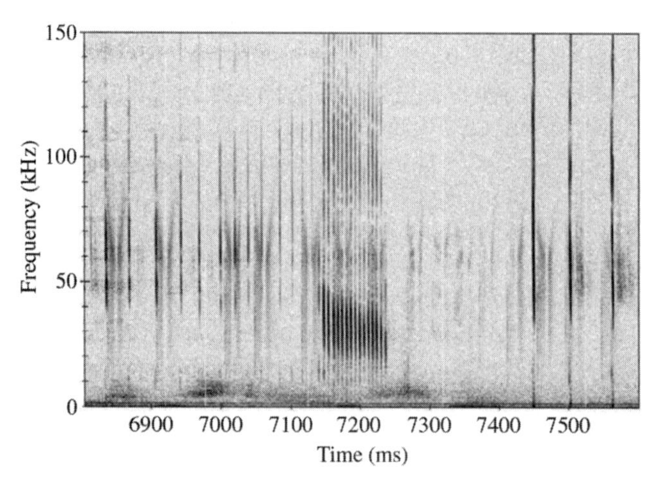

FIGURE 11.8. Spectrograms of *M. septentrionalis* aerial-hawking attacks on intact tethered *C. tenera* (top, aborted attack; bottom, completed attack). From Ratcliffe and Fullard 2005 with the permission of the Company of Biologists.

found that while naive big brown bats, *Eptesicus fuscus,* attacked silent, palatable moths 100% of the time, they aborted attacks on clicking, palatable tiger moths *Euchaetes egle* roughly 20% of the time upon initial presentation but quickly habituated to the sounds. Conversely, naive big brown bats quickly associated clicks with unpalatability when presented with *C. tenera,* aborting attacks ~80% of the time after a few trials. Hristov and Conner's (2005b) study provided the first strong and direct evidence that the clicks of toxic tiger moths function against bats as acoustic aposematic signals; no behavioral evidence so far reported can be interpreted as equally strong evidence for a jamming function. Additional ecologically relevant behavioral experiments using naive bats, high-speed videography, and synchronized high-frequency sound recordings are required to either refute or support the competing, but not mutually exclusive, hypothesis that the clicks of some tiger moths interfere (in some manner) with the processing of echoic information in bats.

Barber and Conner (2007) have recently shown that naive red bats and big brown bats not only readily learn to associate toxic dogbane tiger moths' acoustic signals (on average in less than 7 trials) with their bad taste but generalize this association to other sound-producing and sympatric tiger moth species whether they are toxic or not (*Syntomeida epilais* and *Euchaetes egle* respectively). Muted moths were always attacked; bats therefore did not use visual, echoic, or any other potentially available sensory information for discrimination, and clicks are thus the only signal on which mimicry appears to be based (Barber and Conner 2007). However, red bats, which dietary analyses have shown prey heavily on moths, investigated *E. egle* after a few encounters and eventually learned that these sound-producing moths were safe to eat; the same was not true for big brown bats, which are thought to take predominantly beetles (Barber and Conner 2007).

C. tenera and many other tiger moths most reliably produce clicks in response to close-range approach phase calls of bats attacking them (Fullard et al. 1994; Ratcliffe and Fullard 2005; Barber and Conner 2006) and, in the case of *C. tenera* at least, appear to click most vigorously at approximately the same moment that the bat decides to attack them, that is, during the approach phase after detection and just as the bat plots its course for interception (fig. 11.3; Ratcliffe and Fullard 2005). Because tiger moths are expected to produce clicks only after the bat has detected them, clicks are unlike continuously displayed visual aposematic signals (which may well be the reason for the brightly colored bearer being detected in the first place). Moreover, the potentially negative effects of tiger moths' clicks on the echoic information processing might make them more readily associable with negative consequences than with innocuous cues (Ratcliffe and Fullard 2005). This unique combination of characteristics predicts that the evolution of the acoustic warning signals of tiger moths should be relatively immune to the (at least theoretical) hurdles facing the initial evolution and fixation of always-conspicuous but innocuous warning signals (Ratcliffe and Fullard 2005; Ratcliffe and Nydam 2008). This combination might also make acoustically aposematic models (*i*) more vulnerable to invasion by, but (*ii*) less susceptible to the effects of dilution through, Batesian mimics. However, these predictions—particularly the last—need to be tested at the behavioral level (e.g., Etscorn 1973) and formalized mathematically (e.g., Servedio 2000) to assess their potential validity.

11.4. Bat detection and the primary and secondary defenses of moths

As just discussed, many moths and some butterflies possess bat-detecting ears (Yack and Fullard 2000), and some answer the calls of bats with sounds of their own (sections 11.3.1 and 11.3.2). However, many thousands of earless moth species also fly at night (Fullard 1998; Soutar and Fullard 2004). Edmunds (1974) classifies antipredator behaviors as either primary (functioning to prevent an encounter) or secondary defenses (functioning to reduce mortality after an encounter). As visual information and nonauditory information in general are not relied upon for the detection of bats (Soutar and Fullard 2004), moths' use of primary and secondary defenses against bats can be considered as follows. Earless moths must rely entirely upon primary defenses. Eared moths, however, reduce their risk of attack from bats' using three strategies: (*i*) by not flying when or where bats are hunting to greatly reduce the possibility of detection (a primary defense), (*ii*) when in flight and after detecting echolocation calls, by flying away from bats ostensibly before bats should have detected them (a primary or secondary defense, depending on whether an encounter is defined from the predator's or the prey's perspective), and (*iii*) by effecting erratic flight or going to ground when echolocation calls indicate a bat may well have detected them and is attacking (secondary). The first class of primary defense is shared with earless moths, and only environmental cues (e.g., season, light level) can be used to assess potential risk from bat predation. Eared moths use predator-generated cues to decide when and where to fly to prevent the bat from detecting them.

To test specific hypotheses about the use of these defenses in both eared and earless moths, the 24-hour and nocturnal flight activity of many moths have been quantified using an easily modified behavioral assay (Fullard and Napoleone 2001; Fullard et al. 2003b, 2004; Soutar and Fullard 2004; Ratcliffe and Nydam 2008; Ratcliffe et al. 2008). For the purposes of simply quantifying species-specific flight activity over a 24-hour period (diel flight periodicity)—or over only those 6 hours of the night that bats are most active—the time undisrupted individual moths' were in flight or resting was scored using either 1-minute or 10-minute time bins. For the purposes of quantifying acoustically mediated flight cessation during peak bat activity, over the course of 36 10-minute sound bins individual moths were exposed to 18 randomly assigned bins of batlike 25 kHz pulses and during the remaining 18 bins to silence.

11.4.1. DIEL FLIGHT PERIODICITY AND NOCTURNAL FLIGHT IN EARED AND EARLESS MOTHS

Eared moths sense bats by listening for their calls, and some tiger moths respond to these calls with sounds of their own; earless moths are not so able. During the hours of darkness, Soutar and Fullard (2004) found that earless moths flew significantly less than did species with bat-detecting ears, and they suggested that earless species exhibit lower nocturnal flight activity as a result of selective pressures to reduce encounters with bats. Conversely, eared species, which reduce flight activity when exposed to batlike ultrasound (Fullard et al. 2003b), appear to actively assess the abundance of bats in a given area and use this information to decide where and when not to fly (Soutar and Fullard 2004). Earless moths also tend to fly closer to the ground and vegetation than do eared species, perhaps rendering them less detectable to bats through echolocation, and typically emerge earlier in the season when bats are foraging less often than during the summer months (Yack 1988; Fullard 1998). Fullard and Napoleone (2001) also found that earless moths flew less than did eared species during daylight hours. While ground-hugging flight may make moths less accessible to bats, such flight may put earless moths at greater risk from birds and thus select for diurnal flight reduction (Fullard and Napoleone 2001). On the other hand, the protective warning coloration of some tiger moths may account for their being more active than cryptic species during the day while insectivorous birds are foraging (Dreisig 1986; Ratcliffe and Nydam 2008). Defensive sound production may similarly explain why some tiger moths do not reduce flight activity as silent species do when exposed to batlike ultrasound (Ratcliffe et al. 2008).

11.4.2. SENSORIBEHAVIORAL INTEGRATION AND DISINTEGRATION IN EARED MOTHS

The argument that ears evolved in moths as a direct result of selective pressure from insectivorous bats is supported by a number of premises (section 11.1.2), and thus far, no compelling alternative argument has been put forward. A powerful, if indirect, means to test this hypothesis is provided by comparative studies of species experiencing differential selective pressure from bats today. In southeastern Ontario, where five resident and three migratory insectivorous bat species are found, there is considerable variation in the sensitivity of the A1 auditory afferent and nocturnal flight activity even within the Noctuoidea (Fullard and Napoleone 2001; Soutar and Fullard 2004). Using phylogenetically independent contrasts based on a composite phylogeny, we have recently shown that nocturnal flight activity in a community of eared moths is positively correlated with overall A1 cell sensitivity to the frequencies of the echolocation calls of sympatric bats (ter Hofstede et al. 2008b). One result of this relationship is that, while moths with less sensitive ears fly less often at night, when aloft and in the presence of bats these moths will be less easily driven off course than moths with more sensitive ears. Another example of an evolutionary trade-off between self-preservation and mating opportunities has been shown in the European noctuids *Agrotis segetum* and *Spodoptera littoralis*. Confronted with both female pheromone and batlike ultrasound, moths react less to these predator cues than in the absence of these pheromones. The higher the quality and/or quantity of pheromone, the less likely were male moths to react to batlike sounds, suggesting moths make informed and ultimately adaptive decisions based on conflicting inputs (G. Svensson et al. 2004; Skals et al. 2005).

Given adequate time, in the absence of stabilizing selective pressure from insectivorous bats, the sensoribehavioral mechanisms responsible for evoking anti-bat behaviors are expected to degenerate and disintegrate (if bats are the sole selective pressure maintaining ears in most moths; see D. Jacobs et al. 2008 for a South African moth that uses its ears for the detection of birds). Eared moths can be considered "bat-released" if they fly only when bats do not or where bats are not or if they stop flying altogether (e.g., flightless female gypsy moths, *Lymantria dispar;* Cardone and Fullard 1988). Some bat-released eared moths exhibit auditory sensitivity, as measured from A1 cell activity, similar to that of eared moths still predated upon by aerial-hawking bats, others exhibit high-frequency deafness or appear to be completely deaf (reviewed in Fullard 1998). The amount of time a moth species has lived without bats and/or the amount of gene flow between populations still under threat may account for these differences (Fullard 1998). Fullard (1994) demonstrated that, though the ears of moths endemic to the mountains of the continuously bat-free habitat of Tahiti exhibit high-frequency deafness at A1, they are as sensitive as recently arrived species from the adjacent island of Mo'orea at frequencies below 30 kHz (fig. 11.6). This result was mirrored in the A2 cell, illustrating again the relationship between these two cells (i.e., that A2 is roughly 20 dB less sensitive than A1 across a broad range of frequencies; fig. 11.6; Fullard et al. 2007b). This relationship holds not only in bat-rich environments but in bat-free ones as well (fig. 11.6b). High-frequency deafness may result, not from changes in the auditory afferents themselves, but from thickening of the tympanal membranes (decreasing vibration transduced by these neurons).

Also on the islands of Tahiti and Mo'orea, we investigated acoustically evoked flight cessation using 25 kHz pulsed stimuli in six endemic and six adventive (i.e., recently arrived) species using the sound/no sound design described above (section 11.4). As reported for noctuid moths from southern Ontario, Canada (Fullard et al. 2003b), all six adventive species significantly reduced their flight times when exposed to ultrasound; only one of six endemic species did the same (Fullard et al. 2004; see also fig. 11.9). Interestingly, endemic species also exhibited reduced maximum A2 firing rates in response to similarly intense pulses of 25 kHz sound. Endemic moths still exhibit erratic flight in response to visual stimuli and thus have not lost the behavior; instead, they have decoupled the link between auditory information and evasive flight. In the context of Roeder's bimodal control theory, it is tempting to suggest that (*i*) A2 activity is responsible for the flight cessation in our behavioral assay and (*ii*) the reduced A2 firing rate in noctuid moths endemic to Tahiti (as a result of neural degeneration and/or tympanal thickening) explains this apparent loss. However, further data are required (Fullard et al. 2004, 2007b). As plausibly, it is degeneration of as yet uninvestigated interneuron/interneurons that explains this sensoribehavioral disintegration in the absence of bats over evolutionary time. In Venezuela some species of day-flying notodontid moths possess relatively insensitive ears (Fullard et al. 1997), while nocturnal hedylid butterflies possess bat-detecting ears (Yack and Fullard 2000), complementary evidence that bats are the primary selective agent for the initial evolution and maintenance of ears sensitive to high frequencies in Lepidoptera.

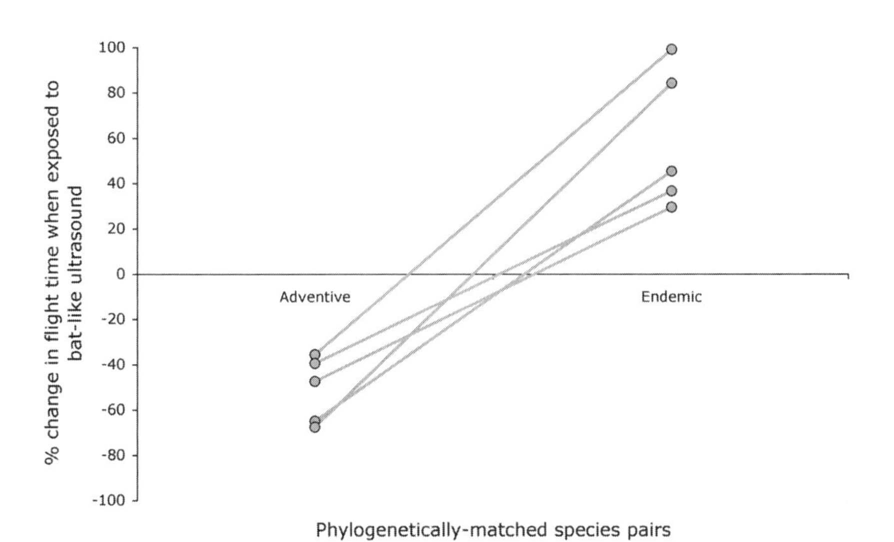

FIGURE 11.9. Phylogenetically matched pairs of 5 adventive and 5 endemic noctuid species ($N = 6$ moths per species) from the islands of Tahiti and Mo'orea (paired *T*-test, $P < 0.05$). Redrawn from Fullard et al. 2004.

11.5. Summary and conclusions

Long after the pioneering work of Griffin, Roeder, and Treat (see D. Griffin 1958; Roeder 1967), bats and moths continue to provide an excellent model system for the study of predation and defense within an auditory world. If the interactions between most bats and moths are indeed confined exclusively to this single sensory modality, this relationship may well prove an exception in a more often than not complex and multimodal world (Ruxton et al. 2004; Hebets and Papaj 2005; Partan and Marler 2005). Predator-prey interactions demand sensory and cognitive systems evolved to acquire and process information with speed and accuracy greater than that required for most other behaviors. Because the interactions of bats and moths are for many species limited to the auditory modality and because the acoustic signals of bats (echolocation calls) and some moths (e.g., tiger moth clicks) can be recorded, they provide a fascinating opportunity to study the information available for processing and to make inferences under ecologically relevant conditions about how that information influences the decisions made and how these decisions impact reproductive success.

While bats and moths appear to detect and communicate with one another almost entirely through sound (see Eklöf et al. 2002 for a rare example of bats using visual information when hunting moths), how the information carried in bats' echolocation calls is processed could hardly be more different. Laryngeal echolocating bats have one of the most sophisticated auditory systems among vertebrates (Popper and Fay 1995). However, relative to the highly variable signals some prey must differentiate (Warkentin and Caldwell, chapter 10 in this volume), at the level of the moth's ear the aerial-hawking attack sequences of distantly related high- and low-duty-cycle bats are encoded, with respect to both individual calls and call emission patterns, in a remarkably similar way (Waters and Jones 1996; D. Jacobs et al. 2008). Moths, which in most species use their ears solely to detect bats, have capitalized on this stereotypy and have invested no more than four auditory afferents per tympanal ear in drepanid and geometrid moths and only one or two per ear in noctuoid moths for initiating anti-bat defensive behaviors. Correspondingly, while long-lived bats have been underestimated with respect to behavioral flexibility and learning (Wilkinson and Boughman 1999; Siemers and Page, in press), moths are short-lived vehicles for reproduction and not expected to be terribly plastic in their sensoribehavioral responses. However, the decisions noctuid moths make when faced with female pheromones and batlike ultrasound (G. Svensson et al. 2004; Skals et al. 2005) and the ability of arctiids to assess the relative risk of hunting bats based on call emission rate alone (Fullard et al.

2007a; Ratcliffe and Fullard, unpublished data)—and the moths' impressive success in species abundance, distribution, and diversity (Spangler 1988)—imply that they too employ evolutionary strategies and cognitive tactics yet to be revealed and investigated.

ACKNOWLEDGMENTS

I thank Reuven Dukas, Brock Fenton, Hannah ter Hofstede, Marie Nydam, and Annemarie Surlykke for helpful comments on earlier versions of the chapter. The Danish Natural Sciences Research Council, the Natural Sciences and Engineering Research Council of Canada, and the National Geographic Society have funded my research.

PART IV

Cognition and Sociality

12 What Do Functionally Referential Alarm Calls Refer To?

MARTA B. MANSER

12.1. Introduction

Nonhuman animals, like humans, categorize their environment according to their requirements and have evolved concepts about objects or events distinguished by specific characteristics. How they categorize their environment depends on the sensory systems and the information processes a species has evolved, shaped by phylogenetic constraints and the habitat they use. In social-living animals, this is influenced not only by ecological factors but also by their conspecifics. Depending on the importance of specific cues or signals in the environment, animals may have evolved concepts of categories that include a broad range of stimuli (e.g., a species preyed on by several species of hawks and eagles may represent all as raptors) or concepts of categories that include highly specific stimuli (e.g., a species preyed on by few species of eagles and not others may distinguish between raptor species). Knowledge about how animals categorize their environment and which characteristics they use in their perception and information processes is crucial. Only the investigation of these processes allows us to understand the evolution of adaptive behaviors, such as antipredator responses, food discrimination, territorial behavior, and navigation, of a specific species in its habitat. One approach to identifying categories and concepts is investigation of the signals animals use and how they transfer this information to others (conspecifics or nonconspecifics). In particular, vocalizations have been used to test the production of different signals as well as the response of receivers to them. This enables us to understand the kinds of signals used and how animals process such information from their environment (for reviews see J. Bradbury and Vehrencamp 1998; Searcy and Nowicky 2005).

To fully understand the meaning of a signal, we have to identify the cognitive mechanisms underlying its production and perception (Marler et al. 1992). Signals will become evolutionary stable only if both the signaler and the receiver benefit from them (Krebs and Dawkins 1984; Maynard Smith and

Harper 2003), yet the meaning and the function of signals from the signaler's and the receiver's perspective are likely to differ (Marler 1961; Seyfarth and Cheney 2003b). In animals, the same call type is given to a wide range of stimuli or in several different contexts (e.g., barks in meerkats; Manser 1998), while other call types are elicited only by specific objects (e.g., aerial alarm calls in meerkats; Manser 2001). Functionally referential calls (Seyfarth et al. 1980) are defined by their high stimulus specificity. Calls are functionally referential only if the receivers use this specificity of the calls in their adaptive responses to them (table 12.1; Marler et al. 1992). However, sometimes it may be difficult to see obvious differences in the responses of the

TABLE 12.1. Alarm calls in meerkats elicited by specific stimuli and context (including predator type and risk) and how the receivers respond to them

Alarm call types		Approach direction	Risk	Stimuli eliciting call	Receiver response
Bark	ba	Nonspecific	High	Raptors perched or circling above, terrestrial predators close by	Move to shelter
Panic	pc	Nonspecific	High	Sudden movements in close proximity and bird alarm calls	Run below ground
Spit	sp	On ground, in holes	High	Mainly snakes, terrestrial animals (also foreign meerkats) within 1–2 m	Approach caller and mobbing behavior
High aerial	ha	Sky	High	Raptors close (<200 m)	Freeze, move to shelter, sometimes below ground
High terrestrial	ht	On ground	High	Terrestrial predators close by (<50 m)	Gather, move together to shelter, bolt-hole, or burrow system
High recruitment	hr	On ground, in holes	High	Snakes, deposits of predators (feces, urine, hair), terrestrial predators in holes	Approach caller with tail and fur erect, inspecting or mobbing behavior
Low aerial	la	Sky	Low	Raptors far away (>200 m)	Vigilant, on guard, move to shelter
Low terrestrial	lt	On ground	Low	Terrestrial predators far away (>50 m)	Vigilant, on guard, Gather, move together to shelter
Low recruitment	lr	On ground, in holes	Low	Deposits from predators or foreign meerkats (feces, urine, hair)	Approach caller with tail erect, inspecting behavior
Alert	al	Mainly sky but also on ground	Low	Nondangerous birds close by, raptors far away	Vigilant, on guard, seldom running to shelter
Animal moving	am	Nonspecific	Low or high	Animals moving: raptors, mammals (also foreign meerkats), nondangerous birds	Vigilant, on guard, gather together, seldom move to shelter

receiver to different call types because the displayed behavioral responses may be very similar (Fischer and Hammerschmidt 2001; Frederiksen and Slobodchikoff 2007). Similarly, on the production side, we may be limited by our current methodological approaches (in particular, measuring the biologically relevant acoustic parameters and their relation to one another) in correctly identifying crucial differences in the structure of the calls, differences that might be informative to animals (Fischer and Hammerschmidt 2001).

Animal vocalizations were long considered to be purely expressions of the caller's emotion (e.g., fear, contempt) without a referential component (Darwin 1872; Myers 1976; Bickerton 1990); however, empirical evidence has accumulated recently that supports the position that animal calls include both components (Manser et al. 2002; Seyfarth and Cheney 2003a). Premack (1972) pointed out that a purely affective communication system could function in a symbolic way. He argues that, even if all the calls given by an individual are the expression of emotions, they may enable receivers to gain specific information. This is possible as long as specific call types are emitted only in narrowly defined contexts (referential specificity). Thereby, a high association exists between a specific context (e.g., predator vs. food encounter) and a call type (e.g., alarm call vs. food call), and most of the time that specific context elicits that call type (information value).

Functionally referential calls have been described within the context of predator alarm calls and food calls (for review, see Searcy and Nowicki 2005). Studies on functionally referential alarm calls have mainly addressed the production and perception specificity, and only in a few species have the underlying cognitive mechanisms been investigated. Diana monkeys, *Cercopithecus diana* (Zuberbühler et al. 1999), and chickens, *Gallus gallus domesticus* (C. Evans and Evans 2007), have been tested on receivers' mental representation of the information they gain from functionally referential calls. In playback experiments subjects presented with a specific call type respond differently and in accordance with the information received from previous vocalizations or the general context they experience at that moment. There is still a lack of knowledge about other fundamental questions on the production of functionally referential calls. For instance, it has not been tested yet whether functionally referential calls are denotative (i.e., as labels for stimulus categories with information on the predator) or imperative (i.e., as instructions describing appropriate responses) (Cheney and Seyfarth 1990; Marler et al. 1992; C. Evans 1997; Palleroni et al. 2005) or refer to the escape behavior of the caller (W. Smith 1981). Furthermore, only recently have investigations been conducted in species in their natural habitat to correlate the specificity of these calls with the

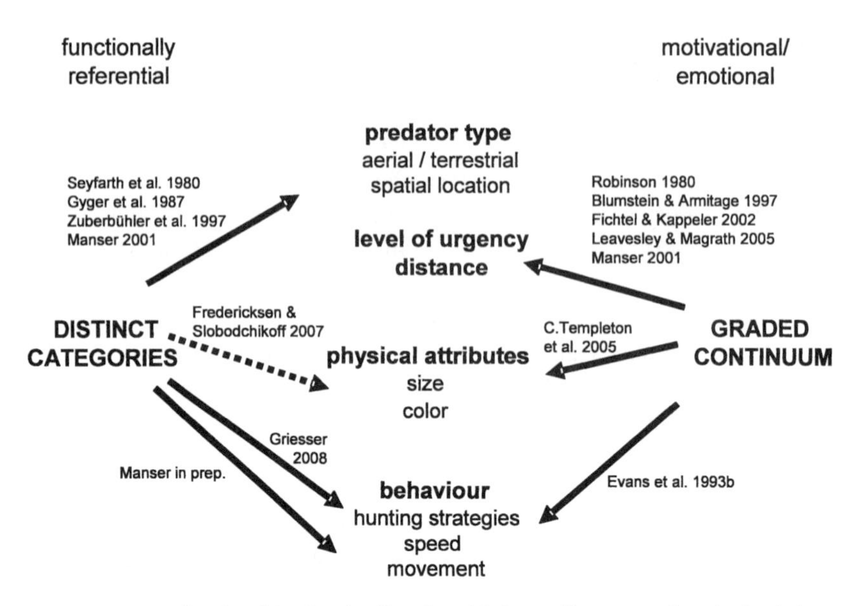

FIGURE 12.1. Studies describing functionally referential alarm call systems and motivational alarm call systems in different contexts (solid arrows indicate studies based on acoustic analysis and playback experiments; dashed arrow indicates acoustic analysis only).

physical properties, behavior, or spatial location of the approaching predator (fig. 12.1; Manser and Fletcher 2004; Palleroni et al. 2005; C. Templeton et al. 2005; Griesser 2008).

Only by identifying in detail what cues in the environment elicit specific call types and what information the receivers use will we understand the cognitive mechanisms underlying functionally referential call systems and thus the evolution of some aspects of categorizations and concepts in animals. The context at the moment a signaler emits a call and a receiver responds to it has to be taken into account, as it will provide crucial information for understanding their respective behaviors. In my discussion on functionally referential alarm calls, I address the following questions: (*i*) Are the calls denotative, are they imperative, or do they refer to the subsequent behavior of the caller? (*ii*) Do the calls refer to specific attributes of the approaching predator (the spatial area the predator is approaching from, physical properties, or behavior)? (*iii*) Why are some calls considered functionally referential and not others? (*iv*) Is the variation of the acoustic structure, though having a referential function, based purely on the expression of the emotion of a caller? I will focus on meerkat alarm calls and then compare this system with others.

12.2. Meerkat alarm calls

Meerkats, a cooperatively breeding mongoose species, have evolved a sophisticated system of predator-specific and nonspecific alarm calls (Manser 2001; Manser et al. 2001). Meerkats live in open, semidesert habitat in the southern part of Africa (Skinner and Chimimba 2006) and experience high predation risk from a range of aerial and terrestrial predators (Clutton-Brock et al. 1999a). They live in groups of 3–50 individuals, wherein mainly the dominant pair breed and adult daughters and sons help to raise their offspring (Clutton-Brock et al. 1998). The groups therefore consist of full and half siblings, cousins, and sometimes one or a few unrelated immigrant males (A. Griffin et al. 2003), which help in the same way as natal subordinate members (Clutton-Brock et al. 2001). They forage together as a cohesive group, each of them digging for their own prey (insect larvae, small reptiles, scorpions) in the sand and dispersed a few meters from each other (Doolan and MacDonald 1996). Although they have a coordinated vigilance system with a sentinel looking out for predators at an exposed position in a bush or tree (Manser 1999; Clutton-Brock et al. 1999b), they have never been observed to climb upward when escaping from a predator or responding to an alarm call. They, like ground-dwelling sciurids (Macedonia and Evans 1993), run to the nearest bolt-hole or move together to other areas in their territory in response to predator encounters (Manser et al. 2001; Manser and Bell 2004).

The variation in the acoustics of meerkat alarm calls can be explained by four main factors: (*i*) type of stimulus, (*ii*) distance to stimulus, (*iii*) risk posed by stimulus, and (*iv*) behavior of stimulus (fig. 12.2 and table 12.1; Manser 2001). Analyzing the elicited alarm calls by predator type and level of urgency revealed distinct categories of calls for aerial and terrestrial predators and a graded variation in acoustic parameters corresponding to the distance to the predator. An additional aspect influencing acoustic structure of a call is stimulus risk. Depending on whether an approaching animal is innocuous or a predator of meerkats, calls differ and this also relates to the distance of the animal. For example, a vulture does not pose a threat to adult meerkats and can approach a group of meerkats to within 200 meters without eliciting even a nonspecific alert call. The same nonspecific alert call may be elicited by a martial eagle (*Polemaetus bellicosus*), the raptor posing the most danger to meerkats, more than 2 kilometers away and turn into a predator-specific aerial alarm call from low to medium urgency as the eagle approaches (Manser, unpublished data). In addition, one of the nonspecific alarm calls, the "animal-moving" call is assigned to a variety of different predator types and also to nondangerous species and appears to refer to the behavior of the other animal rather than any specific physical characteristic (Manser et al., in preparation).

In meerkats, the acoustic structure of the predator-specific calls relates to characteristics of the predator and not the current context of the signaler during the encounter. Meerkats emit the same alarm call types in response to the same predators when they are at their sleeping burrow next to shelter and when they are foraging away from shelter, and they do so independently of whether they are foraging or on sentinel duty (figs. 12.3 and 12.4; Manser, in preparation). At their sleeping burrow, where they are within a few meters of shelter or sometimes directly in the burrow entrance (fig. 12.3b), receivers often do not move and only observe the behavior of the predator. In contrast, while foraging (fig. 12.3a) meerkats often have to run more than 50 meters to

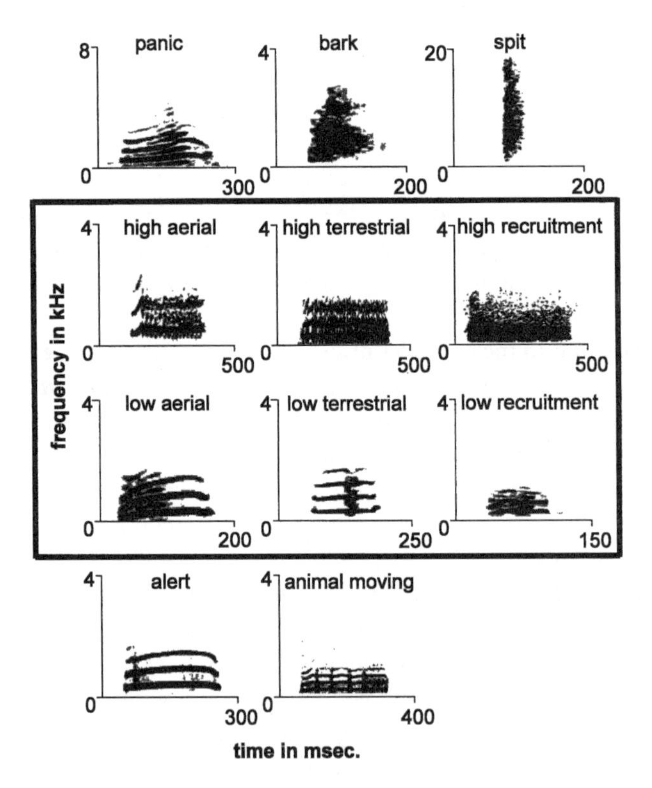

FIGURE 12.2. Spectrograms of the different meerkat alarm calls. Top row shows the highest-urgency calls; bottom row, the lowest-urgency calls. Middle rows (framed) show the predator-specific calls, whereas the calls in the top and bottom rows are not predator type specific.

A

C

B

FIGURE 12.3. Meerkats in different contexts emit alarm calls: (a) away from shelter, foraging; (b) at shelter, here at sleeping burrow; and (c) sentinel on an exposed position.

reach shelter (Manser and Bell 2004). Most of the alarm calls are given by the sentinel (fig. 12.3c), if one is on duty (20%–60% of the foraging time; Manser 1999; Clutton-Brock et al. 1999b), and otherwise by any of the foraging group members older than 6 months (Hollén et al. 2008). Meerkat sentinels climb bushes or trees typically within a few meters of a bolt-hole (Clutton-Brock et al. 1999b). Foraging individuals usually run back to the shelter when alarm calls are given that refer to predators close by (high and medium urgency) and also frequently when the predator is still far away (low urgency) (Manser, in preparation). Sentinels seldom move from their position. They run down to the nearby bolt-hole only when the predator is proximate and poses high risk (Manser, in preparation).

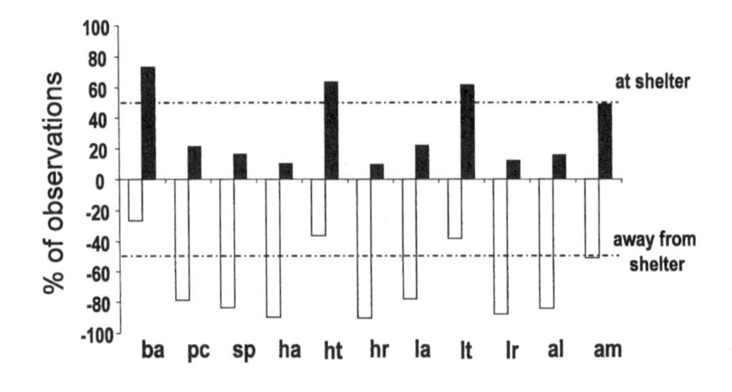

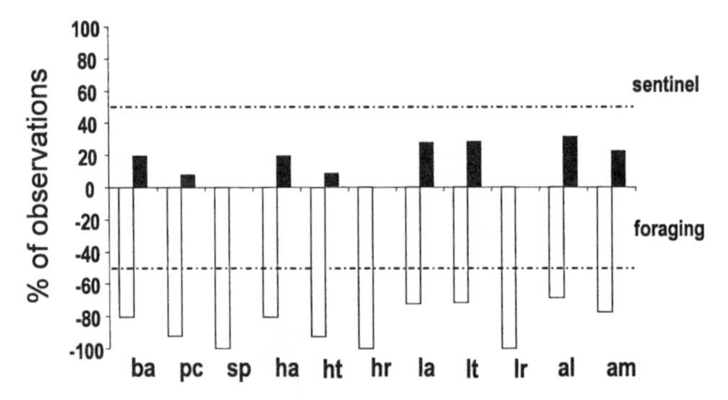

FIGURE 12.4. The proportion (%) of alarm calls observed depending on whether (a) the group was at shelter versus foraging and away from shelter, and (b) the caller was a sentinel versus a foraging individual (ba = bark; pc = panic call; sp = spit call; ha = aerial high urgency; ht = terrestrial high urgency; hr = recruitment high urgency; la = aerial low urgency; lt = terrestrial low urgency; lr = recruitment high urgency; al = alert; am = animal moving).

12.3. What do functionally referential alarm calls refer to?

12.3.1. ARE THE CALLS DENOTATIVE OR IMPERATIVE, OR DO THEY REFER TO THE ESCAPE BEHAVIOR OF THE CALLER?

Whether functionally referential alarm calls are denotative, are imperative, or refer to the escape behavior of the caller is of interest because these options suggest different underlying cognitive mechanisms for the production of calls (Cheney and Seyfarth 1990; Marler et al. 1992; C. Evans 1997; Palleroni et al. 2005). If calls function denotatively, it means they represent labels for stimuli properties (table 12.2). If they function imperatively, it means that the signaler instructs other group members on the appropriate response at the given moment. If the calls refer to the subsequent behavior of the caller, the acoustic structure should relate to the escape mode of that individual (W. Smith 1981). In the case of denotative function the call may express a simple association to the assessed threat. If calls refer to the escape mode of the signaler, the caller must assess the nature of the threat and respond appropriately based on context. Imperative calls may be cognitively more challenging, because they require the signaler to assess both the threat and the appropriate response for the receivers, thereby taking the location and the behavior of the audience into account. Several vertebrate species do adjust their behavior to the presence of the audience (e.g., Marler et al. 1986; Le Roux et al. 2008). Provided information on stimulus properties, receivers are able to assess the danger of the situation and adjust their response according to what they are doing at the moment of the alarm. In contrast, if the calls are imperative, receivers have to rely on the assessment of the caller for their escape.

Distinguishing whether calls are denotative, are imperative, or refer to the subsequent behavior of the caller may be difficult in most systems, as functionally referential alarm calls appear to have evolved in species where predator-specific calls are linked with the escape mode of the animals (Macedonia and Evans 1993). Yet, if the animals gave the same alarm call to the same stimulus independent of their momentary behavior/situation (foraging and away from shelter vs. close to a shelter and without the need to escape), this would clearly suggest that these calls are denotative. If calls varied depending on the location of the animals (caller or receiver) in relation to shelter, this would suggest that they function in an imperative way. However, this pattern could also be explained by the expression of emotions (e.g., fear), with discrete changes depending on the distance to shelter. Yet, this would indicate, not functionally referential properties of the calls, but rather the level of urgency, similar to the calls described for sciurids. Only in this case it would relate, not to the distance of the predator, but to the distance to shelter, which may then also correlate with the escape behavior of the caller.

To identify whether the acoustic structure relates to the subsequent behavior of the signaler, we would have to observe situations in which signaler

TABLE 12.2. Predictions on cognitive mechanisms underlying call production depending on whether they are denotative, are imperative, or refer to the escape mode of the caller

	Representation by signaler	Assessment of context	Receiver	Acoustic structure relates to
Denotative	Label for stimuli	Threat	Own assessment	Properties of stimulus
Imperative	Instruction	Threat and context of audience	Rely on assessment of caller	Properties of stimulus and context of audience
Escape mode	Own behavior in relation to threat	Threat and own context	Rely on assessment of caller	Properties of stimulus and own context

and receiver experience different conditions during the predator encounter that require different escape strategies. Such situations may be given when signaler and receiver are dispersed and located at different distances to shelter, or if they display different activities. Several mammal and bird species have a sentinel on an exposed position while the rest of the group is foraging (vervet monkeys, *Cercopithecus aethiops:* Horrocks and Hunte 1986; meerkats: Manser 1999; Florida scrub jays, *Aphelocoma coerulescens*: McGowan and Woolfenden 1989). In meerkats, foraging group members immediately run back to shelter, while sentinels often do not leave their guarding position in response to an approaching predator (Manser, in preparation) because they typically guard next to shelter (Clutton-Brock et al. 1999b). This allows sentinels to respond only at the very last moment if they need to take cover; otherwise, they stay on guard duty, continuously tracking the location and behavior of the predator. As a consequence, the escape behavior of a sentinel and of a foraging member emitting the alarm call often differs, since a foraging member is farther away from shelter and responds earlier by running for shelter than the sentinel. If the calls referred to the behavior of the caller, we would predict that calls would vary according to the escape behavior of the caller, and that sentinels and foraging individuals would use different call designs for the same predators. If the call given by a foraging group member moving toward shelter was the same as that of a sentinel keeping its position, this would indicate that the call referred to properties of the predator.

That meerkat alarm calls vary with properties of the predator approach independently of spatial position in relation to shelter or the behavior of the signaler or receiver (Manser, in preparation) supports the idea that functionally referential alarm calls in this system are denotative. Considering the different escape modes in this species, it is not immediately obvious why they do not use their alarm calls in an imperative way. In the case of calls for aerial predators, individual meerkats immediately run for the shelter closest to them (Manser 2001; Manser and Bell 2004), but they typically have more time to respond to an alarm call referring to a terrestrial predator or a recruitment call elicited by snakes or deposits. A terrestrial alarm call causes the receivers to gather together and then move as a unit to shelter (Manser et al. 2001; Furrer and Manser, in press). With recruitment calls, a caller recruits others to his location, and they then either mob a predator or inspect the scent and/or scat that elicited the call. In both cases, the behavior of the caller indicates to the receivers the location to gather at. Therefore, it seems that an imperative system in the instance of a predator encounter may be sufficient, as it could clearly cause the different escape mode rather reliably.

Denotative calls may be more informative than imperative calls to the receivers when individuals are dispersed while foraging and when the trade-off between predation risk and energy gain from foraging differs among individuals. The receivers are likely to be in contexts that vary in specific attributes, such as distance to the shelter, distance to the approaching predator, location within the group, or costs of escape due to foraging investment. In meerkats, in particular, which depend on mobile prey, the trade-off between risk of predation and food intake may play an important role in their response to alarm calls. Therefore, receivers may benefit more by information from the signaler referring to the stimulus than information on how to escape. This allows receivers to adjust their own escape strategy in a given moment according to the information on the approaching stimulus in relation to their own occupation, rather than on the assessment of the signaler as to which escape behavior to employ. No study of the meaning of alarm calls has suggested that callers adjust their calls in relation to the distance to shelter or the behavior of themselves or their intended receivers. It is therefore likely that species other than meerkats use primarily denotative alarm calls.

Nevertheless, for some other species it may not be as important for each receiver to assess its own situation individually and adjust its escape strategy to its own situation at the moment of the alarm call. In species where all group members forage in the same patches (e.g., some primates moving from one tree to the other to feed) and very close together, all of them profit from the same escape behavior. In a similar way, for herbivores, rather than carnivores with mobile prey, the investment in pursuing a specific prey item does not apply, and the trade-off between time and energy invested in the prey (here, stationary plant food) and the risk of being predated may not be as important, and it may not be necessary for each receiver to adjust its own escape at the moment of the alarm. To generalize that functionally referential alarm calls are denotative has first to be tested in more species in which group members forage as cohesive units from patch to patch and take stationary food items (e.g., plant products), and not only in carnivores with mobile prey.

12.3.2. WHAT ATTRIBUTES OF THE PREDATOR CAUSE THE ACOUSTIC VARIATION OF FUNCTIONALLY REFERENTIAL ALARM CALLS?

Functionally referential calls are defined to correlate with specific attributes of an object or an event in the environment. In several monkeys (Seyfarth et al. 1980; Zuberbühler 2000), other mammals (Manser 2001), and birds (Gyger et al. 1987), specific alarm call types are elicited by specific predator types (e.g., aerial or terrestrial predator). In birds, calls have been described to correlate with the size of the predator species (C. Templeton et al. 2005). In prairie dogs (*Cynomys gunnisoni*) calls appear to relate to specific predator species within terrestrial predators and also to different characteristics of humans (Frederiksen and Slobodchikoff 2007). Siberian jays signal information about the behavior of the predator (hawks perched, searching for prey, or attacking; Griesser 2008). A recent study on meerkats has suggested that one of the nonspecific alarm call types refers to the movement of a variety of different animal species other than that of the caller (Manser et al., in preparation). This variation in the attributes of predators eliciting specific call types (fig. 12.1) brings up the following questions: What do the different call types really refer to? Do these different types of calls all fit the definition of being functionally referential?

Although in most studies functionally referential calls have been described as referring to aerial or terrestrial predators (Seyfarth et al. 1980; Gyger et al. 1987; Manser 2001), it is not clear whether these calls denote the spatial area from which the predator is approaching, its physical properties, or its behavior (C. Evans et al. 1993b; Manser and Fletcher 2004; C. Templeton et al. 2005). Since aerial and terrestrial predators hunt in different ways, and also from different spatial areas—raptors approach with high speed out of the air, while terrestrial animals approach with moderate speed on the ground—the variation of these call types may refer to the ways and means of approach or the current location of the predator (Manser and Fletcher 2004; C. Templeton et al. 2005). To distinguish among these options, calls for the same predator species or category approaching from the air and on the ground have to be compared. It seems likely that animals recognize a bird species when it is in the air and when it is on the ground, as the sight of a raptor flying from a tree occurs frequently, and the prey will have learned about the different shapes (spread wings vs. tucked wings) of the same predator during flight and when perched in a tree or located on the ground. If aerial predators, typically approaching from the air, elicited different calls when on the ground, this would indicate that calls varied depending on predator location and not on specific predator category. If aerial predators, however, elicited the same calls in the different locations, this would suggest that the calls refer to the species or animal category (e.g., raptors) rather than to the hunting strategy.

For bird and mammal species it has been shown that they emit different call types in response to the same species of predator depending on its spatial location. Chickadees (*Poecile atricapilla*) produce "seet" calls in response to raptors in flight and mobbing calls in response to perched raptors (C. Templeton et al. 2005). Siberian jays also emit different calls for the same species of raptor depending on whether it is flying or perched (Griesser 2008). In the same way, meerkats bark at perched raptors (Manser 2001) and do not emit the predator-specific aerial alarm calls given in response to raptors in the air. Also, a terrestrial predator encountered lying in a bolt-hole elicits recruitment calls for mobbing behavior, while the same species walking or standing elicits terrestrial alarm calls (Manser 2001). Therefore, predator-specific calls in meerkats also refer to the spatial location of the approaching animal. My statement that meerkat alarm calls refer to predator types has to be understood with this additional specification, and this is likely also to be the case for the other species producing predator-specific aerial and terrestrial calls (but see Zuberbühler 2003).

Call specificity has also been suggested to refer to the behavior of the encountered predator (Griesser 2008; Manser et al., in preparation). One could argue it is not the spatial location of the predator causing the use of different call types but rather its hunting strategy or behavior (C. Templeton et al. 2005). However, studies investigating the influence of the spatial position and the speed of predators conducted under controlled conditions on chickens in captivity (C. Evans et al. 1993b) revealed that predators approaching from the air or on the ground elicited acoustically distinct call types, whereas speed of approach influenced call rate but not call structure. This result is supported by the pattern observed in meerkats that circling raptors at some distance elicit aerial alarm calls of the same general type as do raptors approaching them directly and at high speed. The urgency is reflected in call structure and rate, as predicted by Morton's structural rules (Morton 1977), from tonal calls for predators far away (more relaxed context) to harsh calls for predators approaching fast (tense or hostile context). Predator speed appears to be expressed in acoustic structure along a graded continuum and increasing call rate (C. Evans et al. 1993b) rather than through distinct call types. The only clear evidence for a call that refers to the behavior of a predator has recently been demonstrated in Siberian jays (Griesser 2008). The alarm call types in this species were specific to the behavior (perched, prey search, and attack) of different species of hawks and elicited distinct escape responses.

One of the nonspecific alarm calls, the "animal-moving" call in meerkats, appears to refer to the behavior of an animal moving in the caller's vicinity (Manser et al., in preparation). Animal-moving calls are elicited by almost any category of animal in the surrounding area: close, low-flying, nondangerous birds, such as vultures or southern yellow-billed hornbills (*Tockus leucomelas*); raptors such as eagles close on the ground or flying far away; perched raptors moving their wings in preparation for flight; nondangerous terrestrial animals, such as herbivores; terrestrial predators, especially when moving again after an interrupted approach; and foreign meerkats nearby. The consistent aspect is that in all of these contexts, when the call is emitted, the animal of interest is moving. That the movement rather than any other factor is of importance has been shown in experiments where live and stuffed terrestrial predators, as well as stuffed perched raptors, were moved, paused, and moved again. Very often the moving-animal call was given in response to the movement after the pause, before signalers resumed emitting predator-specific calls (Manser et al., in preparation).

Physical attributes, such as size, color, or shape, that correlate to the risk of the encountered stimulus appear to determine the call specificity in a variety of species. Chickadees, when mobbing a perched raptor, vary the number of repetitions of the last call element in their chickadee call depending on the size of the raptor (C. Templeton et al. 2005). Prairie dogs vary the acoustic structure of their alarm calls elicited by humans depending on the color of the person's clothes and their previous experience with him or her (Frederiksen and Slobodchikoff 2007). Observations of naturally occurring predator approaches indicate that meerkats respond more strongly to the most dangerous raptor, the martial eagle, by emitting higher-urgency calls than when detecting a bateleur eagle, *Terathopius ecaudatus* (M. B. Manser, personal observation). As suggested for chickadees and prairie dogs, this may be related to the greater risk posed by this species.

Information on the direction from which and how a predator is approaching is probably of more benefit to the receiver than information on the species. Although more detailed investigations on species specificity may reveal subtle differences in the calls elicited by different raptor species, it is possible that these differences stem more directly from risk posed than from species identification. In an environment where specific predator types typically approach from either the air or the ground, a different response strategy is beneficial, and it is likely that calls referring primarily to the location, rather than to the species, evolved. To a flying raptor approaching at high speed, which also disappears very fast again, the best escape response is to run immediately for the closest shelter, from where they can resume foraging immediately after the encounter. In contrast, for a terrestrial predator, for which a slower response is usually possible and a different escape strategy is appropriate, it is clearly advantageous to have different alarm calls according to spatial location. Other attributes, such as size or color, are of secondary importance and, most of the time, serve only as indicators of urgency and risk, which are reflected more in the continuum of a changing acoustic structure according to Morton's structural rules or call repetition rate (but see Frederiksen and Slobodchikoff 2007).

12.4. Why are some alarm calls considered functionally referential and not others?

Typically, predator-specific calls have been considered to be functionally referential, while nonspecific alarm calls, given to a variety of different predator types, have been considered to be expressions of emotional state with little referential specificity (Macedonia and Evans 1993). The explanation is that receivers do not gain much information from the nonspecific alarm calls and they do not respond in obviously different ways to them. Such calls have been described for many sciurid species (S. Robinson 1980; Blumstein and Armitage 1997) and also birds (Leavesley and Magrath 2005), where individuals use different call types depending on the level of urgency to respond to the approaching predator, regardless of whether it is an aerial or a terrestrial animal (fig. 12.1). This begs the question as to whether these calls might not be referential as regards the level of contextual urgency, a low-urgency situation eliciting a different call type than a high-urgency situation. Similarly, it has been suggested that for meerkat predator-specific calls, the variation of acoustic structure according to animal distance (level of urgency, risk) should not be considered functionally referential. The difference in acoustic structure between the low- and the high-urgency calls in aerial and terrestrial calls is greater than between the low- and the high-urgency calls of the predator-specific calls (fig. 12.2). Yet on the production side, predator specificity is expressed in very distinct acoustic categories, while on the dimension of level of urgency the acoustic structure of the calls grade into each other (Manser 2001). In the same way, receivers show clearly different responses to predator-specific calls, whether they are of low or high urgency; along the dimension of urgency level, the only difference is the strength of the response (Manser et al. 2001).

In contrast, the animal-moving call, although not a predator-specific call, does fit the definition of being functionally referential. It refers to an external event, shows a high referential specificity to the behavior of an animal other than the caller, and receivers exhibit the appropriate response. Despite the fact that this call type is given to a broad variety of different animal species and categories, it has high specificity because it refers to a specific, temporary action of another animal. The animal-moving call indicates that referential specificity is not limited to physical attributes of specific objects or events but that animal calls can refer to a specific action displayed by a variety of external stimuli.

This brings up the question of whether this kind of vocalization represents an additional dimension of information in animal vocalizations, or whether it is just a specific case of the referential and emotional components described for other vocalizations. It is unlikely that this call type is purely an emotional expression, since it is given in extreme situations from very relaxed to highly stressful and yet the call structure stays consistently the same. Whatever these calls express (Zuberbühler 2003), they show that animal vocalizations refer not only to external objects or contexts but also to the behaviors of other animals. This is confirmed by the referential calls of the Siberian jays that indicate whether a raptor is perched, searching for prey, or attacking (Griesser 2008). These recent studies should motivate us to investigate in more detail the acoustic variation of alarm calls and whether the same mechanisms underlie these different types of referentiality to physical properties, spatial locations, and behavior.

12.5. Can functionally referential calls be explained by emotional expression of the signaler?

12.5.1. ACOUSTIC STRUCTURE

Functionally referential alarm call systems may be fully explained by variation in the emotional states of the caller as expressed in the acoustic structure of the calls (Premack 1972). The predator-specific calls of meerkats convey referential and emotional information, whereby the referential component relates to predator type and the emotional component relates to level of urgency (given by risk and distance) as well as other individual characteristics (Manser 2001; Manser et al. 2002). Nevertheless, it has been suggested that the component referential to the receiver may also be an expression of emotion, representing different fear levels of the signaler (Seyfarth and Cheney 2003a). According to this hypothesis, recruitment calls are predicted to induce the lowest level of fear; terrestrial calls, a medium level; and aerial calls, the highest level (fig. 12.5a). In other words, the response urgency explains the different types of calls. Meerkats have a lot of time to recruit, less time to respond when a terrestrial predator approaches, and the least time when an aerial predator approaches. For these different situations we have a graded system of low to high urgency, relating to distance and risk in the case of the terrestrial and aerial predators and to the type of stimulus in the case of the recruitment calls.

The fact that the acoustic structures of the meerkat calls elicited by the different predator types are very similar within the same level of urgency (Manser 2001; Manser et al. 2002; fig. 12.2) supports the idea that the variation of calls

recruitment terrestrial aerial
low med high low med high low med high
Increase of arousal

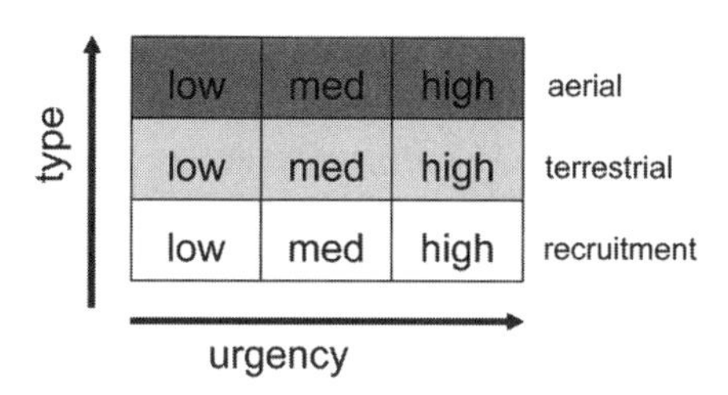

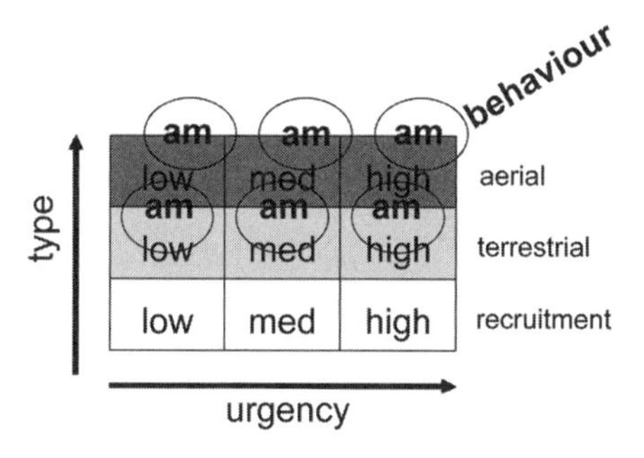

FIGURE 12.5. Models to explain the production of the different alarm calls (i.e., the expression of different emotional states) and how the different types might relate to each other taking context and acoustic structure into account: (a) one-dimensional model; (b) two-dimensional model; (c) three-dimensional model with animal-moving (am) calls.

represents the expression of emotions, not in a single dimension, but rather in at least a two-dimensional way (fig. 12.5b). The information about urgency is encoded in specific rules with the same acoustic parameters changing across aerial, terrestrial, and recruitment alarm calls. However, no consistent acoustic change underlies the differences among the three alarm call types, and the parameters explaining the variation between predator types and urgency level differ (Manser 2001). This pattern of changing acoustic parameters suggests an even more complicated relation not reflected in a simple two-dimensional model. Considering the animal-moving call, no simple model expressing the emotion of a caller could explain the consistent acoustic structure of this call type (fig. 12.5c). The moving-animal call can be given in a relaxed context (encounter of herbivores or nondangerous birds), as well as in highly stressful predator encounters (with dogs or raptors), and yet its acoustic structure is very similar in these different situations. This suggests a different underlying mechanism than the expression of an additional dimension of emotions.

12.5.2. REPETITION RATE

One parameter, the repetition of the call, appears consistent across all different call types and contexts with a linear change in the arousal of an animal. The general pattern described by Morton's acoustic structural rules, with more tonal calls in relaxed situations grading into harsh and noisy calls in tense, unfriendly situations, is reflected in the different alarm calls in meerkats (Manser 2001; Manser et al., in preparation). However, another general pattern observed to correlate with the arousal of the caller seems to be call repetition rate. In several studies, calls are more frequently and more quickly repeated with increased urgency of the situation (Blumstein and Armitage 1997; Manser 2001; Schehka et al. 2007). This is particularly well documented in chickens presented with videos simulating the approach of predators at different speeds (C. Evans et al. 1993b). Similarly, with increased risk reflected by the size of perched raptors, chickadees repeat the last element in their mobbing calls more often (C. Templeton et al. 2005). Furthermore, although the acoustic structure of the meerkat predator-specific call types varies in a linear way with the arousal of the caller, the structure of the animal-moving call does not. Yet, for this call type as well, repetition increases with the tension of the situation (Manser et al., in preparation). Only for the call types emitted under immediate threat and time constraints to respond (high-urgency aerial call and panic call) do meerkats (Manser 2001) and marmots (Blumstein and Armitage 1997) emit only a single or a few calls. This raises the question of why changes in acoustic structure and call rate do not fully correlate with each other, and whether this means that these two parameters express different dimensions of emotions.

12.6. Conclusions

By definition, functionally referential calls refer to specific stimuli in the external environment of the caller and cause receivers to show an adaptive response to them. Yet, the acoustic variation of these calls can vary in relation to many different attributes of a stimulus, as shown for alarm calls that refer to species, size, spatial area, direction, distance, and also behavior (fig. 12.1). What attribute a call refers to is expected to depend on the selective advantage of reliable specific information, which will vary with the ecological environment of a species. This may also explain why, at least in meerkats, functionally referential calls have a denotative function, rather than an imperative one or one referring to the escape behavior of the caller. The information value from denotative calls for receivers dispersed in the environment is likely to be much higher than that from imperative calls or information on the behavior of the caller, as it allows each individual to adjust its response to its own context at the given moment. While the acoustic variation relating to specific attributes may in some cases be explained by different emotional states of the caller, the structure of the predator-specific calls in meerkats and also the call referring to the action of another animal do not predict a simple correlation between arousal of the caller and the acoustic structure of the vocalizations emitted. Arousal state of the caller seems to correlate better with the repetition rate of a call than with its acoustic structure.

To understand the meaning and the function of alarm calls (and this is also true for any other call type), it will be important to not only pick out a single or a few specific call types but consider the whole vocal repertoire of a species, or at least all calls emitted in a similar context (e.g., all calls given during predator encounters). To show only that some calls are functionally referential will not increase our insight into the meaning of these calls and their underlying cognitive processes. The example that the animal-moving call in meerkats refers to a specific behavior also shows that animals categorize their environment in ways different from what we had come to expect. This is true even if the categorization is not based on differences in perception or due to specialized sensory organs but due to differences in processing the perceived information. Neurobiological studies will help to identify whether the same regions of the brain are activated during encounters that elicit functionally referential calls and encounters that cause animals to change their call rate or other acoustic characteristics along a continuum. It is also clear that we have to better describe natural and experimental predator encounters and, in particular, note the specific context that the signaler and the receiver experience in that instance. Only with such a detailed approach will we understand what categories and concepts animals use to behave in an adaptive way in their challenging environment.

12.7. Summary

Functionally referential calls refer to specific stimuli in the external environment of the caller and cause receivers to show an adaptive response to them. Here I have addressed some of the questions about the underlying cognitive mechanisms of functionally referential alarm calls, using as examples meerkats and a few other species. Several social species have evolved alarm call systems with either predator-specific or nonspecific alarm calls, or both. Because meerkats emit the same alarm call types in response to different predators, but independent of the location in relation to shelter or the behavior of the caller, these calls are denotative. Whether this is typical of referential calls in other species has still to be tested. In some species, the predator-specific calls appear to refer to a predator species or category, whereas in others they appear to refer to the spatial area from which the predator is approaching or its behavior. The question arises as to whether the same cognitive mechanism underlies calls referring to the physical characteristics, the spatial position, or the behavior of the predator or whether different information processes distinguish these calls. When considering the different alarm call types with respect to the expression of the emotional state of the caller, the acoustic structure appears not to relate to the arousal of the caller in a simple linear way in all of the alarm call types. This suggests that the different referential and emotional components conveyed in the acoustic structure in the different species reflect more complicated mechanisms than a pure linear emotional system can explain.

ACKNOWLEDGMENTS

I would like to thank the editors of this book, John Ratcliffe and Reuven Dukas, for inviting me to take part in the symposium at the 2007 ABS meeting and for their incredible support in writing this chapter. The meeting was stimulating, and in particular the discussions after the talk with Dan Blumstein and Chris Evans improved the content of this chapter. Furthermore, many thanks to all the people at the Kalahari Meerkat Project, especially Tim Clutton-Brock, for their contributions to the studies described here. I am also grateful to Tim Clutton-Brock, Alex Thornton, my research group in Zurich, and several reviewers for their comments on previous versions of the chapter.

13 Adaptive Trade-offs in the Use of Social and Personal Information

RACHEL L. KENDAL, ISABELLE COOLEN & KEVIN N. LALAND

13.1. Introduction

Social learning—the acquisition of knowledge from others—provides naive animals with information relevant to many life skills, including when, where, what, and how to eat (Galef and Giraldeau 2001), with whom to mate (D. White 2004) or fight (Peake and McGregor 2004), and which predators to avoid and how (A. Griffin 2004). Over the last century, a common assumption of behavioral ecologists, ethologists, and anthropologists has been that such copying is inherently adaptive. Animals are deemed to gain fitness benefits by learning from others, since they acquire adaptive information while avoiding some of the costs (predation risk, search costs, etc.) of learning for themselves—the costs of "personal information" acquisition. However, the use of social information does not, in fact, guarantee success (Boyd and Richerson 1985, 1995; Rogers 1988; Giraldeau et al. 2002). Individual animals face evolutionary trade-offs between the acquisition of costly but accurate information and the use of cheap but potentially less reliable information[1] (Boyd and Richerson 1985), here manifest in a trade-off between reliance on personal and social information.

Theoretical models investigating the adaptive advantages of social learning predict that it should not be employed in an indiscriminate manner, but rather, individual animals should have evolved flexible strategies that dictate precise circumstances under which they copy others (Laland 2004). Such analyses

1. Some readers may object to our use of the phrase "unreliable information," on the grounds that the cues that form the bases of social learning are not so much reliable or unreliable as more or less informative. While we are sympathetic to this objection, we persist with the terminology for three reasons. First, whether appropriate or not, use of such terms is common in the literature that we review. Second, there are no obvious alternative expressions that we find entirely satisfactory. For instance, an "uninformative cue" does not distinguish between a signal designed to mislead and a cue that contains no information at all. Third, it is apparent that we are frequently concerned with the reliability and error associated with potential social and asocial sources of information, for which our use of "reliable or unreliable information" can be taken as shorthand.

reveal that social learners adopting a pure strategy of random copying would have higher fitness than asocial learners only when copying is rare, that is, when most potential "demonstrators" of new behavior would be asocial learners who have acquired accurate information by directly sampling the environment (Boyd and Richerson 1985, 1995; Rogers 1988; Giraldeau et al. 2002). As the frequency of social learners increases, the value of using social information typically declines, as the proportion of individuals demonstrating accurate personal information decreases. At the extreme, with all individuals exhibiting random copying, no one would have acquired accurate personal information by sampling the environment. In order for it to be adaptive, individuals must use social learning selectively and engage in the collection of accurate personal information some of the time (Galef 1995; Laland 2004).

Despite rapid growth in the field of animal social learning, the circumstances under which individuals rely on alternative sources of information remain relatively unexplored. What rules have evolved in animals to specify how they should exploit personal and social information? Do animals copy the behavior of others when they are uncertain how to solve a problem, or, perhaps, when it is easy to do so? Do they have rules dictating from whom they should copy—for example, high-status or apparently successful individuals? Is social learning a last resort when asocial learning has failed, or first port of call? Following Laland (2004), the term "social learning strategies" is used here to equate evolved learning heuristics with those strategies commonly analyzed using evolutionary game theory (Maynard Smith 1982) and to encourage theoretical analysis. Naturally, animals need not be aware that they are following a strategy, nor understand why such strategies may work.

In this chapter we review the predictions arising from theoretical models and outline the current empirical support for several social learning strategies, focusing largely on our own experimental studies and other recent work (Laland 2004; Kendal et al. 2005; Galef 2006). We draw attention to adaptive trade-offs in the use of social and personal information. Laland (2004) distinguished between two classes of social learning strategy: "when" strategies, which dictate the circumstances under which individuals copy others, and "who" strategies, which specify from whom individuals learn. We address each in turn.

13.2. "When" strategies

13.2.1. COPY WHEN ASOCIAL LEARNING IS COSTLY

Trial-and-error learning is often both costly and error prone. Direct interaction with the environment may entail fitness costs, such as injury, sickness, and predation, as well as "missed opportunity" costs, such as the loss of time or energy that could be allocated elsewhere. These costs are expected to restrict an animal's investment in asocial learning, resulting in consequent reliance on social learning, which may lead to "errors" such as a failure to perform an adaptive behavior or the retention of a suboptimal variant. Where these costs are substantive, we anticipate that selection may plausibly have favored shortcuts to adaptive solutions, notably copying others (Boyd and Richerson 1985).

13.2.1.1. *Theoretical background*

Several theoretical analyses conclude that reliance upon social information should increase as the costs associated with acquiring personal information increase (Boyd and Richerson 1985, 1988; Feldman et al. 1996; but see section 13.2.2.3). Boyd and Richerson (1985) propose a "costly information hypothesis," which depicts an evolutionary trade-off between acquiring accurate but costly information versus less accurate but cheap information. While this trade-off may manifest itself at different levels, for our purposes it can be summarized as the idea that when information is too costly to acquire or to utilize personally, individuals will take advantage of the relatively cheap information that can be learned from others. A similar argument was proposed by Bandura (1977, 12), who stated that "the more costly and hazardous the possible mistakes, the heavier is the reliance on observational learning from competent examples." Although the costly information hypothesis lays emphasis on the costs of acquiring personal information, the same reasoning holds with respect to the costs of using personal information.

13.2.1.2. *Empirical evidence*

Laland and Williams (1998) provide an experimental example in which fish were seemingly prepared to pay the costs of using suboptimal foraging information provided by conspecifics in order to avoid the risk of predation associated with the isolated learning of a more efficient foraging route. Small groups of "founder" guppies (*Poecilia reticulata*) were trained to take either an energetically costly circuitous route to a feeder or a less costly direct route. In a transmission chain design, these founders were gradually replaced with naive conspecifics, one individual being replaced each day for a week. Three days after all the trained individuals had been removed, the groups of fish whose founders were trained to swim the circuitous route continued often to use this route to reach the feeder, despite its cost relative to the available direct route. In addition, individuals in groups with founders trained to take the circuitous route took significantly longer to switch to the short route than did otherwise-

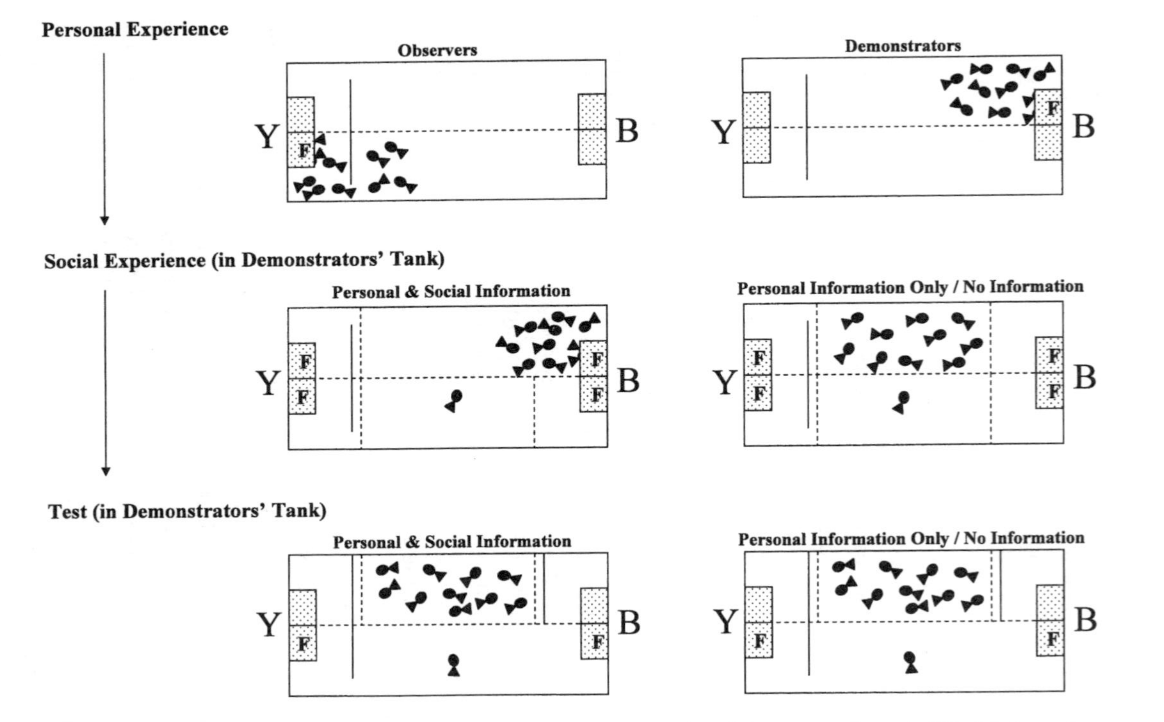

FIGURE 13.1. Experimental procedure, showing the personal experience training of observer guppies to feed (food designated as F) from either blue (B) or yellow (Y) feeders behind an opaque barrier, and demonstrators to feed at the open end of the tank; the social experience procedure for guppies in all conditions (prior personal and social information, personal information only, and no information); and the identical test period for all conditions, in which observers must lose visual contact with the constrained demonstrator shoal in order to feed at the trained feeder. Solid lines indicate opaque partitions and dashed lines indicate transparent partitions. From Kendal et al. 2004.

equivalent solitary fish. It is well established that guppies are reluctant to leave conspecifics and forage alone due to predation risk (see Day et al. 2001); thus, the perpetuation of a suboptimal behavioral tradition in these fish can be explained by the relative cost of acquiring personal information.

Losing visual contact with shoal[2] members is potentially costly to small fish like guppies, since isolation leaves them vulnerable to predation (see Day et al. 2001 and references therein). Kendal et al. (2004) exploited this observation to manipulate the cost of using previously acquired personal information in a social foraging experiment, in the process providing a second illustration of animals utilizing a "copy when asocial learning is costly" strategy. Individuals were allocated to one of three conditions in which they received either (1) prior personal information only, (2) prior personal and social information, or (3) no information (see fig. 13.1). Individuals in the first two conditions had the opportunity to learn through direct experience that food was located in only one of two differently colored feeders at the ends of their tanks. The feeder that contained food was located behind an opaque barrier while the one that did not was in open water. In the next stage of the experiment, one group was then provided with conflicting social information. The fish in the condition that received both personal and social information (2) observed a shoal of demonstrators feed at the feeder in the open water, which their personal experience had indicated never contained food, whereas fish in the other two conditions, (1) and (3), were constrained in the central section of the tank, facing nondemonstrating fish. Following this observation period, there was a test in which the demonstrator shoal was restricted to the center of the tank, both feeders were baited with food, and the fish were released to investigate where each fed. Fish with both sources of information faced a choice between using personal information (i.e., feeding at the feeder that had consistently contained food but that necessitated losing visual contact with conspecifics) and using the social information (i.e., feeding at a feeder that had never previously contained food but did not necessitate loss of contact with conspecifics). Fish in all conditions fed at the feeder in the open water rather than the one behind the opaque partition, supporting the assumption that swimming behind the opaque barrier to feed represented a cost that guppies would avoid if possible. However, fish with both sources of information ignored their personal information and fed at the feeder in the open water more rapidly, and with less variability, than did fish with personal information alone (fig. 13.2);

2. A shoal of fish is a loose aggregation of individuals formed largely for social reasons. When members of a shoal school, they move in a highly coordinated and synchronized fashion. The experiments discussed in this chapter all involve loosely aggregating "shoaling" fish.

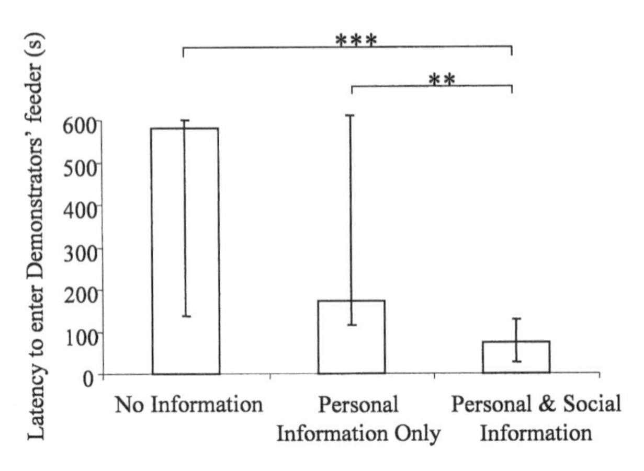

FIGURE 13.2. The latency (median and interquartile range) of guppies to enter the demonstrators' (no barrier) feeder in the "no information," "prior personal information only," and "prior personal and social information" conditions (**$P < 0.01$, ***$P < 0.001$). From Kendal et al. 2004.

hence, it would appear that the former used the social information provided in preference to their personal information. An otherwise-equivalent prior experiment with no visual barrier, in which the use of personal information did not necessitate loss of contact with conspecifics, found that, at test, fish with both sources of information, but not those in other conditions, continued to use their personal information and ignored the conflicting social information (Kendal et al. 2004). As social information outweighed contradictory personal information only where the latter was costly to use, it appears that the guppies were employing a "copy others when asocial learning is costly" strategy.

Further support for this strategy is provided by Coolen et al. (2003), who examined the propensity of wild-caught three-spined *Gasterosteus aculeatus* and nine-spined *Pungitius pungitius* sticklebacks to use public information about the profitability of food patches. In a laboratory test, individual fish were restricted to a central compartment of an aquarium from where they could see two equivalent-sized shoals of conspecifics each feeding at one of two identical but spatially separate feeders. The feeders were designed such that observers could not directly see the food, which was dispensed at different rates, but could use cues indicating the feeding rates (here, the frequency with which individuals pecked at the feeder as the food sank through it) (fig. 13.3). Following observation, the demonstrators and all food were removed from the tank, and the observer was released and its choice of feeder monitored. Solely on the basis of the demonstrators' success, observers were required to choose the richer of the two feeders. At test, nine-spined sticklebacks preferentially chose the goal zone that had formerly held the rich feeder, suggesting that they were able to exploit public information, an interpretation supported by Coolen et al.'s other experiments. However, three-spines swam with equal frequency to the former locations of rich and poor patches (fig. 13.4). This reluctance or inability of three-spines to use prior public information was confirmed by Webster and Hart (2006) in a study involving the acquisition of subhabitat preferences. In Coolen et al.'s final experiment, observers were provided during demonstration with optional use of vegetative cover, which nine-spines, but not three-spines, used. The collection of personal information in open water is costlier for nine-spines than for three-spines because nine-spines have inferior structural antipredator defenses and are consumed preferentially by piscivorous fish (Hoogland et al. 1957). Because of these costs, nine-spines may forgo the opportunity to collect reliable personal information and favor vicarious assessment of foraging opportunities through observational learning. Thus, public information use in sticklebacks can be regarded as an adaptive specialization in learning, reflecting the differential costs of personal information acquisition.

In the first explicit test of flexibility in the social learning strategies of Norway rats (*Rattus norvegicus*), Galef and Whiskin (2006) assessed reliance on social information while foraging in risky situations. Following a 30-minute interaction with a demonstrator rat that imparted cues on its breath as to whether it had been eating either cinnamon- or cocoa-flavored food, naive rats were provided with both cinnamon- and cocoa-flavored food in a single location that afforded them little cover/refuge and entailed traversing an open space (from the safety of the nest box) to reach it. The potential cost of consuming

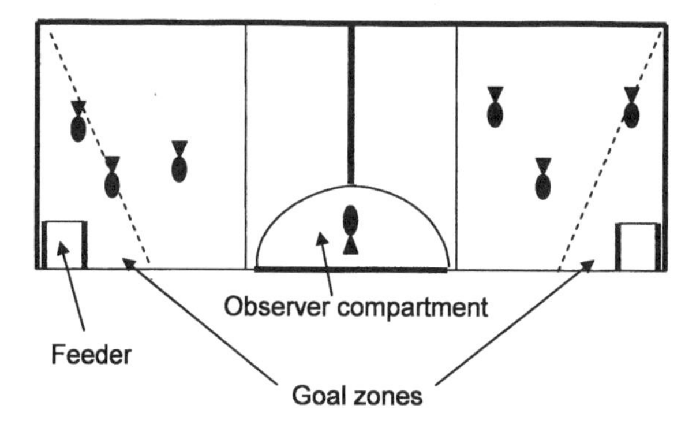

FIGURE 13.3. Diagram of the experimental tank setup allowing single sticklebacks to observe conspecifics feeding at two feeders. Thick lines represent opaque partitions, thin lines represent transparent partitions, and dashed lines represent goal zone virtual delimitations. From Coolen et al. 2003.

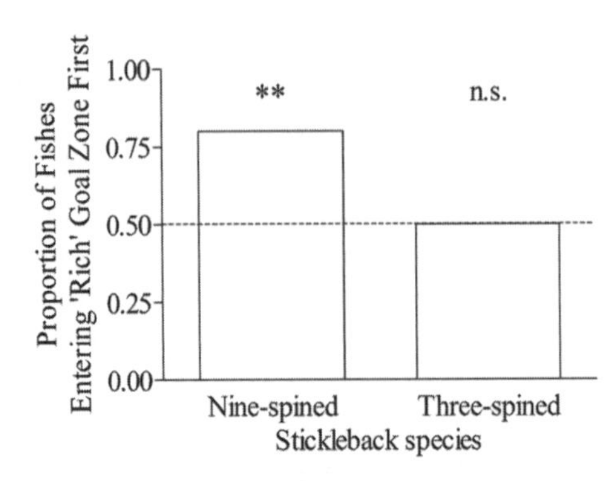

FIGURE 13.4. The proportion of three-spined and nine-spined sticklebacks that entered first the goal zone corresponding to the feeder that appeared "rich" rather than "poor" during the demonstration period ($n = 20$ for each species). The dashed line indicates the proportion expected at random (**$P < 0.01$, n.s. = not significant). From Coolen et al. 2003.

these unfamiliar foods was increased for the rats by placing two cats in a cage in the housing room for either (*i*) 4 or (*ii*) 24 hours or by (*iii*) allowing the cats to roam freely. Contrary to Galef and Whiskin's expectation, rats did not eat more of the diet eaten by their demonstrator as potential predation risk increased; indeed, individuals under potential predation risk were less influenced by the social information than controls, who faced no predation risk. Galef and Whiskin interpreted this finding as contrary to the prediction that, as the costs of acquiring personal information increase, so should the reliance on social information. However, this interpretation is open to question. The cost the rats were facing was largely in traveling, over open ground, to the food site (where both foods were presented in close proximity), rather than in acquiring personal information pertaining to the foods. In contrast to control rats, those under stress of potential predation may have chosen a food at random without taking the time to either (*i*) collect personal information or (*ii*) use their previously acquired social information.

The cost of acquiring personal foraging information, represented by the difficulty of asocial learning, was also manipulated in a study of social foraging in callitrichid monkeys, involving three lion tamarins (*Leontopithecus* sp.), two tamarins (*Saguinus* sp.), and two marmosets (*Callithrix* sp.). Day (2003; Day et al. 2003) presented a series of novel artificial-fruit tasks, requiring the extraction of preferred food items, to zoo-housed groups of monkeys. Judging by the time to learn, as well as the amount of food extracted, the tasks varied significantly in difficulty. For each task there were two options (doors or holes) by which monkeys could extract food, with the alternatives being equivalent except in location and color. While the monkeys learned all of the tasks, a detailed statistical analysis revealed that the difficult, but not the easier, tasks were learned socially (Day 2003). For the difficult, but not the easy, tasks there was a significant tendency for individuals within a group to extract food using the same colored option as others, suggesting nonindependent learning.

Presumably, the personal information required to solve the easy tasks could be acquired at little cost, while the solutions of the more complex tasks were associated with a sufficiently large time or energy cost to render social learning adaptive. Similarly, Baron et al. (1996) reported that human subjects were found to imitate more as task difficulty increased.

In summary, although there is little explicit experimentation, that which exists provides strong support for animals exploiting a "copy when asocial learning is costly" strategy. This is probably the best-supported social learning strategy.

13.2.2. COPY WHEN UNCERTAIN

13.2.2.1. *Copy when have no relevant information*

Theoretical background. Boyd and Richerson (1988) considered a model exploring the advantages of reliance on social and asocial learning in a temporally variable environment in which animals have to make decisions as to which of two environments they are in and choose the appropriate behavior. Behavior 1 is appropriate in environment 1, behavior 2 is appropriate in environment 2, while performing the alternate behavior results in a fitness cost. The animals base their decision on the magnitude of a continuous parameter (x) representing the outcome of direct observation. At one extreme, if x has high values, above a threshold value d, the animals "know" they are in environment 1 and perform behavior 1. At the other extreme, if x has low values (below $-d$), they "know" they are in environment 2 and perform behavior 2. However, if x has intermediate values ($-d < x < d$), animals are uncertain as to whether they are in environment 1 or 2, and it is assumed that they will copy the behavior of others.

We note a (*i*) broad and a (*ii*) narrow interpretation of Boyd and Richerson's assumption. Individuals may be predisposed to rely on social information (*i*) if they lack relevant prior knowledge to guide their decision making, or (*ii*) if they are uncertain as to which of several established behavior patterns is appropriate given the information at hand. We are aware of considerable, albeit often inadvertent, empirical evidence for the former, whereby totally naive individuals use social information, for example, pertaining to the location of food, but none for the latter, whereby individuals would be faced with the choice, given some social information, of using previously acquired personal information, for example, pertaining to the location of food depending upon which laboratory "environment" they believed themselves to be in.

Empirical evidence. In an experiment related to that described in section 13.2.1.2, Kendal et al. (2004) tested in three conditions the propensity of guppies to use social information concerning the availability of food at two differentially colored feeders, although this time the use of personal and social information did not differ in cost. One group was provided first with personal and then with conflicting social information, a second group was given solely social information, and a control group was provided with no personal or social information at all. They found that fish that were provided with social information only, and that lacked relevant prior information, fed at the feeder indicated by conspecifics significantly more often than chance expectation. In contrast, individuals with both sources of information ignored the social information and continued to feed according to their personal information. This finding holds even when the order in which personal and social information are experienced is reversed (i.e., social first, personal second) (Laland, unpublished data). Similarly, Coolen et al. (2003; see section 13.2.1.2) found that nine-spined sticklebacks that did not have personal information copied the patch choices of others, whereas van Bergen et al. (2004; see section 13.2.2.2), testing the same species in an identical setup, found that fish would ignore social information when they had relevant personal information.

Social learning may occur as a result of an individual collecting information either directly, by observing a particular behavior in others (public information), or indirectly, by inferring possible causes of a given behavior pattern (social cues). In an extension of the 2003 study, Coolen et al. (2005) examined the use of direct and indirect social foraging information by nine-spined sticklebacks. The number of demonstrators present at each patch varied (two vs. six fish), supposedly indicating a poor and a rich patch, respectively. This indirect information either conflicted with the demonstrators' feeding rate at each patch (feeding respectively six times vs. twice in the 10-minute demonstration) or was the only information available. The sticklebacks were capable of using both direct and indirect information to make choices about where to forage, but when these contradicted each other, they relied on the direct information provided by feeding rate (see fig. 13.5). These findings are consistent with "copy when have no relevant information," since fish copied their demonstrators' patch choices in both conditions, but also indicate what type of social information individuals prefer to use. Direct information tends to be associated with greater accuracy, whereas by copying the decisions of others (indirect information), one may also copy their mistakes. By relying preferentially on direct information, nine-spines may avoid potentially maladaptive informational cascades (Giraldeau et al. 2002).

Galef et al. (2008) also provide support for a "copy when uncertain" strategy being deployed. They created conditions whereby Norway rats were either certain or uncertain as to the causal relationship between ingesting an

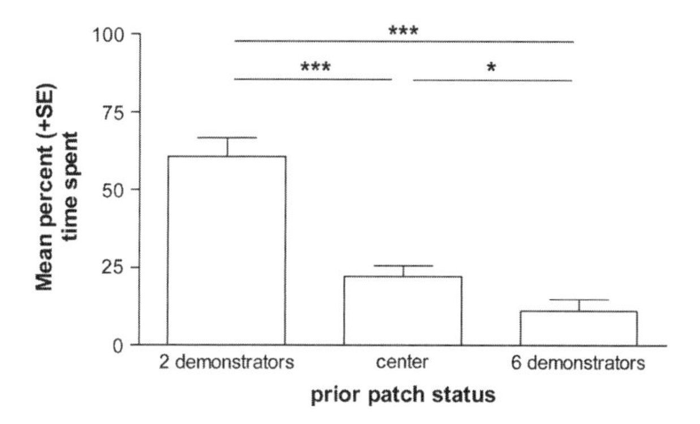

FIGURE 13.5. The mean percentage (+ SE) of time nine-spined sticklebacks spent in each patch ($n = 20$) after they collected both direct (demonstrators' feeding rate) and conflicting indirect information (shoal size) about relative patch quality during the observation period ($^*P < 0.05$, $^{***}P < 0.001$). From Coolen et al. 2005.

unfamiliar flavor and gastrointestinal upset. Thus, "certain" rats were fed cinnamon-flavored food and immediately afterward were injected with lithium chloride, and "uncertain" rats were fed a food containing both cinnamon and cocoa flavors prior to experiencing toxicosis. After a 24-hour recovery period, these rats, having learned that sampling unfamiliar foods can be dangerous, were exposed to a demonstrator that had eaten another unfamiliar food (either anise or marjoram flavor). Following this, when exposed simultaneously to anise- and marjoram-flavored foods for 24 hours, "uncertain" rats ate significantly more of the food eaten by their demonstrator than did "certain" rats.

Galef et al. interpret this study as a test of a response to having "ambiguous prior personal information." However, given that the social information individuals received pertained to different flavors than those experienced during personal information, we posit that the rats were not faced with "ambiguous personal information" but rather lacked personal information relevant to the social information received. While the elevated copying in the "ambiguous" social information condition supports the "copy when uncertain" interpretation, the observation that copying occurs in both conditions to some degree might also be interpreted as consistent with a "copy when asocial learning is costly" strategy; rats may have learned as a result of their poisoning that asocial learning is dangerous in this environment. However, a previous study (Galef and Whiskin 2006) showed that rats experiencing toxicosis after ingesting variously 1, 2, and 4 different foods did not differ in their subsequent use of social information pertaining to unfamiliar foods.

13.2.2.2. *Copy when prior information is unreliable*

Theoretical background. We now consider scenarios where, for whatever reason, personal information is actually less reliable than social information. Boyd and Richerson (1985, 1988) modeled the use of social information in a spatially heterogeneous environment where individuals experience different environments. The average quality of information available from demonstrators enables individuals to weight their use of asocial and social learning according to the likelihood of acquiring erroneous information from each source. As environmental heterogeneity increases and personal information becomes more error prone, the optimal amount of social learning from local residents increases, while as the rate of dispersal between environments increases, social information becomes increasingly unreliable (since individuals will increasingly copy "outsiders"), and the optimal amount of social learning decreases.

Giraldeau et al. (2002; see Bikhchandani et al. 1992, 1998) proposed that individuals may use social information not only because their personal information is in itself unreliable but because the accumulated knowledge of conspecifics potentially represents a source of information with even greater reliability. In spite of this, they predict that, in specific instances, reliance on the decisions of others can lead to arbitrary or even maladaptive traditions in animals (Giraldeau et al. 2002; see also section 13.2.2.1). Theoretical work regarding reliability and the value of information in communication and mating systems (Sirot 2001; Koops 2004) suggests that, even if the costs of misinformation are high, animals should still use information, provided that it is usually reliable. This requires animals to be able to assess the relative reliability of personal versus social information correctly.

Empirical evidence. In a study of nine-spined sticklebacks, van Bergen et al. (2004) manipulated the reliability of personal information concerning the profitability of two foraging patches, using a similar experimental design to that of Coolen et al. (2003; see section 13.2.1.2 and fig. 13.3). Fish were allocated to one of three conditions, where they received 100%, 78%, or 56% reliable personal information as to which of two feeders was "rich" and which "poor." Following this training period, fish were tested individually for their feeder preference. Those in the 100% and 78% reliable conditions significantly preferred the "rich" feeder. Subsequently, the profitability of the two feeders was reversed, and fish were presented with (now conflicting) public information in which they observed demonstrators feeding at the two feeders, with what was according to their earlier sampling the poor feeder now the rich feeder, and vice versa. Following this demonstration, only fish in the 100% reliable condition continued to prefer the feeder that was "rich" according to their

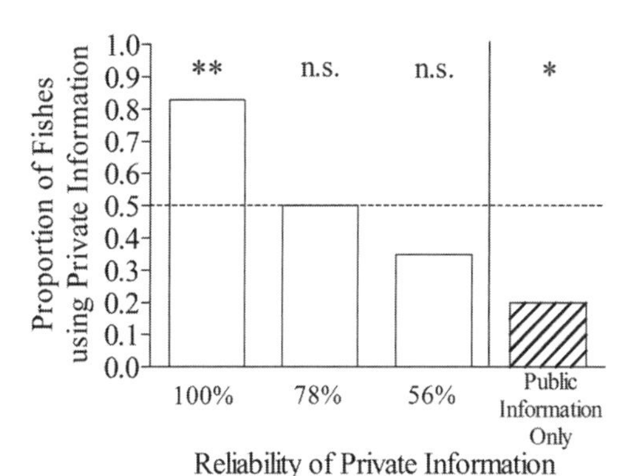

FIGURE 13.6. The proportion of nine-spined sticklebacks that, after receiving personal information of varying reliability followed by conflicting public information, entered first the goal zone of the feeder that was "rich" according to personal information. The dashed line indicates the proportion expected at random, and the hatched bar represents data from Coolen et al. 2003 (*P < 0.05, **P <0.005, n.s. = not significant). From van Bergen et al. 2004.

personal information; fish in the other conditions exhibited no preference (fig. 13.6). Since fish with 56% reliable information probably had not acquired private information (they did not prefer the rich feeder immediately after their training period), this experiment does not provide unequivocal evidence that fish increasingly relied on the social information provided by their demonstrators as the reliability of their personal experience diminished, although it is consistent with this interpretation. It does, however, demonstrate that fish with completely reliable personal information will ignore conflicting social information.

13.2.2.3. *Copy when prior information is outdated*

Theoretical background. Boyd, Richerson, and colleagues (Boyd and Richerson 1985, 1988; Henrich and Boyd 1998) have modeled the use of social information in temporally fluctuating environments, finding that intermediate levels of fluctuation will be most likely to favor social learning. Here individuals can acquire relevant information without bearing the costs of direct interaction with the environment associated with asocial learning but with greater phenotypic flexibility than if the behavior was unlearned (Boyd and Richerson 1985, 1988). Consequently, as socially transmitted information becomes increasingly outdated, we might expect individuals to become less likely to rely on it.

In other theoretical analyses, Doligez et al. (2003) predicted that strategies based on public information use (here the breeding success of conspecifics on particular patches) perform best when fluctuation in patch quality is of intermediate or high temporal predictability. Similarly, Moscarini et al. (1998) have looked at the effect of a changing world on the likelihood of informational cascades and predict that blind copying may occur for some limited time if the state of the world changes stochastically but will not happen anymore when the environment changes too unpredictably (or randomly). Kameda and Nakanishi (2002) reported that in a fluctuating environment, increasing costs of asocial learning initially result in an increase in social learning, a concomitant reduction in fresh information, and thus an outdated "cultural knowledge pool." They predict that natural selection will act against reliance on social learning based on such flawed information, ensuring cultural knowledge tracks environmental change.

These models predict that individuals should acquire personal information and ignore social information when the latter is likely to be outdated. Equally, individuals should opt to frequently update information if the use of their current information, whether acquired asocially or socially, is likely to be outdated.

Empirical evidence. Van Bergen et al. (2004) manipulated the degree to which personal information regarding the relative profitability of two foraging patches was outdated, and they explored how this prior experience affected individuals' subsequent acquisition of public information. Nine-spined sticklebacks were allocated to one of four conditions, where they received personal information as to which of two feeders was "rich" and which "poor" 1, 3, 5, or 7 days prior to receiving conflicting public information and a test of preference. Fish with only a 1-day delay between receiving personal and public information ignored the social information and first visited the feeder that was "rich" according to their personal information. Fish with delays of 3 and 5 days showed no feeder preference, and those experiencing a 7-day delay preferred the feeder that was "rich" according to the public information (see fig. 13.7). Accepting van Bergen et al.'s arguments that personal information was not forgotten after 7 days, comparison with results from Coolen et al. (2003), where fish received public information only, appeared to indicate that

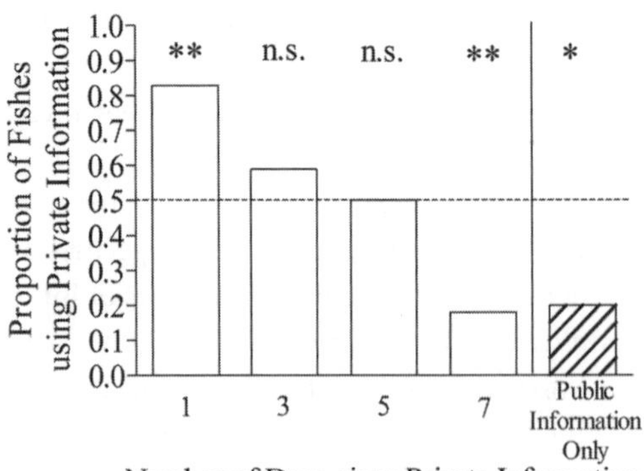

FIGURE 13.7. The proportion of nine-spined sticklebacks that, after receiving personal information followed at varying time lags by conflicting public information, entered first the goal zone of the feeder that was "rich" according to personal information. The dashed line indicates the proportion expected at random, and the hatched bar represents data from Coolen et al. 2003 ($^*P < 0.05$, $^{**}P < 0.005$, n.s. = not significant). From van Bergen et al. 2004.

fish in the 7-day condition ignored their personal information in favor of the public information. As personal information becomes increasingly outdated, nine-spined sticklebacks become increasingly reliant upon socially acquired information.

In summary, experimental data support the prediction that animals utilize a "copy when uncertain" strategy, although this strategy can be interpreted in various ways.

13.2.3. COPY WHEN DISSATISFIED

13.2.3.1. *Theoretical background*

Schlag (1998) explored an imitation rule that he termed "proportional reservation" but that might be called "copy when dissatisfied," finding that it is a highly effective learning rule. Here, the size of the payoff to its current behavior determines an individual's satisfaction, as a linear function. The individual retains its current behavior with a probability equal to this satisfaction and otherwise copies the action of a randomly chosen demonstrator. This strategy has the advantage that it is potentially simple to implement, because it does not require individuals to assess the payoff to a demonstrator or to make any judgment as to the relative profitabilities of alternative behavior patterns. Because "copy when dissatisfied" requires evaluation only of one's own success (e.g., in terms of satiation), not of the relative success of oneself and others, "copy when dissatisfied" seems more likely to have evolved than "copy when better" (Laland 2004), which Schlag found to be equally effective.

13.2.3.2. *Empirical evidence*

Galef et al. (2008) manipulated rats' dissatisfaction with their diet (experiment 1) or environment (experiment 2). In the first experiment, rats were maintained for 1 week on either an unpalatable low-caloric diet, requiring increased handling time to maintain their health (dissatisfied condition), or a palatable relatively high caloric diet (satisfied condition). Following this, the subjects were exposed for 30 minutes to a demonstrator rat carrying cues on the breath as to whether it had eaten either a marjoram- or an anise-flavored diet. In the following 24 hours the subjects were exposed to both diets, and "dissatisfied" rats ate more of the food that their demonstrator had eaten than did "satisfied" rats. In the second experiment, rats, essentially crepuscular burrowing animals, were housed either on a hard substrate with no cover, constant light, and an overly warm ambient temperature (1: dissatisfied condition) or with bedding, cover, a 12:12 light:dark cycle, and appropriate ambient temperature (2: satisfied condition). Again, "dissatisfied" subjects ate more of the diet indicated by their demonstrator than did "satisfied" rats.

In summary, Galef et al.'s experiments are consistent with the strategy of "copy when dissatisfied," although the latter experiment exemplifies "dissatisfaction" more broadly than the former. In experiment 2 rats appear to be generally dissatisfied with their lot, while in experiment 1 they are specifically dissatisfied with their feeding regime—that is, dissatisfied in the same domain as the social information. There is, perhaps, a case for characterizing these as two separate strategies, although it is interesting to note that both types of dissatisfaction translate into the same result with respect to socially induced food preferences.

13.3. "Who" strategies

13.3.1. THEORETICAL BACKGROUND

Boyd and Richerson (1985) proposed a number of cultural transmission biases. For example, a "frequency-dependent bias," in which the most popular variant in a population is disproportionately more likely to be adopted, may be translated into the social learning strategy of "copy the majority" ("conformist bias"). The opposite, negative frequency-dependent bias for rare variants can likewise be characterized as "copy if rare." Henrich and Boyd (1998) established that the range of conditions that favor conformist transmission is likely broader than that for cultural transmission. However, Eriksson et al. (2007) have challenged the theoretical grounds and the empirical evidence for a conformist bias; they claim that the assumptions of previous models may be unrealistic. Hence, whether animals are expected to use a "copy the majority" strategy is very much a controversial point among theoreticians and awaits relevant experimentation.

With "direct bias," individuals adopt cultural variants based upon an assessment of the value of the trait itself, a bias that could be translated into a strategy such as "copy if better" (requiring an individual to evaluate the relative payoffs of the actions of itself and others; Laland 2004). In contrast, with "indirect bias," traits exhibited by successful or high-status individuals may be preferentially adopted, even where the trait in question has absolutely nothing to do with the individuals' success or status. Indirect bias then can be portioned into strategies such as "copy individuals that are successful, high-status (prestige bias), older, or good social learners" (Laland 2004). Indeed, the strategy of "copy older individuals" has been assumed in theoretical analyses (M. Kirkpatrick and Dugatkin 1994) on the basis of reports of mate choice copying in female guppies. Finally, where the adoption of a trait is random with respect to its apparent utility transmission is said to be "unbiased." This is different from "indirect bias," as there is no assumed correlation between the "type" of individual (e.g., high-status) and the adaptiveness of the trait he or she exhibits. Here, individuals would randomly choose a "role model" and copy the trait he or she exhibits. Note, the existence of "random copying" at a population level is perfectly consistent with individuals pursuing directed strategies, provided individuals do not all exhibit the same behavior such that they sum up to generate a population-level bias. Hence, "unbiased transmission" processes may be translated into strategies such as "copy kin" or "copy friends," where the specifics of who is copied vary from one individual to the next. Any kin bias may reflect nothing more than the fact that individuals spend considerably more time in the presence of kin than non-kin. However, there are two reasons to suspect that selection may have favored a disproportionate degree of learning from kin. First, social learning, particularly in a variable environment, is of use only to the extent that demonstrator and observer experience the same environment and reap the same rewards (Boyd and Richerson 1985, 1988). This may be more likely among kin than non-kin because, by virtue of their shared genes (and other sources of similarity), kin may be more likely than non-kin to reproduce similar behavior variants and to experience the same affective sensations in reinforcement. Second, in situations where information transmission is costly, kin may have more to gain by providing reliable information to fellow kin, and less to gain from deceiving them, than non-kin. Witness, for example, teaching among worker ants (Franks and Richardson 2006) and bees (von Frisch 1967). Similarly, if friends are regarded as individuals with whom one trades altruistic acts (Trivers 1971), by similar lines of reasoning we might expect more social learning among friends than nonfriends, in a "copy friends" strategy.

In accordance with mathematical theory, Coussi-Korbel and Fragaszy (1995), after observing the relation between social dynamics and social learning in nonhuman primates, stressed directed social learning. Here, the social rank, sex, age, patterns of association, and other characteristics of demonstrator and observer critically affect the probability of social learning. As a result, information may be transmitted through subsections of animal societies at different rates.

13.3.2. EMPIRICAL EVIDENCE

13.3.2.1. *Frequency-dependent bias*

Following Boyd and Richerson (1985), we treat a conformist bias as entailing a disproportionate tendency to copy the majority. We do not review here recent empirical studies pertaining to a broader interpretation of "conformity" (see Whiten and van Schaik 2007; Galef and Whiskin 2008).

Day et al. (2001) found evidence for conformity in a series of experiments investigating how shoal size affects foraging efficiency in fish. Hidden food was presented to shoals of guppies, and large shoals were found to locate food faster than small shoals, consistent with similar findings in other fishes. This is probably because fish in large shoals have more shoal mates from whom to acquire information, and in open water large numbers of individuals at a food site will attract conspecifics more rapidly than small aggregations. However, in a second experiment, the fish had to swim through a hole in an opaque maze partition to get to a food source, and in this situation the opposite result was found: smaller shoals located food faster than larger shoals. The seemingly conflicting findings of these experiments make sense in the light of the observation that guppies prefer to join large, rather than small, shoals, which implies that individuals ought to be more willing to leave smaller shoals than larger shoals. Swimming through an opaque partition to locate food involved breaking visual contact with, and hence effectively leaving, the shoal. Under such circumstances, conformity, or a strategy of "copy the majority," resulting from the natural shoaling tendency of these fish, results in greater reluctance to acquire a novel behavior in large shoals than in small shoals, at least early in the diffusion process. This interpretation was supported by the findings of a third experiment, which replicated the second except for using a transparent partition: individuals in large shoals once again located the food faster than those in small shoals. Here visual contact between fish was maintained because the partition was transparent, so fish passing through it behaved as if they were not leaving the shoal. Here, social transmission of foraging information was not hindered by conformity, and large shoals were again advantaged.

McElreath et al. (2005) had groups of people repeatedly play a computer-based task (planting one of two crops that gave different yields), either (*i*) where asocial learning alone was possible, (*ii*) with the opportunity to view the previous choice of one randomly selected group member (allowing social learning), or (*iii*) with the opportunity to view the previous choices of all group members (allowing conformity). The use of social information increased when individual learning was relatively inaccurate, thus confirming the strategy of "copy when uncertain" (see section 13.2.2). Similarly, although models indicated that conformity was the better strategy under all conditions, it was used only when the environment fluctuated—hence, when there was a cost to individual learning. Similarly, Efferson et al. (2008), in another computer-based study, found considerable, as yet unexplained, heterogeneity in the extent to which humans follow a strategy of "copy the majority," even when doing so would be in their interests. In accordance with these findings are those of

Eriksson et al. (2007) indicating that the likelihood of conforming is highly context dependent. People reported that, when presented with vignettes pertaining to either a scenario of novel food choice (different soups) or one of punishment of social defectors (defection being the undertaking of large print jobs on a communal machine), they would conform in regard to the former but not the latter.

In summary, empirical evidence for utilization of a "copy the majority" strategy is surprisingly weak, given the emphasis on conformity in the social science literature.

13.3.2.2. *Indirect bias*

Mesoudi and O'Brien (2008) provide the first attempt to simulate the cultural transmission of prehistoric artifacts in order to determine whether theoretical predictions can explain patterns observed in the archeological record. They examined a previously observed pattern in which the attributes (e.g., length and width) of arrowheads dating to AD 300–600 were poorly correlated with each other in eastern California but were well correlated in central Nevada. The pattern in California was thought to be due to cultural transmission involving "guided variation" (individual trial-and-error experimentation of attributes) and that in Nevada due to "indirect bias" (wholesale copying of a successful individual's design) (Bettinger and Eerkens 1999). Mesoudi and O'Brien's study involved participants playing a computer game whereby they tested "virtual projectile points" in "virtual hunting environments." Consistent with "indirect bias" and a strategy of "copy the most successful individual" they found that individuals chose to copy the arrowhead design of the single most successful demonstrator. In addition, where individuals were allowed to adapt their culturally acquired arrowheads via individual trial-and-error learning ("guided variation"), the resulting arrowheads were less uniform in design than those of individuals allowed only to choose which individual to copy arrowhead attributes from (their hunting success being indicated on screen). Consistent with the adaptiveness of a strategy of "copy if asocial learning is costly" (see section 13.2.1), participants who could engage in indirectly biased cultural transmission outperformed those engaging in individual learning, particularly when the latter was costly (modifications incurred a caloric penalty).

13.3.2.3. *Unbiased transmission*

Many fish express preferences for shoaling with familiar individuals (Griffiths 2003), and we may interpret a strategy of "copy familiar individuals" in the same light as "copy friends" or "copy kin." Guppies have been reported to acquire foraging information faster when in groups of familiar individuals than when in groups of unfamiliar individuals (Swaney et al. 2001). A meta-analysis by Reader (2000) found disproportionate numbers of reports of learning from mothers in nonhuman primates. Galef and Whiskin (unpublished data) have explicitly attempted to examine the influence of both familiarity and kinship on the transmission of food preferences in rats. Following a similar protocol to that outlined above, they found that both familiar-kin and familiar-non-kin had an equivalent social influence as unfamiliar-kin and unfamiliar-non-kin, respectively. However, when considering only studies where observer rats interacted simultaneously with familiar and unfamiliar individuals, regardless of relatedness, they found unfamiliar individuals to have the greatest social influence over subsequent food preferences. Galef (personal communication) suggests that the peculiarities of rat social life, where interaction with a strange conspecific takes precedence over social interaction with a known individual, may override any strategy of "copy kin or familiar individuals."

13.4. Evolutionary implications

An individual animal's use of social learning strategies may generate specific population-level effects on evolutionary processes. For instance, theoretical analyses suggest that a strategy of "copy the majority" may underlie widespread cooperation generated by cultural group selection, while a strategy of "copy successful individuals" may generate runaway prestige markers that become exaggerated in a similar manner to runaway sexual selection (Boyd and Richerson 1985). Maladaptive cultural traditions may result from the (asocial learning) costs of breaking a convention or from informational cascades (Giraldeau et al. 2002) where individuals base behavioral decisions on a prior decision of others without observing the cue upon which that decision was based (akin to personal information). When "asocial learning is costly" or "uncertainty" reigns, animals and humans may be more likely to engage in erroneous or maladaptive informational cascades (e.g., Cavalli-Sforza and Feldman 1981; Boyd and Richerson 1985; Laland and Williams 1998).

Consideration of social learning strategies may explain why social learning abilities appear to reflect ecological (environmental and social) rather than taxonomic affinities among species. There may be a greater reliance on social information in species that use complex foraging skills or must overcome challenging prey defenses than in species that do not (e.g., folivorous vs. frugivorous species: Fragaszy and Visalberghi 1996; extractive vs. nonextractive foragers: Day et al. 2003; Zentall 2004); but see Lefebvre and Giraldeau 1996 for caution regarding inferences based on comparative studies of social learning. Although our findings cast doubt on the widespread belief that social learning is particularly important to large-brained species, Sol (chapter 7 in this volume) argues for a positive correlation between innovation, the starting point of any socially learned trait, and brain size.

In fact, the accumulating evidence for social learning strategies may guide researchers in the difficult task of assessing the plausibility of social learning being involved in any observed patterns of behavioral variation in the wild (see Laland et al., in press). For example, given the theoretical and empirical support for a "copy when asocial learning is costly" strategy, researchers might reasonably question the plausibility of putative cultural traits that are relatively simple and cheap to acquire asocially. Similarly, a growing understanding of the contexts in which social learning is used may enhance our understanding of factors that may have influenced the evolution of cultural capacities. For example, several empirical findings (e.g., Coolen et al. 2003; Kendal, unpublished data) indicate that the costs associated with acquiring or using personal information may promote the evolution of increasingly complex social learning processes. If replicated in our closest ancestors, these findings may shed light on the evolution of the potent cultural capacity of humans. Theoretical studies indicate that an individual's ability to adopt a strategy of "copy if better" may be more important for the evolution of a cumulative culture than either conformist transmission or imitation (Eriksson et al. 2007; Enquist and Ghirlanda 2007). Thus, a "cognitive deficit" in evaluating the relative payoff of actions performed by the self and by others, preventing animals from discriminating between cultural variants (or "adaptive filtering"), may explain why cumulative culture appears to be virtually unique to humans. Incidentally, a similar line of reasoning, regarding cognitive abilities (e.g., theory of mind) required to cope with increasingly complex social life, may be important in the apparent distribution of social intelligence and deception in human and nonhuman animals (see Federspiel, chapter 14 in this volume). Indeed, for some, a theory of mind is deemed necessary for both imitation and cumulative culture (Tomasello 1990).

In contrast, several of the proposed social learning strategies may actually hinder the cultural transmission of information. As we have seen, a strategy of "copy the majority" may hinder the adoption of novel information in large groups (Day et al. 2001), and a recent theoretical study (Eriksson et al. 2007) indicates that it may also hinder the development of cumulative culture. In addition, a strategy of "copy high-status individuals" may inhibit the spread of novel information where the innovator, as is often the case in primates, is of low status (Reader and Laland 2003). The apparent failure of most animal innovations to spread (Reader and Laland 2003) has a profound influence on species' evolution and survival, not least because the ability to adopt innovations is implicated in invasion success and the ability to respond to anthropocentrically altered environments (Lee 1991; Sol et al. 2002; Sol, chapter 7 in this volume).

13.5. Summary and future directions

In summary, we report good support for several "when" strategies, particularly "copy when asocial learning is costly," "copy when uncertain," and "copy when dissatisfied." However, in many cases the precise strategy employed is ambiguous, conditional, and context specific. In contrast, support for the existence of "who" strategies is weaker, and even the comparatively well established "copy the majority" strategy remains contentious in the absence of unambiguous support.

It is early days in the study of social learning strategies, and clearly further research is necessary before a deep understanding is gained of how and when acquired social information is used. We would like to encourage further empirical research in this area, particularly as there is a paucity of direct investigation, in nonhuman animals especially. Of particular interest is investigation into whether there is a hierarchy in the implementation of strategies (Laland 2004). For example, individuals might practice unbiased transmission (e.g., copy kin) unless they have access to relevant biased transmission information (e.g., copy high-status/successful individuals) or can afford to invest in the time, energy, or cognition required for individual learning/direct bias. We note also that many of the studies investigating "who" strategies hint at the existence of conditional strategies. For example, both of the reviewed conformist transmission studies (Day et al. 2001; McElreath et al. 2005; section 13.3.2.1) and, to a certain extent, the results of another study in guppies (Kendal et al. 2004; see section 13.2.1.2) appear to indicate a mixed strategy of "copy the majority when asocial learning is costly." Individual characteristics of observers, favoring the overriding of social learning strategies, and the continued acquisition of personal information (as reported by Efferson et al. 2008 and Kendal et al. 2004 with regard to conformity), may be influential in determining the innovatory capacities of individuals.

In parallel, we argue that there is considerable potential for fruitful integration of empirical and theoretical work, particularly game-theoretical analyses (Laland and Kendal 2003; Laland 2004). We hope that consideration of the trade-offs inherent in the adaptive use of social and asocial learning will contribute to an increased understanding of the observed pattern of social learning and behavioral traditions in the animal kingdom, especially as the use of social information may lead to cultural evolution, which may in turn affect biological evolution (Boyd and Richerson 1985; Feldman and Laland 1996; Danchin et al. 2004).

ACKNOWLEDGMENTS

We would very much like to thank John Ratcliffe and Reuven Dukas for inviting us to submit a chapter and also for their encouragement and helpful suggestions. Thanks too to the past and present members of the Laland lab, who have, over the years, provided fruitful discussions and data upon the topic of social learning strategies.

14 The 3E's Approach to Social Information Use in Birds: Ecology, Ethology, and Evolutionary History

IRA G. FEDERSPIEL, NICOLA S. CLAYTON & NATHAN J. EMERY

14.1. Introduction

The field of social learning has attracted considerable attention over the last few years. Researchers from various fields, including psychologists, zoologists, and ecologists, have been working on both the theoretical framework and the practical methods for studying social information use in animals, from mammals and birds to reptiles and fish. A significant obstacle for learning about how the mechanisms and functions of social learning may interact arises from the lack of communication between these fields, resulting in nonunified definitions, noncomparable results due to different methodologies, and different ways of interpreting results.

Kamil (1998) underlined the importance of integrating the different approaches to the study of animal behavior that developed from two scientific revolutions: the cognitive revolution that stemmed from comparative psychology, and the behavioral ecology revolution that originated in biology. As discussed by Dukas (1998e, 405) in the first edition of this book, the philosophy underlying cognitive ecology is that "cognition must be studied with regard to an animal's ecology and evolutionary history, and that knowledge of cognitive mechanisms can help us explain behavioral, ecological, and evolutionary phenomena."

In this chapter, we shall evaluate the success of this integrative approach, focusing specifically on social information use by birds. We shall argue that investigating the psychological mechanisms underlying social learning processes in the light of an animal's *Umwelt*, that is, in terms of the "3E's"—ecology, ethology, and evolutionary history (as shown in fig. 14.1, but see also section 14.1.2)—is critical for gaining a more complete, unified picture of social information, not only for interpreting existing results, but also in designing new experiments with high ecological validity as well as rigorous experimental control.

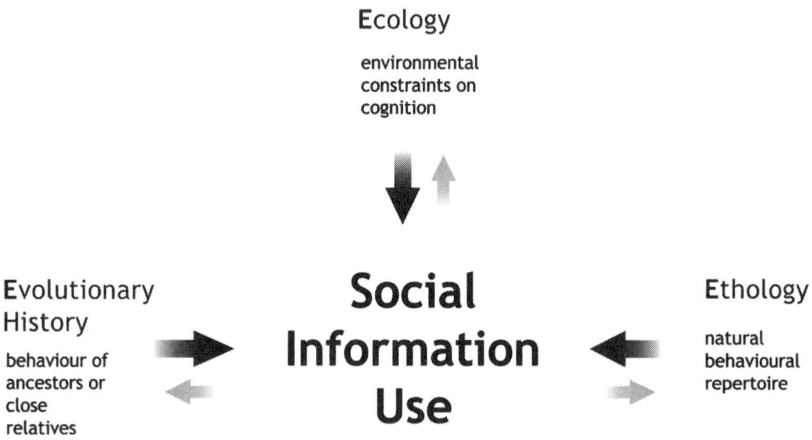

FIGURE 14.1. Diagram of the 3E's that influence social information use: ecology, ethology, and evolutionary history. Black arrows represent the influence of the 3E's on social information use; gray arrows indicate the possible influence of social information use on the ecology, ethology, and evolutionary history of a species.

Studying the psychological mechanisms of social learning in the light of the 3E's and drawing conclusions from such studies may help us to *define* the underlying mechanisms, whereas in reverse it is more difficult, and consequently, explaining an animal's ecology by examining its cognition is more complicated and more speculative. In this chapter, we shall largely concentrate on the former approach, but a good example of how a specific type of social learning can influence a species' ecology is the caching (i.e., food hoarding) behavior of corvids. The sophisticated arms race between cachers and pilferers (i.e., thieves) would not be possible without the basic cognitive ingredient of observational spatial memory, in which pilferers can accurately locate the caches others have made, even when the cacher has left the scene (see section 14.2.1).

As highlighted by Kendal and colleagues (chapter 13, in this volume), the successful use of social information is a matter of gaining benefits (i.e., fitness) and avoiding costs. Being part of a social group or a pair-bond appears to be a prerequisite for exploiting the knowledge of others, since only being in the company of others opens up opportunities to "scrounge" information from knowledgeable conspecifics (Giraldeau et al. 2002; Laland 2004). Different types of social information and a varying number of opportunities to access that information are available, depending on the social and mating system (Lefebvre and Giraldeau 1996). For example, for territorial birds there may be fewer opportunities for picking up social information than for gregarious birds, and for pair-bonded birds there may be different types of information than for lekking birds.

Social learning, defined as "changes in the behavior of one individual that result, in part, from paying attention to another" (Box 1984, 213), involves various factors, ranging from "low-fidelity copying mechanisms" (Whiten et al. 2004), which include mechanisms of social influence, such as social facilitation (i.e., the mere presence of another animal affects the motivation or arousal of the observer) and contagious behavior (i.e., species-typical behavior is released by the sight of others engaged in that activity), to "high-fidelity" social learning, such as imitation and emulation. Animals can learn from others using a variety of different mechanisms, such as local enhancement (i.e., facilitation of learning that results from drawing the observer's attention to a location or object with which the other individual is interacting), stimulus enhancement (i.e., attention is drawn to a certain stimulus), and observational conditioning (learning about the positive or negative reinforcement of an object or event). In Great Britain, a small population of blue tits (*Parus caeruleus*), great tits (*P. major*), and coal tits (*P. ater*) learned how to open milk bottles to get to the cream off the top of the milk. Different bottle-opening techniques were observed, indicating a social learning mechanism other than imitation (i.e., copying the exact action of another), most likely stimulus enhancement (J. Fisher and Hinde 1949; Hinde and Fisher 1951; Sherry and Galef 1984).

Imitation and emulation (i.e., copying the goal or result of an action sequence or learning about the operating characteristics of objects) are often considered more complex forms of social learning than those already discussed, although some authors argue against that view (Heyes 1999; for reviews and definitions see Thorpe 1956; Zajonc 1965; Mineka et al. 1984; Galef 1988; Zentall and Galef 1988; Whiten and Ham 1992; Heyes 1994; Heyes and Galef 1996; Tomasello 1996; Custance et al. 1999; Whiten et al. 2004; Zentall 2004). Well-known examples of imitation include vocal imitation, used to increase the repertoire size (Pepperberg 2007), and the imitation of human's greeting gestures (B. Moore 1992) in African Grey parrots (*Psittacus erithacus*).

In addition to commonly described forms of social learning, the use of other types of social information may play a role in decision making (Bonnie and Earley 2007). Mechanisms for copying another's choice of mate and eavesdropping (the use of information in signals by individuals other than the primary target; Peake 2005), for example, learning the whereabouts of potential predators, are also important information resources. In eavesdropping, some authors distinguish between interceptive and social eavesdropping (Peake 2005). Interceptive eavesdropping is common when the eavesdropper is a different species from the signaler, for example, prey detecting predator cues or individuals from one species picking up information from individuals of another. For example, in a playback experiment it was found that black-casqued hornbills (*Ceratogyma atrata*) responded to alarm calls given by Diana and Campbell's monkeys (*Cercopithecus diana* and *C. campbelli*) and could distinguish between calls that for these monkeys referred to crowned eagles and leopards (Rainey et al. 2004; see Manser, chapter 12 in this volume). Within species, animals of the same sex may pick up information that was intended for the opposite sex (Mennill et al. 2002; Peake 2005). Social eavesdropping takes place within a species when individuals intercept signals that were sent between conspecifics; for example, female great tits (*Parus major*) gain information about potential mates by listening to song interactions between neighboring males (Otter et al. 1999), and domestic fowl (*Gallus gallus domesticus;* Hogue et al. 1996) and pinyon jays (*Gymnorhinus cyanocephalus*) infer their own dominance rank after watching encounters between conspecifics (Bond et al. 2003; see also later in this chapter).

However, there are also good reasons not to use social information. First, although "less expensive" than information acquired via individual trial-and-error learning, information gained through social learning may come at a cost. Observing others performing an action involves forgoing other behaviors that could have been pursued in the meantime, such as watching out for predators, looking for food, or finding a potential mate (McGregor and Dabelsteen 1996). Second, there is the possibility of learning incorrect or inefficient behaviors. Therefore, an animal should employ social learning only if it contributes to its survival or reproductive success (see Kendal et al., chapter 13 in this volume).

Another reason for suggesting that social learning might be costly is that social species often have large brains relative to their body size (Dunbar 1992), although this does not hold for all social species, and there are a number of positive correlations between brain size and various indices of sociality, such as grooming (Kudo and Dunbar 2001), forming coalitions (Shultz and Dunbar 2007), and deceiving others (Byrne and Corp 2004). Indeed, according to the social function of intellect hypothesis (Humphrey 1976; see also Jolly 1966), it is the ability to survive the political dynamics of a complex social world that has been the primary driving force shaping primate intelligence. Keeping track of others' interactions and relationships in addition to their own in large social groups may be beneficial in future interactions when it comes to deceiving others, knowing whom to ask for support in a fight, or climbing up the dominance hierarchy. This imposes an additional burden, since a large amount of social information has to be processed every day, and may have led to the development of social intelligence in animals living in large groups (Humphrey 1976).

In primates, relative neocortex volume (neocortex volume/brain volume remainder) increases with increasing group size, indicating an effect of social complexity on the brain (Dunbar 1992). But other indications of social complexity—such as size of grooming networks (Kudo and Dunbar 2001), whether the species forms coalitions (Dunbar and Shultz 2007b), and the rate of tactical deception (Byrne and Corp 2004)—are all positively correlated with relative neocortex volume in many social primates.

Striking similarities between apes and corvids suggest that these social skills may not be unique to primates: for example, to stabilize their bonds in a group, birds engage in allopreening bouts, similar to grooming in apes (Emery et al. 2007). Furthermore, there is evidence for postconflict affiliation (Seed et al. 2007) and deception in corvids (Bugnyar and Kotrschal 2004). The different types of mating systems, such as monogamy or promiscuity, seem to also have an impact on brains in birds (Dunbar and Shultz 2007a). Birds that form lifelong pair-bonds or are cooperative breeders were found to have the largest relative brain size (Emery et al. 2007). Similar to primate alliances, members of lifelong pairs in birds spend a lot of time and energy on maintaining their relationship. The benefits of pair-bonding include, but are not limited to, food sharing, allopreening, support during fights, and reducing stress levels by initiating affiliative postconflict behaviors such as bill twining, the avian equivalent of chimpanzee kissing (Emery et al. 2007; Seed et al. 2007; von Bayern et al. 2007). These skills require high levels of coordination and may have led to a certain form of intelligence, so-called relationship intelligence (Emery et al. 2007).

14.1.1. A COMPARATIVE APPROACH

It has been suggested that social learning may be an adaptive specialization to group living in birds (Klopfer 1959) and that scramble competition in opportunistic species (i.e., simultaneous competition over food in a group) may have pushed the development of learning skills in a "mental arms race" (Lefebvre and Palameta 1988, 155). This was tested using two species of columbid (the pigeons and doves). The investigation compared social information use in the gregarious, group-foraging feral pigeon *Columba livia* and in the territorial tropical dove *Zenaida aurita* (Lefebvre et al. 1996). The animals had to find food and were provided with information about the location of the food by a demonstrator (social condition) or by the apparatus that contained the food itself (nonsocial condition). The pigeons learned much faster than the doves in both conditions. Thus, the species differences appeared to be due to some general difference in learning abilities rather than a specialization for social learning (Shettleworth 1998). Social species seem to be generally better at learning tasks, but this is probably the result of being simply more attentive or better at detecting information.

In a similar study using corvids, social pinyon jays (*Gymnorhinus cyanocephalus*) and territorial Clark's nutcrackers (*Nucifraga columbiana*) were provided with two tasks—a motor task (lifting a lid off a shallow well containing food) and a color discrimination task (lifting a lid of a particular color off a well containing food)—that could be learned individually or socially (J. Templeton et al. 1999). The pinyon jays learned faster in the social condition than in the nonsocial condition, whereas there was no difference in learning rate for the social and nonsocial conditions in Clark's nutcrackers. These results run contrary to Lefebvre and Palameta's (1988) study, because the pinyon jays were significantly worse at the individual condition than the pigeons, which displayed a more general learning ability. Although the authors suggest that their results support the idea of an adaptive specialization for social learning, we believe these studies highlight the importance of taking other aspects of an animal's natural history into consideration when designing cognitive ecological tasks and interpreting the findings from them. Emery and Clayton (Emery et al. 2004; Emery and Clayton 2008) have recently suggested the importance of the 3E's approach for studying comparative cognition, extending Kamil's (1988) synthetic approach to animal intelligence to integrate information about the evolutionary history and ethology of an animal as well as its ecology.

14.1.2. THE 3E'S APPROACH AND WHY IT IS IMPORTANT

At the outset it is important to make a distinction between the ecological factors that affect an animal's behavior, namely its diet, the habitat in which it lives, and its social system and mating system, and the ethology, by which we mean the natural history of a species-specific behavior, which also contributes to "skillful" social information use. For example, knowing who associates with whom may play a major role for social birds, such as colonial, cooperatively breeding (ecology) pinyon jays (Marzluff and Balda 1992). As intense social cachers (ethology), they need to know whom to protect caches from and who is safe as an observer, which they can infer from watching interactions between conspecifics coupled with an understanding of their social relationships. Transitive inference in social scenarios is defined as the ability to infer the relative dominance status of an individual based on observed interactions, and it should be an essential skill in this type of complex social environment. In a laboratory experiment designed to test this ability, three groups of pinyon jays with a linear hierarchy were formed: group 1 with birds A to F, group 2 with birds 1 to 6, and group 3 with birds P to S. Group 1 was dominant to group 2, and group 2 was dominant to group 3. An observer, bird 3, was allowed to compete with bird B. Bird 3 had never met bird B before, but he had watched encounters between bird A and B and between 2 and B. Also, bird 2 was part of 3's group and therefore bird 3 would have information that bird 2 was dominant to him from their previous interactions. In the observed encounters, bird 3 could watch bird B being submissive to bird A but dominant to bird 2. Bird 3 was then allowed to interact with bird B, and during their encounter, bird 3 demonstrated a greater number of submissive displays to bird B, suggesting that bird 3 had formed a representation of the relative dominance of those birds from its previous observations. All of the tested birds showed similar appropriate behaviors across different combinations of birds (Paz-y-Miño et al. 2004). It would be most informative to know whether these birds could also extrapolate this information about dominance relationships in order to determine from whom they should protect their caches and when.

In tests of this transitive inference ability using arbitrary stimuli, pinyon jays outperformed the less social western scrub jays; however, the scrub jays did learn (Bond et al. 2003). In an initial experiment, learning to discriminate between successive color pairs that were implicitly ordered was tested. Social pinyon jays were faster at learning the dyadic relationships and made fewer errors. Whereas both species learned the first pair without problems, pinyon jays adapted more rapidly to reversals and thus made fewer errors in subsequent reversal trials, in which the previously rewarded pair was incorrect. When more pairs were included in the tests, pinyon jays learned and improved faster than the less social scrub jays. In a second experiment, the birds were tested for transitive inference by intermixing familiar pairs with novel, nonadjacent pairs. Both species showed high accuracy and, thus, transitive inference. However, differences were found in the responses to the position of the stimulus color pair in the implicit rank order. Pinyon jays responded more slowly to low-ranking pairs, although no effect was found with the highest-ranking pair, whereas western scrub jays displayed a first-item accuracy with almost no effects on latency. The authors concluded that the two species may have used different methods for representing the rank order, with pinyon jays using relational representations in which novel pairs can be inserted into a preexisting structure, and western scrub jays building a series of associative representations (Bond et al. 2003).

It therefore appears that differences in socioecology between these species may have driven them to develop different social information use skills.

Pinyon jays are colonial birds, living in large groups, that breed in aggregated pairs of 50 and show cooperative breeding. By contrast, western scrub jays are semiterritorial birds that breed in single pairs (Clayton and Emery 2007). Inferring the dominance status of conspecifics therefore seems to be much more important for the pinyon jays, since using social information to gain knowledge about one's own and others' relationships seems to be an (adaptive) advantage for species living in large, social groups.

Finally, considering the putative evolutionary history of the species under consideration is also important. Although a specific aspect of an animal's ethology or ecology may not be present in the extant species, it may have been present in the common ancestor of the group in question. For example, there is no evidence that jackdaws cache in the wild, but there are anecdotes that jackdaws do display proto-caching, by which we mean they are sometimes seen to place food in nooks and crannies without ever hiding it or leaving it for any length of time. A reconstruction of the evolution of caching in corvids has suggested that the common ancestor of corvids was a caching species (de Kort and Clayton 2006). Therefore, taking these points into consideration may aid in interpreting the results of cognitive ecological studies and help in the design of experiments comparing closely related species that appear to differ in ecology and ethology, that is, in the challenges they face in their given environment and their natural behavioral repertoire.

In justifying the 3E's approach we have described a few studies on a small number of different species. In the following sections, we will discuss further why the 3E's approach, combining ecology, ethology, and evolutionary history, is important, but will focus on social information use in birds (fig. 14.1). By using detailed case studies of a small number of species, we hope to show how the 3E's determine when the solution to a problem is learned socially and what mechanism may be employed.

14.2. Case studies

14.2.1. CORVIDAE

14.2.1.1. *Western scrub jays and ravens*

Many birds, including western scrub jays and ravens, cache temporary surpluses of food for future consumption, which they recover days, if not months, later. Krebs (1990) argued that, for efficient cache recovery, there has likely been considerable selection pressure for them to have highly accurate and long-lasting memories of where they hid the food, and he argued that, as a consequence, food-caching birds had an adaptive specialization in behavior, in terms of this enhanced spatial memory, and an adaptive specialization in the brain, in terms of an enlarged hippocampus relative to the rest of the brain. The most striking example of an adaptive specialization in caching, memory, and the hippocampus comes from two populations of black-capped chickadees that live in very different environmental conditions, one in the harsh climatic conditions in Alaska and one in the milder conditions in Colorado, thereby highlighting the importance of ecology (Pravosudov and Clayton 2002). The Alaskan chickadees cache considerably more food than the Colorado ones, even when housed in identical conditions in the laboratory. Furthermore, the Alaskan birds were much more efficient at cache recovery, and their performance of spatial but not nonspatial memory tasks was much more accurate. They also had much larger hippocampal volumes than the Colorado ones, both in terms of absolute size and in relation to the rest of the brain. Taken together, these findings support the hypothesis that population differences within a species reflect adaptations to ecological conditions.

The abilities needed for recovering food have been investigated in a study combining spatial memory and social learning and have been called observational spatial memory (Balda et al. 1997). Three corvids that differ in levels of sociality and the number of caches made were required to remember where a conspecific had hidden food (Bednekoff and Balda 1996a, 1996b). Clark's nutcrackers, territorial birds that are thought to be able to remember up to 30,000 food caches (Balda et al. 1997), were less accurate in finding another's caches and could not remember for as long as social pinyon jays and Mexican jays (*Aphelocoma ultramarina*), two species that cache much less food than Clark's nutcrackers. However, less social western scrub jays (*Aphelocoma californica*), which also cache much less food than Clark's nutcrackers, remembered the location of the caches almost as accurately as the cachers themselves (Clayton et al. 2001). Ravens (*Corvus corax*) were more successful in raiding another's caches if the caches were made more than 3 meters away from them, suggesting that they were accurate in recovering caches when the cacher was not present to defend them (Bugnyar and Kotrschal 2002).

Although spatial memory is essential for the birds' accurate cache recovery, the birds also need to keep track of not only what they have cached but also what has been recovered, for cache sites may have been emptied, either by themselves or by pilferers. To protect against cache theft, cachers have to employ strategies to either distract others from their caches or to defend them when conspecifics approach them (Dally et al. 2006a). This may be an issue only for corvids, because there is little evidence that other caching species, such as parids (e.g., black-capped chickadees), can remember where another

FIGURE 14.2. A raven caching food in the snow. Photograph by I. G. Federspiel.

individual has cached (Hitchcock and Sherry 1995; but see Pravosudov 2008 for contrary evidence on mountain chickadees).

Most of the studies on social information use in caching experiments have been performed on two species of corvid: common ravens and western scrub jays. These two species have similar ecologies; both live in monogamous, territorial (common ravens) or semiterritorial (western scrub jays) pairs or in flocks that include pairs (Clayton and Emery 2007). Although these species are not particularly social in the traditional sense, social information plays an important role for both species in their caching and pilfering. One striking fact is that each individual can play the role of both cacher and pilferer simultaneously, caching their own food while at the same time watching others cache food and them attempting to pilfer those caches (fig. 14.2).

This "cognitive arms race" between cachers and pilferers seems to have driven the food-caching corvids to excel when it comes to social skills (Bugnyar and Kotrschal 2002) such as keeping track of who was watching when and where. Applying the necessary tactics to the acquired knowledge would seem to be beneficial in such a highly competitive environment.

Western scrub jays, a species also known for its skills in mental time travel (i.e., recollecting what they cached where and how long ago and planning for the future in terms of what to cache for tomorrow's breakfast; Clayton and Dickinson 1998; Clayton et al. 2003; Correia et al. 2007; Raby et al. 2007), are also able to use various cache protection strategies in a flexible manner, choosing the technique most suitable to the context in which the caches were made (fig. 14.3).

When given the opportunity to cache in full view of an observer or behind a barrier, cachers chose to hide more food items behind the barrier

FIGURE 14.3. A western scrub jay (a) carrying worms toward a caching tray and (b) caching a worm. Photograph by I. Cannell.

(fig. 14.4a; Dally et al. 2005). When given the opportunity to either cache close to or further away from an observing bird, they chose to cache at a distance but showed no preference when they were visually isolated from the potential observer (fig. 14.4b; Dally et al. 2005). When caching in view of the observer, they moved the food item around multiple times during the caching process. In recovery sessions, during which the food-hiding birds were allowed to approach the caches without being observed, they tended to retrieve the caches made either close to or in view of the observing conspecifics and recached them in new sites. When provided with the opportunity to choose between "shady" and "well-lit" sites for caching when observed, the jays showed a clear preference for the shady sites, whereas they cached equally at both sites when no observer was present (fig. 14.4c; Dally et al. 2004). When recovering, they tended to recache those food items hidden in the well-lit sites once the observer had left the scene.

The scrub jays therefore appear to use different strategies in different caching contexts. Caching out of view clearly limits the information that can be gained by an observer. In addition to the strategies mentioned, the jays were often observed redistributing the substrate after caching, making it almost impossible to tell where they had hidden their food (Dally et al. 2005). Increasing the number of caches made in view of the observer may have been used as a strategy to offset the risk of cache loss. Confronted by a single trial with a

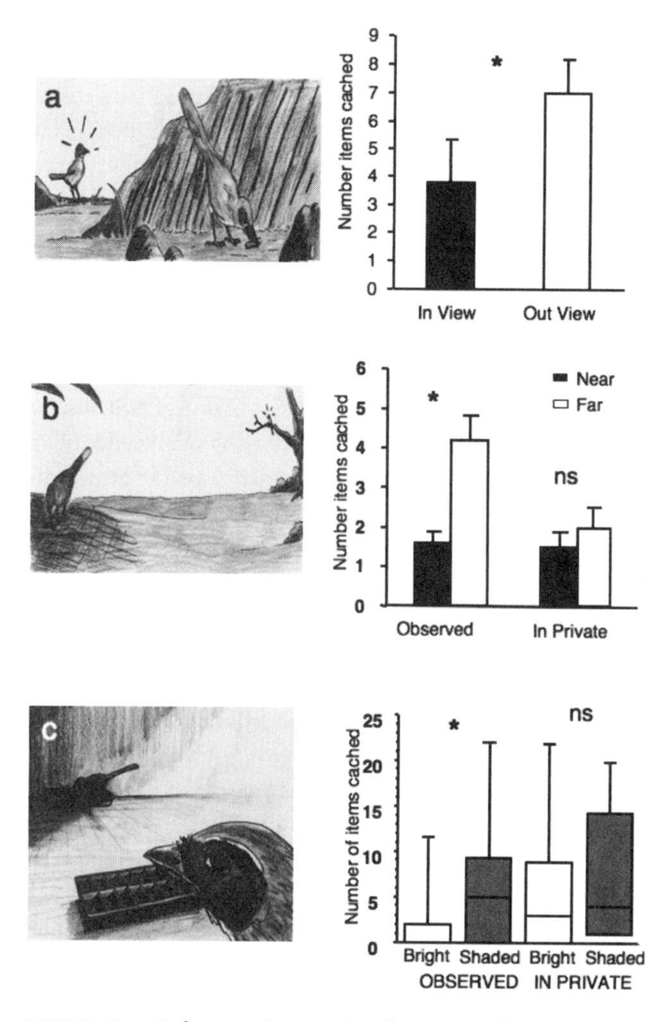

FIGURE 14.4. Cache protection strategies of western scrub jays. a. Drawing of scrub jay caching behind a rock, out of view of a competitor. Number of food items cached either behind a barrier or in the open. $^*P < 0.05$. Adapted from Dally et al. 2005. b. Drawing of scrub jay caching as far as possible from another scrub jay. Number of food items cached in private or when observed by another scrub jay. In each condition the cacher could store food in two trays: one close to the observer, the other farther away. $^*P < 0.05$. Adapted from Dally et al. 2005. c. Drawing of a scrub jay caching in front of another scrub jay in a shady part of their environment rather than in a well-lit part. Number of food items cached in private or when observed by another scrub jay. In each condition, the cacher could store in two trays: one in a darkened part of the cage, the other in a brightly lit part of the cage. $^*P < 0.05$. Adapted from Dally et al. 2004. All illustrations by Scott Stevens.

pilfering conspecific or a human experimenter taking away the caching trays after food had just been hidden, jays switched from checking the caches to predominately eating them and caching in "out-of-tray" sites that the human experimenter or conspecific could not get to (Emery at al. 2004). Cachers also use combinations of at least two strategies; for example, when forced to cache in view of the observer, scrub jays chose to cache at a distance and increase the number of caches, which they later recached if they had not been stolen. Especially the caches that were at high risk (i.e., those cached in the sight of and/or close to the observer) were later recached (Dally et al. 2005). Scrub jays also remembered who was watching them when they made caches and which caches were witnessed by whom (Dally et al. 2006b). They were given the opportunity to recache food in private or in front of one of two observers, one of which had been present during caching. When in the presence of an observer, they moved more caches from the tray in which they had been watched making caches by that same observer. When recaching in private, jays cached in out-of-tray locations, whereas when recaching in front of an observer, the jays seemed to use a confusion tactic, moving the caches twice as often as during private recaching (Dally et al. 2006b).

Experience as a pilferer also appears to play an important role. Birds were allowed to cache either in private or while a conspecific was watching. Individuals given the chance to pilfer others' caches prior to the experiment recached only those caches that they had made in front of an observer. Jays without pilfering experience did not, suggesting some sort of experience projection (Emery and Clayton 2001, 2008). Social information use plays a role not only in the caching context but also during foraging at a new food resource. In a field study, Florida scrub jays (*Aphelocoma coerulescens*) learned to forage at a novel patch where conspecifics had already foraged successfully. Juveniles that watched others digging for peanuts and were able to scrounge learned more about the technique than control individuals. Watching the demonstrations increased the probability that the birds would approach the novel patch, and occasionally, demonstrators modified their behavior "in a way that suggested teaching" (Midford et al. 2000, 1205).

Ravens have also demonstrated impressive performances in social information use, including caching and pilfering tasks. Recently, ravens have been found to learn about their competitors in caching bouts during play caching of objects (Bugnyar et al. 2007). In the wild, ravens were observed to hide themselves when caching food (Heinrich and Pepper 1998) and to protect their caches by retrieving the food or aggressively approaching others who came close to their caches (Heinrich 1999). Controlled experiments revealed that they appeared to differentiate between birds that were present during a

FIGURE 14.5. Ravens during a caching experiment; the observer (left) watches the storer (right) through a window. Photograph by T. Bugnyar.

caching event and those that were not (fig. 14.5; Bugnyar and Heinrich 2005, 2006).

When released back into the caching area, storers retrieved more caches when they were accompanied by a knowledgeable conspecific (present during caching) than they did when they were with ignorant birds (absent at caching) (fig. 14.6a). However, they retrieved the caches only when the conspecific was moving toward them, suggesting that the ravens were acting in response to the behavior of the competitors (fig. 14.6b). To rule out the possibility of the ravens acting on the basis of whom they had seen during a caching event, a second experiment was performed, this time with a human experimenter making the caches. A subject was able to observe the experimenter and could then enter the site, with either a co-observer or a nonobserver or in private. The subjects delayed retrieving the caches when given access to the site with a dominant nonobserver but did not differentiate between dominant and subdominant co-observers, going straight to the cache with either observer. This led to the subjects being first at the cache in all but one of the cases with the nonobservers, even though they delayed approaching the cache with the dominant nonobserver. In cases in which the subject was dominant, it always reached

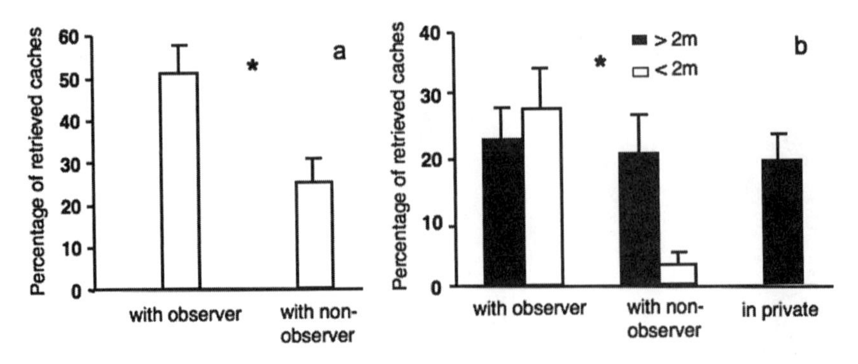

FIGURE 14.6. Mean percentage of caches retrieved by storers (a) with previous observers and nonobservers and (b) when competitors approached the caches (white bars) or did not come near them (black bars). $^*P < 0.05$. Adapted from Bugnyar and Heinrich 2005.

the cache before the co-observer, but with a dominant co-observer, the subjects almost always failed (Bugnyar and Heinrich 2005). Approaching the cache as fast as possible when in the presence of a co-observer therefore seemed to be the only method of getting the food that these ravens employed.

Ravens adjust their behavior in line with the social relationship and knowledge state of their opponent. As storers, they selectively retrieved caches that others were able to watch being made, and as pilferers they hurried to the cache with a co-observer but delayed approaching with a dominant nonobserver (fig. 14.6). Similar to western scrub jays, ravens not only appear to understand that objects such as rocks and tree stumps can degrade visual information but also demonstrate some understanding of another's perspective. Further evidence for this comes from an experiment in which a subordinate began to lead a dominant conspecific away from a food resource that only the subordinate knew about. Subsequently, the dominant bird learned not to stop following the knowledgeable subordinate bird and started searching for the food itself (Bugnyar and Kotrschal 2004).

Ravens feed opportunistically on animal and plant material and are also carrion scavengers, feeding on road kill. Information about new food resources may be shared at so-called information centers (Heinrich 1999) through food calls (Bugnyar et al. 2001) or via visual social information. Individuals accessing new food resources are thus an interesting stimulus for others. In an experiment with a small group of ravens, birds observing others opening a food box tended to open the box using the same technique (Fritz and Kotrschal 1999). A demonstrator was trained to open the box by pulling a flap to open the lid. Nonobserver birds (i.e., the birds in the control group, which did not watch demonstrations of how to open the box by pulling the flap) exclusively opened the box by pecking at a crevice in the middle and then levering the lid open. The observer birds (i.e., the birds that could watch demonstrations) were faster at approaching the box and also at opening it than the nonobservers. They also initially used both possible opening techniques: pecking and pulling. It therefore seems likely that the observing birds were influenced in their choice of technique by the demonstrating birds. Whereas two of the three observers went on to almost exclusively open the box by pecking, the third retained the pulling technique. Due to the two different initial positions a bird had to assume in order to perform one of the two actions, the authors concluded that both imitation and stimulus enhancement could have accounted for the observers' performance (Fritz and Kotrschal 1999).

Social information use seems to be somewhat enhanced between siblings. When they were able to watch a conspecific manipulate an object, ravens manipulated the same object (out of a choice of five) if the conspecific was their sibling (Schwab et al. 2008). This may have been due to socially biased learning (Fragaszy and Visalberghi 2004), with individuals tending to learn from siblings rather than nonsiblings.

Social learning was also found to have an effect on the spread of vocalizations by ravens (Enggist-Dueblin and Pfister 2002). Similar to other songbird species, ravens learn calls from other individuals in their group, and specific dialects spread within subpopulations via social learning (Gwinner 1964). Enggist-Dueblin and Pfister (2002) recorded the vocalizations of free-ranging ravens interacting with a captive pair and analyzed the different types of vocalizations, their distribution, and their differences and similarities. Interestingly, there seemed to be no difference in repertoire size between males and females, which contrasts with the majority of songbirds (Catchpole and Slater 1995). The calls were mainly transmitted within sex and, in a few cases, to the partner, leading to a sex-specific call repertoire. The authors were able to divide the study site into three different parts on the basis of the geographical distribution of the different call types, suggesting a cultural process.

In summary, although territorial, ravens need to be able to use social information to deal with a highly competitive environment. An additional selection pressure is their diet and heterospecific competitors, such as wolves (*Canis lupus*), at carcasses. It was observed that ravens adjust their behavior to the presence of wolves, which occasionally kill ravens when defending food, but do not adjust their behavior to boars (*Sus scrofa*), which do not pose a threat to ravens. When at the carcass with wolves, ravens specialized in scrounging rather than approaching the food resource themselves (Bugnyar and Kotrschal 2002). Finding a new food resource, dealing with competition over food, and also establishing a vocal repertoire and communicating with others require social information use in a raven's life.

For caching species in general, the ability to use social information seems to be vital in order to pilfer another's caches, remember who was watching during a caching event, and protect one's own caches. Furthermore, the evolutionary history of the species must also be taken into account. De Kort and Clayton (2006) concluded that the common ancestor of all corvids was a moderate cacher and that the emergence of specialized cachers evolved independently at least twice. It seems obvious that moderate and specialized species possess the skill of using social information for employing cache protection strategies, and since the ancestors of the western scrub jay were either moderate or specialized cachers and the ancestors of the raven were moderate cachers (de Kort and Clayton 2006), one can legitimately assume that the ancestors of both of these species were able to use social information.

Together these findings indicate that scrub jays and ravens use flexible caching and recovery strategies when hiding and protecting their caches from thieves and also in their role as thieves of others' caches. Their feeding ecology and social system and the evolution of caching behavior have shaped their skills in social information use with respect to storing and stealing.

14.2.1.2. *Jackdaws*

In another corvid species, the jackdaw (*Corvus monedula*), social information may play a role during foraging. Jackdaws are gregarious birds that are especially attentive to their partners' behavior (fig. 14.7; Röell 1978; von Bayern and Emery, in press). However, they are less attentive to conspecifics than ravens (Scheid et al. 2007). Jackdaws feed mainly on invertebrates, including opportunistically catching insects on the wing, but they also forage in flocks on the ground for seeds, etc., with single birds joining these flocks when searching for a food resource (Wechsler 1988a).

Although jackdaws are a noncaching corvid species, they can remember the location of food on the basis of spatial and object-specific information, whereas food-caching Eurasian jays (*Garrulus glandarius*) respond preferentially to spatial cues (Clayton and Krebs 1994b). In a field study on the social influences on foraging, a jackdaw colony was presented with food hidden in nine containers (Röell 1978). The alpha male had been trained to find the food, and which individuals approached and explored the containers was also recorded. The jackdaws hardly ever approached the containers in the absence of others. However, when the alpha male was present and gaining food, other jackdaws immediately joined him, exploring the containers and so learned

A

B

FIGURE 14.7. A jackdaw foraging (left) and a conspecific (right) paying close attention (a), then approaching and exploring at the same location (b). Photographs by I. G. Federspiel.

how to open them. Most of the birds learned by supplanting a bird exploring the food source, indicating local enhancement followed by individual learning as the underlying learning mechanism. Similar results were found in a study using two different food dispensers (Wechsler 1988b). The birds had to either press a lever or pull a plastic disk to gain food. The behavior spread through the group within two months, and 22/28 birds learned to press the lever, 23/28 to pull the plastic disk; however, the birds did not preferentially choose to copy the method of their social partner. These results could not rule out individual trial-and-error learning, and it appears that there was some general enhancement effect, because the birds' attention was drawn to the dispensers, but in the beginning of the study they did not seem to know exactly how to operate the mechanisms. A similar result was found in a recent experiment in which jackdaws that had the opportunity to watch a conspecific opening a food box and feeding from it approached the box faster, stayed close to the box for longer, and showed a higher persistency in exploring the box than birds who were tested without demonstrations (Federspiel and Emery, in preparation).

In conclusion, social information seems to play a less vital role in jackdaws than ravens and western scrub jays. Although jackdaws are a more social corvid species, their ecology does not require them to use social information to the same extent as scrub jays and ravens. They feed on abundant types of food; jackdaws may therefore not need to learn much about food via social information. However, the fact that they learn where to find food from their conspecifics in an experimental context shows that they have the capacity to do so. Since their diet does not include hard-to-access or -process foods (no evidence for extractive foraging), they need to learn only the location of the food but not how to process it. Therefore, the relatively simple mechanism of local enhancement seems to be adequate for their requirements.

14.2.2. PSITTACIDAE

14.2.2.1. *Keas*

In contrast to jackdaws, some birds that feed on hard-to-access foods may need more complex social learning mechanisms, such as imitation or emulation, to learn how to access the palatable part of the fruit, nut, meat, etc. One example of the influence of ecology on social information use is the kea (*Nestor notabilis*), New Zealand's mountain parrot. Little is known about the mating system of this species, but certain occasions, such as feeding, appear to bring them together in large gatherings. As juveniles, keas form large flocks and travel around exploring their environment together (Diamond and Bond 1999). A complex dominance network, a relatively long developmental period of the young, and various forms of social and object play may have contributed to their social intelligence (Diamond and Bond 1999; Keller 1975). Furthermore, the lack of predators and the patchy distribution of food in the winter are thought to have led to their extreme neophilia (Diamond and Bond 1999). A group of captive keas demonstrated flexible use of different social techniques to gain food in cooperative tasks (Federspiel 2006; Werdenich 2006), and dominant birds were even found to manipulate lower-ranking birds to coerce them into producing a food reward (Tebbich et al. 1996). In social learning tasks, ambiguous results have been found. In some field studies (tube lifting: Gajdon et al. 2004; rubbish bin opening: Gajdon et al. 2006) and laboratory studies (social learning apparatuses: Pesendorfer 2007), there is little or no evidence of social learning, but in others, indicators of imitation and emulation have been reported. In a task with an artificial fruit and three locking devices (a bolt, a pin, and a screw), birds that observed a conspecific performing the opening sequence displayed shorter latencies to approach the apparatus, a greater persistence in manipulating the devices, and a higher success rate at opening the apparatus than those that did not observe a demonstrator (fig. 14.8; Huber et al. 2001).

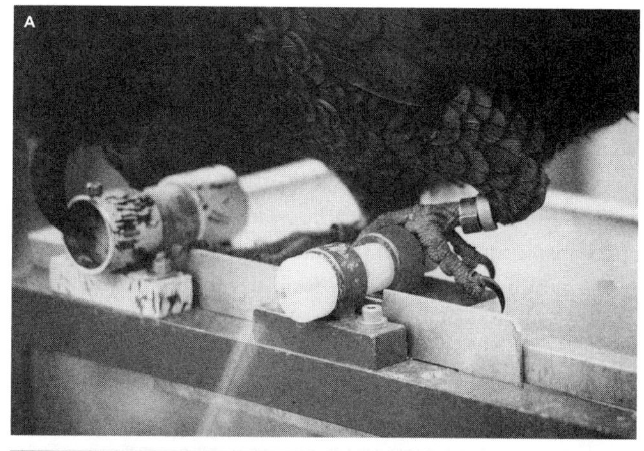

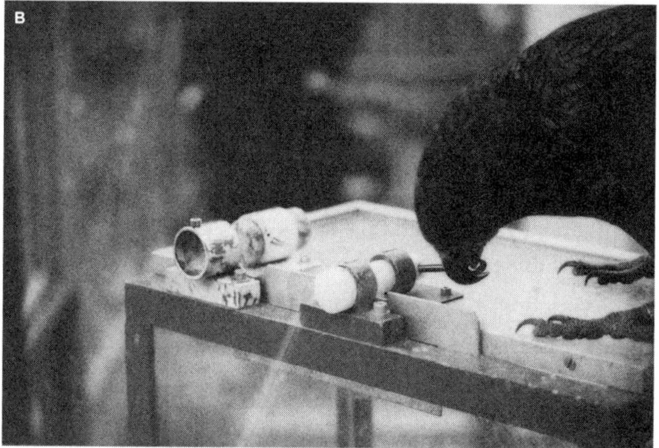

FIGURE 14.8. A kea performing two of the three required actions: (a) poking out the bolt and (b) pulling out the pin. Photographs by L. Huber.

A second experiment with just one opening mechanism revealed keas' use of imitation; a crank had to be rotated repeatedly to get to the food reward, and two of the six observers copied the model's actions rather than just the outcome (L. Huber, personal communication). Although generalization across the whole species cannot be drawn from just two individuals, there does seem to be an indication of the capacity to imitate. Why the keas did not use imitation in the more complex task with the three locking devices remains unclear. There may be an urge to employ individual learning once they have learned about the affordances of an object, that is, what can be done with/to the object.

Ecology must be a strong factor in the keas' use of social information. With no predators around and the urge to always search for food, especially in winter, when the surroundings are covered in snow (B. Campbell 1976), keas have become a flexible, opportunistic, and very curious species (Diamond and Bond 1991). They live in the Southern Alps of New Zealand between elevations of 700 and 2000 meters and in winter fly down to the coast to find food. They feed on at least 89 plant species (including nectar and fruit) and 9 animal species, turning over stones to get to bugs and larvae (Breejaart 1988). Foraging on hard-to-access foods is thought to result in the development of larger brains in primates (K. Gibson 1986), and so the same argument could be made for large-brained birds, such as keas, that feed on similar food types. In general, habitat opportunists show lower levels of neophobia (Greenberg 1990), and island populations show more exploratory behavior (Mettke-Hofmann et al. 2002). This may have led to keas' well-developed manipulation skills and their persistence on instrumental tasks (Kubat 1992), influencing their approach to solving technical problems and the development of an advanced form of technical intelligence. During object play, they train their manipulatory skills and gather information about the affordances of an object (Inglis 1983). They are also well known for their playfulness and destructiveness around human settlements (Forshaw 1977), exploring and tearing apart everything they can get to.

As mentioned earlier, the social component may have also contributed to the keas' skills in social information use. They experience an extended post-fledging period of dependency on the parents and join juvenile flocks in their second summer. Under these conditions there are plenty of opportunities to utilize social information (Lint 1958; Jackson 1963; Porter 1947). Furthermore, there is an intricate dominance hierarchy in kea groups (Diamond and Bond 1999), which resembles a network rather than a linear hierarchy such as found in chickens. Keeping track of relationships in the group may have further driven their skills for picking up social information, similar to pinyon jays.

Evolution also suggests the presence of social information use in the ancestors of the kea. The order Psittaciformes includes roughly 350 species that are commonly grouped into two families, namely cockatoos (Cacatuidae) and true parrots (Psittacidae), although different systematics exist. One tribe of the subfamily of the typical parrots (Psittacinae) is the Nestorini, consisting of three species: the Norfolk kaka (*Nestor productus*), which became extinct in the first half of the nineteenth century, the kaka (*Nestor meridionalis*), and the kea (Pies-Schulz-Hofen 2004). The ancestor of the three species was a "Proto-kaka," which is thought to have lived 15 million years ago, when New Zealand was still a single, large island. During the Pleistocene, New Zealand was dissected into two smaller islands, the climatic conditions changed, and the differences between the northern and southern environments became extreme. The population living under harsh conditions in the south became keas; the population in the north became kakas, which specialized on fruits and insects of the rain forest (Diamond and Bond 1999). During the Holocene, there was an increase in the growth of forests, and the kaka returned to the South Island of New Zealand, pushing the keas out of their foraging niche and thus forcing them into the alpine habitat. Although little is known about the behavior of the ancestors of the kea, their habitat and the extreme changes in temperature and environment they experienced suggest that the use of social information would have been advantageous for them. The kaka's varied diet suggests something similar to the kea. Similarities in the complexity and diversity of social play in keas and kakas (Diamond and Bond 2004) also indicate pronounced social skills in the kaka, a close relative of the kea. The relatively solitary lifestyle and the less complex play of the kakapo (*Strigops habroptilus;* Diamond et al. 2006), another close relative of the kea belonging to the Psittacinae and the tribe Strigopini, indicate that the kea's social information use is a relatively novel phenomenon in evolutionary terms in this group of birds (kea, kaka, and kakapo). However, this is necessarily speculative as it is constrained by information on the number of species.

14.2.3. ANATIDAE

14.2.3.1. *Geese*

Another gregarious species, greylag geese (*Anser anser*), live in large flocks of families and form lifelong pairs in their third or fourth year. Their diet mainly consists of roots, fruits, flowers, leaves, and stems (Cramp 1977). Certain behaviors related to feeding, such as adopting a novel food into the diet, are transmitted via social information use within groups (fig. 14.9).

One year after some individuals were first observed to bite through the stems of butterbur leaves (*Petasites hybridus*), almost all the individuals of a

FIGURE 14.9. A goose foraging in the grass (front) and another watching (back). Photograph by I. Nedelcu.

semi-tame flock of geese displayed that behavior (Fritz et al. 2000). The behavior spread particularly quickly between birds that spent most of their time together, such as related animals, suggesting social influence on the learning process (Coussi-Korbel and Fragaszy 1995). It was even observed that ganders left stems to goslings if the goslings approached them while the gander was biting on the stems. To gain an insight into the processes involved, Fritz and colleagues (2000) conducted an experiment with a food box that could be opened by sliding a lid open. A human demonstrator showed seven geese how to open the box, and all seven observers learned the action, whereas only one of the seven control animals that had not received a demonstration managed to solve the task. The observers showed no imitation but explored more often the location at which the demonstrator had touched the box, which facilitated individual trial-and-error learning. Thus, observing an experienced conspecific chewing through the stems of butterbur leaves may have accounted for the spread of the behavior through the group (Fritz et al. 2000; Fritz and Kotrschal 2002).

The social lifestyle of the geese provides them with numerous opportunities to use social information. They are gregarious, except when nesting, live in flocks made up of families, and come together during molting and migration. Individuals in a group support one another actively (participation of a social ally) and passively (mere presence of an ally reduces stress) during agonistic encounters (Scheiber et al. 2005; Weiss and Kotrschal 2004). Lifelong monogamy is the rule, with males and females associating all year round (Cramp 1977). Nevertheless, their diet suggests no particular need to learn from others. They feed on abundant food resources, such as plants on the ground or on the surface of water, flowers, and fruits. It seems that geese will pick up social information if they are able to but do not rely on it. Similar to jackdaws, local enhancement may therefore be sufficient for their requirements.

Although numerous fossil species have been suggested as ancestral to this genus (Brodkorb 1964; Short 1970; Livezey 1986), little is known about the social life of the ancestors of geese, and living close relatives share a similar ecology. Species of the genus *Anser* are largely herbivorous wetland species (Cramp 1977). Given the similar diet and environment, it is unlikely that the relatives and the common ancestor of the genus *Anser* employed more complex forms of social information use.

14.2.4. TETRAONIDAE

14.2.4.1. *Black grouse*

A different form of social information use is employed by some families of the order Galliformes, such as certain game birds of the family Phasianidae (for

Japanese quail, *Coturnix japonica,* see Galef and White 2000; D. White 2004) and grouse (family Tetraonidae). One example is the European black grouse (*Tetrao tetrix*). Black grouse tend to be gregarious throughout the year with stable lekking groups in the spring and less stable groups in the autumn and winter (Cramp 1977). At leks (i.e., mating areas), males gather within sight of each other to court and compete at the outset of the breeding season. They perform mating displays and thus attract females they subsequently mate with (Tomkins et al. 2004).

At the leks, it was found that, after being chosen by one female, a male is generally more likely to be chosen by other females. Further experiments with model females revealed that, rather than just seeing a male close to other females, watching a male actually copulating with a female made him more attractive to other females (Höglund et al. 1995). Young, inexperienced animals may benefit from copying others' choices. Copying the choice of others seems to be a process additional to the normal mate choice process of females (Höglund et al. 1995); since some females may not have the opportunity to copy others, mate choice copying may not always be employed. Although a well-known phenomenon, it seems that the ultimate reasons for mate choice copying are not yet clear. Theories range from avoiding the costs of independently sampling mates, managing information and filtering out the unimportant information, to simply reducing the errors made during mate choice and learning from knowledgeable conspecifics about the quality of potential mates (D. White 2004). Although the costs and benefits of mate choice copying are not clear, using the social information of a conspecific about a male's fitness seems to have an advantage over individual learning. Furthermore, local adaptations of the population in terms of heritably transmitted mate preferences could lead to a change in reproductive patterns and genetic changes (Freeberg 1998, 2004).

The diet of European black grouse consists predominantly of abundant plants, such as berries and grasses (Cramp 1977), suggesting no particular requirement for social information use when it comes to accessing or processing food. Their usage of social information may therefore be restricted to the context of mate choice, but more experimental work needs to be performed to draw firm conclusions.

In terms of the influence of evolutionary history on the existence and the type of social information use, the close relatives of the black grouse may provide an insight. The birds of the genus *Tetrao,* such as the Caucasian black grouse (*Tetrao mlokosiewiczi*), the capercaillie (*Tetrao urogallus*), and the black-billed capercaillie (*Tetrao parvirostris*), all form leks to display and choose mates (Madge and McGowan 2002). Even earlier in the genealogy, within the family Tetraonide, the males are polygamous in all but one species, the willow grouse (*Lagopus lagopus*). This suggests similar social information use in the common ancestor.

14.3. Conclusions

In this chapter we have argued that knowledge of the ecology (environmental pressures), ethology (natural behavior), and evolutionary history of a species is essential for understanding how the mechanisms responsible for social information use have been shaped. However, to understand these mechanisms and to define in what ways, if any, they differ between species we must adopt an integrative approach, combining the 3E's with knowledge of experimental psychology, in order to obtain rigorous experimental validity. Only then can we hope to understand why two behaviors that, when taken at face value, appear identical are in fact examples of convergent evolution resulting from similar ecological, but not evolutionary, histories. Similarly, it is only by taking this integrated approach that we can hope to show that two "different" behaviors share an evolutionary antecedent and components of the same underlying mechanisms. Instead of only analyzing the psychological processes involved in social learning mechanisms, taking these other components into consideration provides us with a more complete picture. There are still obvious gaps in our knowledge that restrict our use of this approach when it comes to investigating the evolutionary history of social information use. For these cases, the 3E's approach may nevertheless be a useful tool for making predictions about how to direct future research. More comparative work is clearly needed to determine the influence of ecology and evolutionary history on social information use, and experiments using different tasks should shed light on whether the capacity to use social information is domain specific.

For example, rooks and jackdaws both pair for life, yet rooks engage in extrapair copulations (Røskaft 1983) and jackdaws do not (Henderson et al. 2000). We might predict that rooks will copy another's choice of sexual partner during acts of promiscuity (because this is within the capacity of the species), whereas we might also predict that it is futile to test mate choice copying in jackdaws because they do not appear to possess the capacity for infidelity. Jackdaws' social information use may be restricted to certain domains (e.g., behavioral coordination within the pair-bond, learning about the location of good food patches), excluding others (e.g., choice of partners during extrapair relations, extractive foraging techniques). We can therefore use knowledge of how the 3E's have influenced a particular species' life history not only to drive the development of appropriate research questions and methodology but also to provide post hoc explanations of successes and failures in psychological experiments. This makes the 3E's approach a very powerful research tool for cognitive ecology—but one that is only as good as the information provided by ecologists, ethologists, evolutionary biologists, and comparative psychologists.

ACKNOWLEDGMENTS

Ira G. Federspiel is partially funded by the Cambridge European Trust, and Nathan J. Emery is funded by a Royal Society University Research Fellowship. We would like to thank Ivo D. Federspiel for constructing the diagram in figure 14.1, and everyone who sent photographs of their research animals.

15 Prospects

REUVEN DUKAS & JOHN M. RATCLIFFE

> In the future I see open fields for far more important researchers. Psychology will be securely based on the foundation already well laid by Mr. Herbert Spencer, that of the necessary acquirement of each mental power and capacity by gradation.
>
> CHARLES DARWIN, *The Origin of Species* (1859)

As all chapters in this volume superbly illustrate, Darwin's 150-year-old prediction quoted above (Darwin 1859) has materialized, and current cognitive research is indeed securely based on the foundations laid by evolutionary research. Darwin's forecast of open fields has materialized as well, and researchers from many of the disciplines that emerged during the twentieth century have produced an enormous amount of information relevant to cognitive ecology. Contributors to this book have exhibited a remarkable ability for combining pertinent data across many of these disciplines.

The chapters in this volume deal with a rich variety of topics in a broad assortment of animals. Learning, memory, and overall behavioral flexibility are analyzed from a variety of angles primarily in insects and birds (chapters 2–7). The two chapters devoted to mate choice elegantly synthesize broad ideas across all animals (chapter 8) and over a few levels of analysis from neural substrates to field behavior in a mammal (chapter 9). Antipredatory behavior is discussed for the embryonic stage, which is relevant for most animals (chapter 10), and for the insightful acoustic domain of insects and their mammalian predators (chapter 11). Finally, the use of social information is examined in a single mammal (chapter 12), in birds (chapter 14), and in a broad review with a focus on several vertebrates (chapter 13). The variation in topics and taxa, however, does not obscure the clear thread tightly linking the chapters, which is the underlying approach of relating cognitive mechanisms to animal ecology and evolution.

For a variety of reasons, including the cognitive limitations of the editors, authors' availability, and space constraints, some relevant established topics and exciting new developments in cognitive ecology are not included here. Several areas have been admirably covered in other recent books. For example, the recent volume on foraging by Stephens, Brown, and Ydenberg (2007) provides thorough analyses of the proximate and ultimate mechanisms determining animal feeding and of the importance of foraging for animal population dynamics and community ecology. The major advances in our understanding of the interactions between predator cognition and prey camouflage, warning signals, and mimicry are reviewed in a few books (Ruxton et al. 2004; Enquist and Ghirlanda 2005; Searcy and Nowicki 2005). North and Greenspan's volume on invertebrate neurobiology (2007) remarkably relates neurobiological findings to whole-animal behavior. Finally, the emerging discipline of neuroeconomics integrates economic and cognitive analyses with neurobiological information for deciphering key human economic decisions (Glimcher and Rustichini 2004; Sanfey et al. 2006).

This volume makes it clear that cognitive ecology is a vibrant discipline, which has significantly advanced since the publication of the first *Cognitive Ecology* (Dukas 1998a). Many future directions in the discipline have been clearly identified in all chapters of the current volume and hence will not be repeated here. Overall, we believe that the major challenge facing future work in cognitive ecology is the further integration of knowledge on the genetic, neurobiological, and endocrinological mechanisms underlying cognition with information about animal ecology and evolution. For example, we can go a long way by examining how a cognitive constraint influences behavior and fitness (Dukas 1998b). However, constraint is defined as anything that prevents, delays, or increases the cost of attaining a certain ability (Dukas 2002). This means that we should study what mechanism generates a given constraint. Is it merely lack of genetic variation? Is it some fundamental physical property of either individual neurons or neuronal networks? Or does it simply reflect energetic costs (Clark and Dukas 2003)? Ultimately, we require this level of fundamental mechanistic understanding, combined with detailed knowledge of ecology and evolutionary history, for deciphering the evolutionary biology of cognition.

A few chapters in this volume have addressed essential issues related to development. Whereas the importance of developmental mechanisms for behavior has been recognized for decades (Tinbergen 1963), new tools from the emerging field of evolutionary developmental biology (Raff 1996; Wilkins 2002) can help us understand the underlying developmental pathways shaping cognition. Somewhat related to development is the topic of phenotypic plasticity. Whereas the term "plasticity" is widely used in a variety of disciplines, it has a specific meaning in evolutionary biology (Futuyma 1998;

Pigliucci 2001). Evolutionary biologists have studied phenotypic plasticity mostly within the morphological domain, which is easier to quantify than that of cognition. We can rely on this coherent body of literature for addressing key questions about cognitive plasticity. For example, is adaptive phenotypic plasticity of cognitive traits prevalent in nature? Is there genetic variation for reaction norms of cognitive traits (Dukas 2004a)?

In closing, the historical traditions that maintained cognitive and evolutionary research on parallel, distinct paths are rapidly vanishing. Remaining, however, is the sheer challenge of integrating large amounts of information across disciplines and levels of analysis. This volume, however, vividly indicates that such work is both feasible and enlightening.

REFERENCES

Abisgold, J. D., and S. J. Simpson. 1988. The effect of dietary-protein levels and hemolymph composition on the sensitivity of the maxillary palp chemoreceptors of locusts. *Journal of Experimental Biology* 135:215–229.

Aboitiz, F. 2001. What determines evolutionary brain growth? *Behavioral and Brain Sciences* 24:278.

Abrams, P. A. 1993. Does increased mortality favor the evolution of more rapid senescence? *Evolution* 47:877–887.

Aceves-Pina, E. O., and W. G. Quinn. 1979. Learning in normal and mutant *Drosophila* larvae. *Science* 206:93–96.

Acharya, L., and M. B. Fenton. 1992. Echolocation behavior of vespertilionid bats (*Lasiurus cinereus* and *Lasiurus borealis*) attacking airborne targets including arctiid moths. *Canadian Journal of Zoology—Revue Canadienne de Zoologie* 70:1292–1298.

———. 1999. Bat attacks and moth defensive behaviour around street lights. *Canadian Journal of Zoology—Revue Canadienne de Zoologie* 77:27–33.

Acharya, L., and J. N. McNeil. 1998. Predation risk and mating behavior: The responses of moths to bat-like ultrasound. *Behavioral Ecology* 9: 552–558.

Aiello, L. C., and P. Wheeler. 1995. The expensive-tissue hypothesis: The brain and the digestive system in human and primate evolution. *Current Anthropology* 36:199–221.

Airey, D. C., K. L. Buchanan, T. Szekely, C. K. Catchpole, and T. J. DeVoogd. 2000. Song, sexual selection, and a song control nucleus (HVc) in the brains of European sedge warblers. *Journal of Neurobiology* 44:1–6.

Allen, C., and C. F. Stevens. 1994. An evaluation of causes for unreliability of synaptic transmission. *Proceedings of the National Academy of Sciences of the United States of America* 91:10380–10383.

Allman, J. M. 1977. Evolution of the visual system in early primates. In J. M. Sprague and J. M. Epstein, eds., *Progress in Psychobiology, Physiology, and Psychology* 7:1–53. New York: Academic Press.

———. 2000. *Evolving Brains*. New York: Scientific American Library.

Allman, J. M., T. McLaughlin, and A. Hakeem. 1993. Brain-weight and life-span in primate species. *Proceedings of the National Academy of Sciences of the United States of America* 90:118–122.

Alloway, T. M. 1972. Learning and memory in insects. *Annual Review of Entomology* 17:43–56.

Alvarez-Buylla, A., and J. R. Kirn. 1997. Birth, migration, incorporation, and death of vocal control neurons in adult songbirds. *Journal of Neurobiology* 33:585–601.

Alvarez-Buylla, A., J. R. Kirn, and F. Nottebohm. 1990a. Birth of projection neurons in adult avian brain may be related to perceptual or motor learning. *Science* 249:1444–1446.

Alvarez-Buylla, A., M. Theelen, and F. Nottebohm. 1988. Birth of projection neurons in the higher vocal center of the canary forebrain before, during, and after song learning. *Proceedings of the National Academy of Sciences of the United States of America* 85:8722–8726.

——. 1990b. Proliferation "hot spots" in adult avian ventricular zone reveal radial cell division. *Neuron* 5:101–109.

Amdam, G. V., K. Norberg, R. J. Page, J. Erber, and R. Scheiner. 2006. Downregulation of vitellogenin gene activity increases the gustatory responsiveness of honey bee workers (*Apis mellifera*). *Behavioral Brain Research* 169:201–205.

Amlacher, J., and L. A. Dugatkin. 2005. Preference for older over younger models during mate-choice copying in young guppies. *Ethology, Ecology and Evolution* 17:161–169.

Anderson, P. A. V. 1989. *Evolution of the First Nervous Systems*. New York: Plenum Press.

Anderson, R. C., W. A. Searcy, and S. Nowicki. 2005. Partial matching in an eastern population of song sparrows, *Melospiza melodia*. *Animal Behaviour* 69:189–196.

Andersson, M. 1986. Evolution of condition-dependent sex ornaments and mating preferences: Sexual selection based on viability differences. *Evolution* 40:804–816.

——. 1994. *Sexual Selection*. Princeton: Princeton University Press.

Anisman, H., M. D. Zaharia, M. J. Meaney, and Z. Merali. 1998. Do early-life events permanently alter behavioral and hormonal responses to stressors? *International Journal of Developmental Neuroscience* 16:149–164.

Arlettaz, R., G. Jones, and P. A. Racey. 2001. Effect of acoustic clutter on prey detection by bats. *Nature* 414:742–745.

Aron, A., H. Fisher, D. J. Mashek, G. Strong, H. F. Li, and L. L. Brown. 2005. Reward, motivation, and emotion systems associated with early-stage intense romantic love. *Journal of Neurophysiology* 94:327–337.

Arrow, K. J. 1962. The economic implication of learning by doing. *Review of Economic Studies* 29:155–173.

Ash, J., and G. G. Gallup Jr. 2007. Paleoclimatic variation and brain expansion during human evolution. *Human Nature* 18:109–124.

Ashburner, M., H. L. Carson, and J. N. Thompson. 1976. *The Genetics and Biology of* Drosophila. London: Academic Press.

Aubin, T., and P. Jouventin. 1998. Cocktail-party effect in king penguin colonies. *Proceedings of the Royal Society of London Series B* 265:1665–1673.

Avery, M., and L. W. Oring. 1977. Song dialects in bobolink (*Dolichonyx oryzivorus*). *Condor* 79:113–118.

Bacher, S., J. Casas, and S. Dorn. 1996. Parasitoid vibrations as potential releasing stimulus of evasive behaviour in a leafminer. *Physiological Entomology* 21:33–43.

Baker, M. C. 1983. The behavioral response of female Nuttall's white-crowned sparrows to male song of natal and alien dialects. *Behavioral Ecology and Sociobiology* 12:309–315.

Baker, M. C., P. K. McGregor, and J. R. Krebs. 1987. Sexual response of female great tits to local and distant songs. *Ornis Scandinavia* 18:186–188.

Balda, R. P., and A. C. Kamil. 1989. A comparative study of cache recovery in three corvid species. *Animal Behaviour* 38:486–495.

Balda, R. P., A. C. Kamil, and P. A. Bednekoff. 1996. Predicting cognitive capacity from natural history: Examples from four species of corvids. In V. Nolan Jr. and E. D. Ketterson, eds., *Current Ornithology*, 33–66. New York: Plenum Press.

Balda, R. P., A. C. Kamil, P. A. Bednekoff, and A. G. Hile. 1997. Species differences in spatial memory performance on a three-dimensional task. *Ethology* 103:47–55.

Balda, R. P., I. M. Pepperberg, and A. C. Kamil. 1998. *Animal Cognition in Nature*. London: Academic Press.

Baldwin, J. M. 1893. *Elements of Psychology*. New York: Holt.

Ballentine, B., J. Hyman, and S. Nowicki. 2004. Vocal performance influences female response to male bird song: An experimental test. *Behavioral Ecology* 15:163–168.

Bamshad, M., M. A. Novak, and G. J. Devries. 1994. Cohabitation alters vasopressin innervation and paternal behavior in prairie voles (*Microtus ochrogaster*). *Physiology and Behavior* 56:751–758.

Bandura, A. 1977. *Social Learning Theory*. Englewood Cliffs, NJ: Prentice-Hall.

Banks, B., and T. J. C. Beebee. 1988. Reproductive success of natterjack toads *Bufo calamita* in two contrasting habitats. *Journal of Animal Ecology* 57:475–492.

Baptista, L. F., and M. L. Morton. 1988. Song learning in montane white-crowned sparrows: From whom and when. *Animal Behaviour* 36:1753–1764.

Bar, M. 2004. Visual objects in context. *Nature Reviews Neuroscience* 5:617–629.

Barber, J. R., and W. E. Conner. 2006. Tiger moth responses to a simulated bat attack: Timing and duty cycle. *Journal of Experimental Biology* 209:2637–2650.

———. 2007. Acoustic mimicry in a predator-prey interaction. Proceedings of the National Academy of Sciences of the United States of America 104:9331–9334.

Barclay, R. M. R., and R. M. Brigham. 1994. Constraints on optimal foraging: A field test of prey discrimination by echolocating insectivorous bats. *Animal Behaviour* 48:1013–1021.

Bargmann, C. I. 1998. Neurobiology of the *Caenorhabditis elegans* genome. *Science* 282:2028–2033.

Barnea, A., and F. Nottebohm. 1994. Seasonal recruitment of hippocampal neurons in adult free-ranging black-capped chickadees. *Proceedings of the National Academy of Sciences of the United States of America* 91:11217–11221.

Baron, R. S., J. A. Vandello, and B. Brunsman. 1996. The forgotten variable in conformity research: The impact of task importance on social influence. *Journal of Personality and Social Psychology* 71:915–927.

Barton, R. A. 2004. Binocularity and brain evolution in primates. *Proceedings of the National Academy of Sciences of the United States of America* 101:10113–10115.

Barton, R. A., A. Purvis, and P. H. Harvey. 1995. Evolutionary radiation of visual and olfactory brain systems in primates, bats and insectivores. *Philosophical Transactions of the Royal Society of London Series B* 348:381–392.

Barton, R. A., and P. H. Harvey. 2000. Mosaic evolution of brain structure in mammals. *Nature* 405:1055–1058.

Basil, J. A., A. C. Kamil, R. P. Balda, and K. V. Fite. 1996. Differences in hippocampal volume among food storing corvids. *Brain, Behavior and Evolution* 47:156–164.

Basolo, A. L. 1990. Female preference for male sword length in the green swordtail, *Xiphophorus helleri* (Pisces: Poeciliidae). *Animal Behaviour* 40:332–338.

Bateman, A. J. 1948. Intra-sexual selection in *Drosophila*. *Heredity* 2:349–368.

Bateson, M., and S. D. Healy. 2005. Comparative evaluation and its implications for mate choice. *Trends in Ecology and Evolution* 20:659–664.

Bateson, M., S. D. Healy, and T. A. Hurly. 2002. Irrational choices in hummingbird foraging behaviour. *Animal Behaviour* 63:587–596.

———. 2003. Context-dependent foraging decisons in rufous hummingbirds. *Proceedings of the Royal Society of London Series B* 270:1271–1276.

Baugh, A. T., K. L. Akre, and M. J. Ryan. 2008. Categorical perception of a natural, multivariate signal: Mating call recognition in túngara frogs. *Proceedings of the National Academy of Sciences of the United States of America* 105:8985–8988.

Beardsley, P. 2006. Behavioural flexibility in the mating system buffers population extinction: Lessons from the lesser spotted woodpecker *Picoides minor*. *Journal of Animal Ecology* 75: 540–548.

Bedi, K. S. 1992. Spatial leaarning ability of rats undernourished during early postnatal life. *Physiology and Behavior* 51:1001–1007.

———. 2003. Nutritional effects on neuron numbers. *Nutritional Neuroscience* 3:141–152.

Bednekoff, P. A. 2007. Foraging in the face of danger. In D. W. Stevens, J. S. Brown, and R. C. Ydenberg, eds., *Foraging: Behavior and Ecology,* 305–330. Chicago: University of Chicago Press.

Bednekoff, P. A., and R. P. Balda. 1996a. Observational spatial memory in Clark's nutcrackers and Mexican jays. *Animal Behaviour* 52:833–839.

———. 1996b. Social caching and observational spatial memory in pinyon jays. *Behaviour* 133: 807–826.

Bee, M. A., and G. M. Klump. 2004. Primitive auditory stream segregation: A neurophysiological study in songbird forebrain. *Journal of Neurophysiology* 92:1088–1104.

———. 2005. Auditory stream segregation in the songbird forebrain: Effects of time intervals on responses to interleaved tone sequences. *Brain, Behavior and Evolution* 66:197–214.

Bee, M. A., and C. Micheyl. 2008. The "cocktail party problem": What is it? How can it be solved? And why should animal behaviorists study it? *Journal of Comparative Psychology* 122:235–251.

Beebee, M. D. 2002. Song sharing by yellow warblers differs between two modes of singing: Implications for song function. *Condor* 104:146–155.

Beecher, M. D. 1996. Birdsong learning in the laboratory and field. In D. E. Kroodsma and E. H. Miller, eds., *Ecology and Evolution of Acoustic Communication in Birds,* 61–78. Ithaca: Cornell University Press.

Beecher, M. D., and E. A. Brenowitz. 2005. Functional aspects of song learning in songbirds. *Trends in Ecology and Evolution* 20:143–149.

Beecher, M. D., and J. M. Burt. 2004. The role of social interaction in bird song learning. *Current Directions in Psychological Science* 13:224–228.

Beecher, M. D., J. M. Burt, A. L. O'Loghlen, C. N. Templeton, and S. E. Campbell. 2007. Bird song learning in an eavesdropping context. *Animal Behaviour* 73:929–935.

Beecher, M. D., and S. E. Campbell. 2005. The role of unshared songs in singing interactions between neighbouring song sparrows. *Animal Behaviour* 70:1297–1304.

Beecher, M. D., S. E. Campbell, and J. M. Burt. 1994a. Song perception in the song sparrow: Birds classify by song type but not by singer. *Animal Behaviour* 47:1343–1351.

Beecher, M. D., S. E. Campbell, J. M. Burt, C. E. Hill, and J. C. Nordby. 2000a. Song type matching between neighbouring song sparrows. *Animal Behaviour* 59:21–27.

Beecher, M. D., S. E. Campbell, and J. C. Nordby. 1998. The cognitive ecology of song communication and song learning in the song sparrow. In R. Dukas, ed., *Cognitive Ecology,* 175–199. Chicago: University of Chicago Press.

———. 2000b. Territory tenure in song sparrows is related to song sharing with neighbours, but not to repertoire size. *Animal Behaviour* 59:29–37.

Beecher, M. D., S. E. Campbell, and P. K. Stoddard. 1994b. Correlation of song learning and territory establishment strategies in the song sparrow. *Proceedings of the National Academy of Sciences of the United States of America* 91:1450–1454.

Beecher, M. D., J. C. Nordby, S. E. Campbell, J. M. Burt, C. E. Hill, and A. L. O'Loghlen. 1997. What is the function of song learning in songbirds? In D. H. Owings, M. D. Beecher, and N. S. Thompson, eds., *Perspectives in Ethology,* vol. 12, *Communication,* 77–97. New York: Plenum Press.

Beecher, M. D., P. K. Stoddard, S. E. Campbell, and C. L. Horning. 1996. Repertoire matching between neighbouring song sparrows. *Animal Behaviour* 51:917–923.

Behrends, A., R. Scheiner, N. Baker, and G. V. Amdam. 2007. Cognitive aging is linked to social role in honey bees (*Apis mellifera*). *Experimental Gerontology* 42:1146–1153.

Beletsky, L. D., S. Chao, and D. G. Smith. 1980. An investigation of song-based species recognition in the red-winged black bird. *Behaviour* 73–74:189–203.

Bell, G. P. 1982. Behavioral and ecological aspects of gleaning by a desert insectivorous bat, *Antrozous pallidus*. *Behavioral Ecology and Sociobiology* 10:217–223.

Bell G. P., and M. B. Fenton. 1984. The use of Doppler-shifted echoes as a flutter-detection and clutter-rejection system: The echolocation and feeding behavior of *Hipposideros ruber* (Chiroptera: Vespertilionidae). *Behavioral Ecology and Sociobiology* 15:109–114.

Beltman, J. B., P. Haccou, and C. ten Catea. 2004. Learning and colonization of new niches: A first step toward speciation. *Evolution* 58:35–46.

Beltman, J. B., and J. A. J. Metz. 2005. Speciation: More likely through a genetic or through a learned habitat preference? *Proceedings of the Royal Society of London Series B* 272:1455–1463.

Benney, K. S., and R. F. Braaten. 2000. Auditory scene analysis in estrildid finches (*Taenopygia guttata* and *Lonchura striata domestica*): A series advantage for detection of conspecific song. *Journal of Comparative Psychology* 114:174–182.

Berger, J., J. E. Swenson, and I.-L. Persson. 2001. Recolonizing carnivores and naive prey: Conservation lessons from Pleistocene extinctions. *Science* 291:1036–1039.

Bernays, E. A., and W. T. Wcislo. 1994. Sensory capabilities, information-processing, and resource specialization. *Quarterly Review of Biology* 69:187–204.

Bernstein, C., and H. Bernstein. 1991. *Aging, Sex, and DNA Repair*. San Diego: Academic Press.

Berthold, P., and S. B. Terrill. 1991. Recent advances in studies of bird migration. *Annual Review of Ecology and Systematics* 22:357–378.

Bessman, A. N., N. Muzyczka, M. Goodman, and R. Schnaar. 1974. Studies on the biochemical basis of spontaneous mutation. II. The incorporation of a base and its analogue into DNA by wild-type, mutator, and anti-mutator DNA polymerase. *Journal of Molecular Biology* 88:409–421.

Bester-Meredith, J. K., L. J. Young, and C. A. Marler. 1999. Species differences in paternal behavior and aggression in *Peromyscus* and their associations with vasopressin immunoreactivity and receptors. *Hormones and Behavior* 36:25–38.

Bettinger, R. L., and J. Eerkens. 1999. Point typologies, cultural transmission, and the spread of bow-and-arrow technology in the prehistoric Great Basin. *American Antiquity* 64:231–242.

Bicker, G. 1999. Histochemistry of classical neurotransmitters in antennal lobes and mushroom bodies of the honeybee. *Microscopy Research and Technique* 45:174–183.

Bickerton, D. 1990. *Language and Species*. Chicago: University of Chicago Press.

Bielsky, I. F., S. B. Hu, X. H. Ren, E. F. Terwilliger, and L. J. Young. 2005. The V1a vasopressin receptor is necessary and sufficient for normal social recognition: A gene replacement study. *Neuron* 47:503–513.

Bikhchandani, S., D. Hirshleifer, and I. Welch. 1992. A theory of fads, fashion, custom, and cultural change as informational cascades. *Journal of Political Economy* 100:992–1026.

———. 1998. Learning from the behavior of others: Conformity, fads, and informational cascades. *Journal of Economic Perspective* 12:151–170.

Bingman, V. P., A. Gagliardo, G. E. Hough, P. Ioale, M. C. Kahn, and J. J. Siegel. 2005. The avian hippocampus, homing in pigeons and the memory representation of large-scale space. *Integrative and Comparative Biology* 45:555–564.

Bingman, V. P., G. E. Hough, M. C. Kahn, and J. J. Siegel. 2003. The homing pigeon hippocampus and space: In search of adaptive specialization. *Brain, Behavior and Evolution* 62:117–127.

Bingman, V. P., J. J. Siegel, A. Gagliardo, and J. T. Erichsen. 2006. Representing the richness of avian cognition: Properties of a lateralized homing pigeon hippocampus. *Reviews in the Neurosciences* 17:17–28.

Birkhead, T. R., F. Fletcher, and E. J. Pellatt. 1999. Nestling diet, secondary sexual traits and fitness in the zebra finch. *Proceedings of the Royal Society of London Series B* 266:385–390.

Bishop, P. J., M. D. Jennions, and N. I. Passmore. 1995. Chorus size and call intensity: Female choice in the painted reed frog, *Hyperolius marmoratus*. *Behaviour* 132:721–731.

Bitterman, M. E., R. Menzel, A. Fietz, and S. Schäfer. 1983. Classical conditioning of proboscis extension in honey bees. *Journal of Comparative Psychology* 97:107–119.

Black, J. M., and M. Owen. 1995. Reproductive performance and assortative pairing in relation to age in barnacle geese. *Journal of Animal Ecology* 64:234–244.

Blaustein, L. 1997. Non-consumptive effects of larval *Salamandra* on crustacean prey: Can eggs detect predators? *Oecologia* 110:212–217.

Blaustein, L., and M. Spencer. 2001. Hatching responses of temporary pool invertebrates to signals of environmental quality. *Israel Journal of Zoology* 47:397–417.

Blest, A. D., T. S. Collett, and J. D. Pye. 1963. The generation of ultrasonic signals by a New World arctiid moth. *Proceedings of the Royal Society of London Series B* 158:196–207.

Blumstein, D. T., and K. B. Armitage. 1997. Alarm calling in yellow-bellied marmots. 1. The meaning of situationally variable alarm calls. *Animal Behaviour* 53:143–171.

Bohus, B., G. L. Kovacs, and D. Dewied. 1978. Oxytocin, vasopressin and memory: Opposite effects on consolidation and retrieval processes. *Brain Research* 157:414–417.

Boisvert, M. J., A. J. Veal, and D. F. Sherry. 2007. Floral reward production is timed by an insect pollinator. *Proceedings of the Royal Society of London Series B* 274:1831–1837.

Bolhuis, J. J., and E. M. Macphail. 2001. A critique of the neuroecology of learning and memory. *Trends in Cognitive Sciences* 5:426–433.

Bond, A. B., A. C. Kamil, and R. P. Balda. 2003. Social complexity and transitive inference in corvids. *Animal Behaviour* 65:479–487.

Bonnie, K. E., and R. L. Earley. 2007. Expanding the scope for social information use. *Animal Behaviour* 74:171–181.

Borror, D. J. 1965. Song variation in Maine song sparrows. *Wilson Bulletin* 77:5–37.

Bottjer, S. W., S. L. Glaessner, and A. P. Arnold. 1985. Ontogeny of brain nuclei controlling song learning and behavior in zebra finches. *Journal of Neuroscience* 5:1556–1562.

Bottjer, S. W., E. A. Miesner, and A. P. Arnold. 1984. Forebrain lesions disrupt development but not maintenance of song in passerine birds. *Science* 224:901–903.

Bouchard, J., W. Goodyer, and L. Lefebvre. 2007. Innovation and social learning are positively correlated in pigeons. *Animal Cognition* 10:259–266.

Box, H. O. 1984. *Primate Behaviour and Social Ecology*. London: Chapman and Hall.

Boyd, R., and P. J. Richerson. 1985. *Culture and the Evolutionary Process*. Chicago: University of Chicago Press.

———. 1988. An evolutionary model of social learning: The effects of spatial and temporal variation. In T. R. Zentall and B. G. Galef Jr., eds., *Social Learning: Psychological and Biological Perspectives*, 29–48. Hillsdale, NJ: Lawrence Erlbaum.

———. 1995. Why does culture increase human adaptability? *Ethology and Sociobiology* 16:125–143.

Bradbury, I. R., S. E. Campana, P. Bentzen, and P. V. R. Snelgrove. 2004. Synchronized hatch and its ecological significance in rainbow smelt *Osmerus mordax* in St. Mary's Bay, Newfoundland. *Limnology and Oceanography* 49:2310–2315.

Bradbury, J. W., and S. L. Vehrencamp. 1998. *Principles of Animal Communication*. Sunderland, MA: Sinauer.

Brandon, J., and R. Coss. 1982. Rapid dendritic spine shortening during one-trial learning: The honeybee's first orientation flight. *Brain Research* 252:51–61.

Breed, M. D. 1983. Nestmate recognition in honey bees. *Animal Behaviour* 31:86–91.

Breed, M. D., M. F. Garry, A. N. Pearce, B. E. Hibbard, L. B. Bjostad, and J. R. E. Page. 1995. The role of wax comb in honey bee nestmate recognition. *Animal Behaviour* 50:489–496.

Breed, M. D., K. R. Williams, and J. H. Fewell. 1988. Comb wax mediates the acquisition of nestmate recognition cues in honey bees. *Proceedings of the National Academy of Sciences of the United States of America* 85:8766–8769.

Breejaart, R. 1988. Diet and feeding behaviour of the kea (*Nestor notabilis*). Thesis, Lincoln College, Canterbury.

Bregman, A. 1990. *Auditory Scene Analysis*. Cambridge: MIT Press.

Bremond, J. C. 1968. Recherches sur la semantique et les elements vecteurs d'information dans les signaux acoustiques du rouge-gorge (*Erithacus rubecula* L.). *Terre Vie* 2:109–220.

Brenner, S. 1974. The genetics of *Caenorhabditis elegans*. *Genetics* 77:71–94.

Brenowitz, E. A. 2004. Plasticity of the adult avian song control system. *Annals of the New York Acedemy of Sciences* 1016:560–585.

Brenowitz, E. A., and D. E. Kroodsma. 1996. The neuroethology of bird song. In D. E. Kroodsma and E. H. Miller, eds., *Ecology and Evolution of Acoustic Communication in Birds*, 285–304. Ithaca: Cornell University Press.

Brenowitz, E. A., K. Lent, and D. E. Kroodsma. 1995. Brain space for learned songs in birds develops independently of song learning. *Journal of Neuroscience* 15:6281–6286.

Brenowitz, E. A., B. Nalls, J. C. Wingfield, and D. E. Kroodsma. 1991. Seasonal changes in avian song nuclei without seasonal changes in song repertoire. *Journal of Neuroscience* 11:1367–1374.

Breuner, C. W., and M. Orchinik. 2002. Plasma binding proteins as mediators of corticosteroid action in vertebrates. *Journal of Endocrinology* 175:99–112.

Breuner, C. W., S. E. Lynn, G. E. Julian, J. M. Cornelius, B. J. Heidinger, O. P. Love, R. S. Sprague, H. Wada, and B. A. Whitman. 2006. Plasma-binding globulins and acute stress response. *Hormone and Metabolic Research* 38:260–268.

Briggs, S. E., J. G. J. Godin, and A. Dugatkin. 1995. Mate-choice copying under predation risk in the Trinidadian guppy (*Poecilia reticulata*). *Behavioral Ecology* 7:151–157.

Brigham, R. M., S. D. Grindal, M. C. Firman, and J. L. Morissette. 1997. The influence of structural clutter on activity patterns of insectivorous bats. *Canadian Journal of Zoology—Revue Canadienne de Zoologie* 75:131–136.

Briskie, J. V. 1999. Song variation and the structure of local song dialects in the polygynandrous Smith's longspur. *Canadian Journal of Zoology—Revue Canadienne de Zoologie* 77:1587–1594.

Brodin, A., and Lundberg, K. 2003. Is hippocampal volume affected by specialization for food hoarding in birds? *Proceedings of the Royal Society of London Series B* 270:1555–1563.

Brodkorb, P. 1964. Catalogue of fossil birds: Part 2 (Anseriformes through Galliformes). *Bulletin of the Florida State Museum* 8:195–335.

Brooker, M. L., N. B. Davies, and D. G. Noble. 1998. Rapid decline of host defences in response to reduced cuckoo parasitism: Behavioural flexibility of reed warblers in a changing world. *Proceedings of the Royal Society of London Series B* 265:1277–1282.

Brown, J. H., and R. M. Sibly. 2006. Life-history evolution under a production constraint. *Proceedings of the National Academy of Sciences of the United States of America* 103:17595–17599.

Brown, R. M., and D. T. Iskandar. 2000. Nest site selection, larval hatching, and advertisement calls of *Rana arathooni* from southwestern Sulawesi (Celebes) Island, Indonesia. *Journal of Herpetology* 34:404–413.

Brua, R. B. 2002. Parent-embryo interactions. In D. C. Deeming, ed., *Avian Incubation: Behaviour, Environment, and Evolution,* 88–99. Oxford: Oxford University Press.

Brumm, H. 2004. The impact of environmental noise on song amplitude in a territorial bird. *Journal of Animal Ecology* 73:434–440.

Brumm, H., and H. Slabbekoorn. 2005. Acoustic communication in noise. *Advances in the Study of Behavior* 35:151–209.

Brumm, H., and D. Todt. 2002. Noise-dependent song amplitude regulation in a territorial songbird. *Animal Behaviour* 63:891–897.

Brzek, P., and M. Konarzewski. 2007. Relationship between avian growth rate and immune response depends on food availability. *Journal of Experimental Biology* 210:2361–2367.

Buchanan, K. L. 2000. Stress and the evolution of condition-dependent signals. *Trends in Ecology and Evolution* 15:156–160.

Buchanan, K. L., S. Leitner, K. A. Spencer, A. R. Goldsmith, and C. K. Catchpole. 2004. Developmental stress selectively affects the song control nucleus HVC in the zebra finch. *Proceedings of the Royal Society of London Series B* 271:2381–2386.

Buchanan, K. L., K. A. Spencer, A. R. Goldsmith, and C. K. Catchpole. 2003. Song as an honest signal of past developmental stress in the European starling (*Sturnus vulgaris*). *Proceedings of the Royal Society of London Series B* 270:1149–1156.

Bugnyar, T., and B. Heinrich. 2005. Ravens, *Corvus corax*, differentiate between knowledgeable and ignorant competitors. *Proceedings of the Royal Society of London Series B* 272:1641–1646.

———. 2006. Pilfering ravens, *Corvus corax*, adjust their behaviour to social context of competitors. *Animal Cognition* 9:369–376.

Bugnyar, T., M. Kijne, and K. Kotrschal. 2001. Food calling in ravens: Are yells referential signals? *Animal Behaviour* 61:949–958.

Bugnyar, T., and K. Kotrschal. 2002. Scrounging tactics in free-ranging ravens, *Corvus corax*. *Ethology* 108:993–1009.

———. 2004. Leading a conspecific away from food in ravens (*Corvus corax*)? *Animal Cognition* 7:69–76.

Bugnyar, T., C. Schwab, C. Schloegl, K. Kotrschal, and B. Heinrich. 2007. Ravens judge competitors through experience with play caching. *Current Biology* 17:1804–1808.

Buonomano, D. V., and M. M. Merzenich. 1998. Cortical plasticity: From synapses to maps. *Annual Review of Neuroscience* 21:149–186.

Burger, J. 1998. Antipredator behaviour of hatchling snakes: Effects of incubation temperature and simulated predators. *Animal Behaviour* 56:547–553.

Burkhardt, J., and W. Richard. 2005. Patterns of Behavior: Konrad Lorenz, Niko Tinbergen, and the Founding of Ethology. Chicago: University of Chicago Press.

Burt, J. M., S. C. Bard, S. E. Campbell, and M. D. Beecher. 2002. Alternative forms of song matching in song sparrows. *Animal Behaviour* 63:1143–1151.

Burt, J. M., S. E. Campbell, and M. D. Beecher. 2001. Song type matching as threat: A test using interactive playback. *Animal Behaviour* 62:1163–1170.

Burt, J. M., A. L. O'Loghlen, C. N. Templeton, S. E. Campbell, and M. D. Beecher. 2007. Assessing the importance of social factors in bird song learning: A test using computer simulated tutors. *Ethology* 113:917–925.

Byrne, R. W., and N. Corp. 2004. Neocortex size predicts deception rate in primates. *Proceedings of the Royal Society of London Series B* 271:1693–1699.

Byrne, R. W., and A. Whiten, eds. 1988. Machiavellian Intelligence: Social Expertise and the Evolution of Intellect in Monkeys, Apes and Humans. Oxford: Clarendon Press.

Caldwell, H. K., H. J. Lee, A. H. Macbeth, and W. S. Young. 2008. Vasopressin: Behavioral roles of an "original" neuropeptide. *Progress in Neurobiology* 84:1–24.

Caldwell, M. S., J. G. McDaniel, and K. M. Warkentin. 2009. Frequency information in the vibration-cued escape hatching of red-eyed treefrogs. *Journal of Experimental Biology* 212:566–575.

———. Submitted. Is it safe? Characteristics of benign stimuli aid in the discrimination of predator cues.

Campbell, B. A. 1976. Feeding habits of the kea in the Routeburn Basin. Thesis, University of Otago, Dunedin.

Campbell, L. F., and K. S. Bedi. 1989. The effects of undernutrition during early life on spatial learning. *Physiology and Behavior* 45:883–890.

Candolin, U. 1998. Reproduction under predation risk and the trade-off between current and future reproduction in the threespine stickleback. *Proceedings of the Royal Society of London Series B* 265:1171–1175.

Capellán, E., and A. G. Nicieza. 2007. Trade-offs across life stages: Does predator-induced hatching plasticity reduce anuran post-metamorphic performance? *Evolutionary Ecology* 21:445–458.

Capranica, R. R. 1966. Vocal response of the bullfrog to natural and synthetic mating calls. *Journal of the Acoustical Society of America* 40:1131–1139.

Cardinal, R. N., J. A. Parkinson, J. Hall, and B. J. Everitt. 2002. Emotion and motivation: The role of the amygdala, ventral striatum, and prefrontal cortex. *Neuroscience and Biobehavioral Reviews* 26:321–352.

Cardone, B., and J. H. Fullard. 1988. Auditory characteristics and sexual dimorphism in the gypsy moth. *Physiological Entomology* 13:9–14.

Caro, T. M. 1994. Cheetahs of the Serengeti Plains: Group Living in an Asocial Species. Chicago, University of Chicago Press.

Carroll, E. J., and J. L. Hedrick. 1974. Hatching in the toad *Xenopus laevis:* Morphological events and evidence for a hatching enzyme. *Developmental Biology* 38:1–13.

Carter, C. S., L. L. Getz, L. Gavish, J. L. McDermott, and P. Arnold. 1980. Male-related pheromones and the activation of female reproduction in the prairie vole (*Microtus ochrogaster*). *Biology of Reproduction* 23:1038–1045.

Cash, A. C., C. Whitfield, N. Ismail, and G. E. Robinson. 2005. Behavior and the limits of genomic plasticity: Power and replicability in microarray analysis of honeybee brains. *Genes, Brain and Behavior* 4:267–271.

Cassidy, A. L. E. V. 1993. Song variation and learning in island populations of song sparrows. Thesis, University of British Columbia.

Castellanos, I., and P. Barbosa. 2006. Evaluation of predation risk by a caterpillar using substrate-borne vibrations. *Animal Behaviour* 72:461–469.

Catchpole, C. K. 1986. Song repertoires and reproductive success in the great reed warbler *Acrocephalus arundinaceus*. *Behavioral Ecology and Sociobiology* 19:439–445.

Catchpole, C. K., J. Dittami, and B. Leisler. 1984. Differential responses to male song repertoires in female songbirds implanted with oestradiol. *Nature* 312:563–564.

Catchpole, C. K., and P. J. B. Slater. 1995. *Bird Song: Biological Themes and Variations*. Cambridge: Cambridge University Press.

Catts, E. P. 1982. Biology of the new world bot flies: Cuterebridae. *Annual Review of Entomology* 27:313–338.

Cavalli-Sforza, L. L., and M. W. Feldman. 1981. *Cultural Transmission and Evolution: A Quantitative Approach*. Princeton: Princeton University Press.

Chalcraft, D. R., and R. M. Andrews. 1999. Predation on lizard eggs by ants: Species interactions in a variable physical environment. *Oecologia* 119:285–292.

Changizi, M. 2003. Relationship between number of muscles, behavioral repertoire size, and encephalization in mammals. *Journal of Theoretical Biology* 220:157–168.

Charlesworth, B., and J. A. Leon. 1976. The relation of reproductive effort to age. *American Naturalist* 110:449–459.

Cheney, D. L., and R. M. Seyfarth. 1990. *How Monkeys See the World.* Chicago: University of Chicago Press.

Cherry, E. 1953. Some experiments on the recognition of speech, with one and two ears. *Journal of the Acoustical Society of America* 25:975–979.

Cheverud, J. M., D. Falk, M. Vannier, L. Konigsberg, R. C. Helmkamp, and C. Hildebolt. 1990. Heritability of brain size and surface features in rhesus macaques (*Macaca mulatta*). *Journal of Heredity* 81:51–57.

Chivers, D. P., J. M. Kiesecker, A. Marco, J. DeVito, M. T. Anderson, and A. R. Blaustein. 2001. Predator-induced life-history changes in amphibians: Egg predation induces hatching. *Oikos* 92:135–142.

Christy, J. H. 1995. Mimicry, mate choice, and the sensory trap hypothesis. *American Naturalist* 146:171–181.

———. 2003. Reproductive timing and larval dispersal of intertidal crabs: The predator avoidance hypothesis. *Revista Chilena de Historia Natural* 76:177–185.

Clare, A. S. 1997. Eicosanoids and egg-hatching synchrony in barnacles: Evidence against a dietary precursor to egg-hatching pheromone. *Journal of Chemical Ecology* 23:2299–2312.

Clare, A. S., G. Walker, D. L. Holland, and D. J. Crisp. 1985. The hatching substance of the barnacle, *Balanus balanoides* (L.). *Proceedings of the Royal Society of London Series B* 224:131–147.

Clark, C. W., and R. Dukas. 2003. The behavioral ecology of a cognitive constraint: Limited attention. *Behavioral Ecology* 14:151–156.

Clayton, N. S. 1990. Assortative mating in zebra finch subspecies, Taeniopygia guttata guttata and T. g. castanotis. Philosophical Transactions of the Royal Society of London Series B 330:351–370.

———. 1996. Development of food-storing and the hippocampus in juvenile marsh tits (*Parus palustris*). *Behavioral Brain Research* 74:153–159.

———. 2001. Hippocampal growth and maintenance depend on food-caching experience in juvenile mountain chickadees (*Poecile gambeli*). *Behavioral Neuroscience* 115:614–625.

Clayton, N. S., and A. Dickinson. 1998. Episodic-like memory during cache recovery by scrub jays. *Nature* 395:272–278.

Clayton, N. S., and N. J. Emery. 2007. The social life of corvids. *Current Biology* 17:652–656.

Clayton, N. S., D. P. Griffiths, N. J. Emery, and A. Dickinson. 2001. Elements of episodic-like memory in animals. *Philosophical Transactions of the Royal Society of London Series B* 356:1483–1491.

Clayton, N. S., and J. R. Krebs. 1994a. Hippocampal growth and attrition in birds affected by experience. *Proceedings of the National Academy of Sciences of the United States of America* 91:7410–7414.

———. 1994b. Memory for spatial and object-specific cues in food-storing and non-storing birds. *Journal of Comparative Physiology A* 174:371–379.

Clayton, N. S., K. S. Yu, and A. Dickinson. 2003. Interacting cache memories: Evidence of flexible memory use in scrub jays. *Journal of Experimental Psychology: Animal Behavior Processes* 29:14–22.

Clutton-Brock, T. H. 1984. Reproductive effort and terminal investment in iteroparous animals. *American Naturalist* 123:212–229.

Clutton-Brock, T. H., P. N. M. Brotherton, M. J. O'Riain, A. S. Griffin, D. Gaynor, R. Kansky, L. Sharpe, and G. M. McIlrath. 2001. Contributions to cooperative rearing in meerkats. *Animal Behaviour* 61:705–710.

Clutton-Brock, T. H., D. Gaynor, R. Kansky, A. D. C. MacColl, G. McIlrath, P. Chadwick, P. N. M. Brotherton, J. M. O'Riain, M. Manser, and J. D. Skinner. 1998. Costs of cooperative behaviour in suricates (*Suricata suricatta*). *Proceedings of the Royal Society of London Series B* 265:185–190.

Clutton-Brock, T. H., D. Gaynor, G. M. McIlrath, A. D. C. Maccoll, R. Kansky, P. Chadwick, M. Manser et al. 1999a. Predation, group size and mortality in a cooperative mongoose, *Suricata suricatta*. *Journal of Animal Ecology* 68:672–683.

Clutton-Brock, T. H., and P. H. Harvey. 1980. Primates, brains and ecology. *Journal of Zoology* 190:309–323.

Clutton-Brock, T. H., and K. McComb. 1993. Experimental tests of copying and mate choice in fallow deer (*Dama dama*). *Behavioral Ecology* 4:191–193.

Clutton-Brock, T. H., M. J. O'Riain, P. N. M. Brotherton, D. Gaynor, R. Kansky, A. S. Griffin, and M. Manser. 1999b. Selfish sentinels in cooperative mammals. *Science* 284:1640–1644.

Coe, C. L., M. Kramer, B. Czeh, E. Gould, A. J. Reeves, C. Kirschbaum, and E. Fuch. 2003. Prenatal stress diminishes neurogenesis in the dentate gyrus of juvenile rhesus monkeys. *Biological Psychiatry* 54:1025–1034.

Cogley, T. P., and M. C. Cogley. 1989. Morphology of the eggs of the human bot fly, *Dermatorbia homins* (L. Jr.) (Diptera: Cuterebridae) and their adherence to the transport carrier. *International Journal of Insect Morphology and Embryology* 18:239–248.

Cohen, J. A., and L. P. Brower. 1983. Cardenolide sequestration by the dogbane tiger moth. *Journal of Chemical Ecology* 9:521–531.

Collins, S. A. 1995. The effect of recent experience on female choice in zebra finches. *Animal Behaviour* 49:479–486.

Conner, W. E. 1999. "Un chant d'appel amoureux": Acoustic communication in moths. *Journal of Experimental Biology* 202:1711–1723.

Coolen, I., Y. van Bergen, R. L. Day, and K. N. Laland. 2003. Species difference in adaptive use of public information in sticklebacks. *Proceedings of the Royal Society of London Series B* 270:2413–2419.

Coolen I., A. J. W. Ward, P. J. B. Hart, and K. N. Laland. 2005. Foraging nine-spined sticklebacks prefer to rely on public information over simpler social cues. *Behavioral Ecology* 16:865–870.

Cooper, B. G., T. F. Manka, and S. J. Y. Mizumori. 2001. Finding your way in the dark: The retrosplenial cortex contributes to spatial memory and navigation without visual cues. *Behavioral Neuroscience* 115:1012–1028.

Correia, S., P. C., A. Dickinson, and N. S. Clayton. 2007. Western scrub-jays anticipate future needs independently of their current motivational state. *Current Biology* 17:856–861.

Coss, R. G., J. G. Brandon, and A. Globus. 1980. Changes in morphology of dendritic spines on honeybee calycal interneurons associated with cumulative nursing and foraging experience. *Brain Research* 192:49–59.

Coussi-Korbel, S., and D. M. Fragaszy. 1995. On the relation between social dynamics and social learning. *Animal Behaviour* 50:1441–1453.

Cramp, S. 1977. Handbook of the Birds of Europe, the Middle East and North Africa: The Birds of the Western Palearctic. Oxford: Oxford University Press.

Crisp, D. J., and C. P. Spencer. 1958. The control of the hatching process in barnacles. *Proceedings of the Royal Society of London Series B* 149:278–299.

Cristol, D. A., Reynolds, E. B., Leclerc, J. E., Donner, A. H., Farabaugh, and C. W. S., Ziegenfus. 2003. Migratory dark-eyed juncos, *Junco hyemalis,* have better spatial memory and denser hippocampal neurons than nonmigratory conspecifics. *Animal Behaviour* 66:317–328.

Cummings, M. E., F. J. García de León, D. M. Mollaghan, and M. J. Ryan. 2006. Is UV ornamentation an amplifier in swordtails? *Zebrafish* 3:91–100.

Custance, D., A. Whiten, and T. Fredman. 1999. Social learning of an artificial fruit task in capuchin monkeys. *Journal of Comparative Psychology* 113:13–23.

Cynx, J., R. Lewis, B. Tavel, and H. Tse. 1998. Amplitude regulation of vocalizations in noise by a songbird, *Taeniopygia guttata. Animal Behaviour* 56:107–113.

Dall, S. R. X., L.-A. Giraldeau, O. Olsson, J. M. McNamara, and D. W. Stephens. 2005. Information and its use by animals in evolutionary ecology. *Trends in Ecology and Evolution* 20:187–193.

Dally, J. M., N. S. Clayton, and N. J. Emery. 2006a. The behaviour and evolution of cache protection and pilferage. *Animal Behaviour* 72:13–23.

Dally, J. M., N. J. Emery, and N. S. Clayton. 2004. Cache protection strategies in western scrub-jays (*Aphelocoma californica*): Hiding food in the shade. *Proceedings of the Royal Society of London Series B* 271:387–390.

———. 2005. Cache protection strategies by western scrub-jays, *Aphelocoma californica:* Implications for social cognition. *Animal Behaviour* 70:1251–1263.

———. 2006b. Food-caching western scrub-jays keep track of who was watching when. *Science* 312:1662–1665.

Danchin, E., L.-A. Giraldeau, T. J. Valone, and R. H. Wagner. 2004. Public information: From nosey neighbours to cultural evolution. *Science* 305:487–491.

Dantzer, R. 1998. Vasopressin, gonadal steroids and social recognition. *Progress in Brain Research* 119:409–14.

Dantzer, R., G. F. Koob, R. M. Bluthe, and M. Lemoal. 1988. Septal vasopressin modulates social memory in male rats. *Brain Research* 457:143–147.

Darwin, C. 1859. The Origin of Species by Means of Natural Selection. London: John Murray.

———. 1871. The Descent of Man, and Selection in Relation to Sex. London: John Murray.

———. 1872. The Expression of the Emotions in Man and Animals. London: John Murray.

Dauncey, M. J., and Bicknell, R. J. 1999. Nutrition and neurodevelopment: Mechanisms of developmental dysfunction and disease in later life. *Nutrition Research Reviews* 12:231–253.

Daunt, F., V. Afanasyev, A. Adam, J. P. Croxall, and S. Wanless. 2007. From cradle to early grave: Juvenile mortality in European shags *Phalacrocorax aristotelis* results from inadequate development of foraging proficiency. *Biology Letters* 3:371–374.

Davis, R. L. 2005. Olfactory memory formation in *Drosophila:* From molecular to systems neuroscience. *Annual Review of Neuroscience* 28:275–302.

Dawson, D. W., and J. H. Fullard. 1995. The neuroethology of sound production in tiger moths (Arctiidae). II. Location of the thoracic circuitry controlling the tymbal response in *Cycnia tenera* Hubner. *Journal of Comparative Physiology A* 176:541–549.

Day, R. L. 2003. Innovation and social learning in monkeys and fish: Empirical findings and their application to reintroduction techniques. Thesis, Cambridge University.

Day, R. L., R. L. Coe, J. R. Kendal, and K. N. Laland. 2003. Neophilia, innovation and social learning: A study of intergeneric differences in callitrichid monkeys. *Animal Behaviour* 65: 559–571.

Day, R. L., T. MacDonald, C. Brown, K. N. Laland, and S. M. Reader. 2001. Interactions between shoal size and conformity in guppy social foraging. *Animal Behaviour* 62:917–925.

Deaner, R. O., R. A. Barton, and C. P. Svan Schaik. 2003. Primate brains and life histories: Renewing the connection. In P. M. Kappeler and M. E. Pereira, eds., *Primate Life Histories and Socioecology*, 233–265. Chicago: University of Chicago Press.

Deaner, R. O., K. Isler, J. Burkart, and C. P. van Schaik. 2007. Overall brain size, and not encephalization quotient, best predicts cognitive ability across non-human primates. *Brain, Behavior and Evolution* 70:115–124.

Deaner, R. O., and C. L. Nunn. 1999. How quickly do brains catch up with bodies? A comparative method for detecting evolutionary lag. *Proceedings of the Royal Society of London Series B* 266:687–694.

Deaner, R. O., C. P. Van Schaik, and V. Johnson. 2006. Do some taxa have better domain-general cognition than others? A meta-analysis of nonhuman primate studies. *Evolutionary Psychology* 4:149–196.

deBelle, J. S., and M. Heisenberg. 1994. Associative odor learning in *Drosophila* abolished by chemical ablation of mushroom bodies. *Science* 263:692–695.

de Bono, M., and A. V. Maricq. 2005. Neuronal substrates of complex behaviors in *C. elegans*. *Annual Review of Neuroscience* 28:451–501.

de Kort, S. R., and N. S. Clayton. 2006. An evolutionary perspective on caching by corvids. *Proceedings of the Royal Society of London Series B* 273:417–423.

De Roeck, E. R. M., T. Artois, and L. Brendonck. 2005. Consumptive and non-consumptive effects of turbellarian (*Mesostoma* sp.) predation on anostracans. *Hydrobiologia* 542:103–111.

Detto, T., P. R. Y. Backwell, J. M. Hemmi, and J. Zeil. 2006. Visually mediated species and neighbour recognition in fiddler crabs (*Uca mjoebergi* and *Uca capricomis*). *Proceedings of the Royal Society of London Series B* 273:1661–1666.

DeVoogd, T. J., J. R. Krebs, S. D. Healy, and A. Purvis. 1993. Relations between song repertoire size and the volume of brain nuclei related to song: Comparative evolutionary analyses amongst oscine birds. *Proceedings of the Royal Society of London Series B* 254:75–82.

de Vries, G. J., and M. A. Miller. 1998. Anatomy and function of extrahypothalamic vasopressin systems in the brain. *Progress in Brain Research* 119:3–20.

DeWied, D. 1971. Long term effect of vasopressin on maintenance of a conditioned avoidance response in rats. *Nature* 232:58–60.

DeWitt, T. J., and S. M. Scheiner. 2004. *Phenotypic Plasticity: Functional and Conceptual Approaches*. Oxford: Oxford University Press.

Diamond, J., and A. Bond. 1991. Social behaviour and the ontogeny of foraging in the kea (*Nestor notabilis*). *Ethology* 88:128–144.

———. 1999. Kea, Bird of Paradox: The Evolution and Behavior of a New Zealand Parrot. Berkeley and Los Angeles: University of California Press.

———. 2004. Social play in kaka (*Nestor meridionalis*) with comparisons to kea (*Nestor notabilis*). *Behaviour* 141:777–798.

Diamond, J., D. Eason, C. Reid, and A. Bond. 2006. Social play in kakapo (*Strigops habroptilus*) with comparisons to kea (*Nestor notabilis*) and kaka (*Nestor meridionalis*). *Behaviour* 143:1397–1423.

Diekamp, B. M., and H. C. Gerhardt. 1992. Midbrain auditory sensitivity in the spring peeper (*Pseudacris crucifer*): Correlations with behavioral studies. *Journal of Comparative Physiology A* 171:245–250.

DiMichele, L., and M. H. Taylor. 1980. The environmental control of hatching in *Fundulus heteroclitus*. *Journal of Experimental Zoology* 214:181–187.

Dluzen, D. E., V. D. Ramirez, C. S. Carter, and L. L. Getz. 1981. Male vole urine changes luteinizing hormone releasing hormone and norepinephrine in female olfactory bulb. *Science* 212:573–575.

Doligez B., C. Cadet, E. Danchin, and T. Boulinier. 2003. When to use public information for breeding habitat selection: The role of environmental predictability and density dependence. *Animal Behaviour* 66:973–988.

Doody, J. S., A. Georges, J. E. Young, M. D. Pauza, A. L. Pepper, R. L. Alderman, and M. A. Welsh. 2001. Embryonic aestivation and emergence behaviour in the pig-nosed turtle, *Carettochelys insculpta*. *Canadian Journal of Zoology—Revue Canadienne de Zoologie* 79:1062–1072.

Doolan, S. P., and D. W. Macdonald. 1996. Diet and foraging behaviour of group-living meerkats, *Suricata suricatta*, in the southern Kalahari. *Journal of Zoology* 239:697–716.

Draganoiu, T. I., L. Nagle, and M. Kreutzer. 2002. Directional female preference for an exaggerated male trait in canary (*Serinus canaria*) song. *Proceedings of the Royal Society of London Series B* 269:2525–2531.

Drake, J. M. 2007. Parental investment and fecundity, but not brain size, are associated with establishment success in introduced fishes. *Functional Ecology* 21:963–968.

Dreisig, H. 1986. Timing of daily activities in adult Lepidoptera. *Entomologia Generalis* 12:25–43.

Drickamer, L. C. 1996. Intra-uterine position and anogenital distance in house mice: Consequences under field conditions. *Animal Behaviour* 51:925–934.

Drickamer, L. C., F. S. vomSaal, L. M. Marriner, and C. A. Mossman. 1995. Anogenital distance and dominance status in male house mice (*Mus domesticus*). *Aggressive Behavior* 21:301–309.

Dubnau, J., A.-S. Chiang, and T. Tully. 2003. Neural substrates of memory: From synapse to system. *Journal of Neurobiology* 54:238–253.

Duellman, W. E. 2001. *The Hylid Frogs of Middle America: Contributions to Herpetology*. Vol. 18. St. Louis: Society for the Study of Amphibians and Reptiles.

Dugatkin, L. A. 1992. Sexual selection and imitation: Females copy the mate choice of others. *American Naturalist* 139:1384–1389.

———. 1996. Interface between culturally based preferences and genetic preferences: Female mate choice in *Poecilia reticulata*. *Proceedings of the National Academy of Sciences of the United States of America* 93:2770–2773.

———. 2007. Developmental environment, cultural transmission, and mate choice copying. *Naturwissenschaften* 94:651–656.

Dugatkin, L. A., and J. G. J. Godin. 1992. Reversal of female mate choice by copying in the guppy (*Poecilia reticulata*). *Proceedings of the Royal Society of London Series B* 249:179–184.

Dukas, R. 1998a. Cognitive ecology: The evolutionary ecology of information processing and decision making. Chicago: University of Chicago Press.

———. 1998b. Constraints on information processing and their effects on behavior. In R. Dukas, ed., *Cognitive Ecology*, 89–127. Chicago: University of Chicago Press.

———. 1998c. Evolutionary ecology of learning. In R. Dukas, ed., *Cognitive Ecology*, 129–174. Chicago: University of Chicago Press.

———. 1998d. Introduction. In R. Dukas, ed., *Cognitive Ecology*, 1–19. Chicago: University of Chicago Press.

———. 1998e. Prospects. In R. Dukas, ed., *Cognitive Ecology*, 405–409. Chicago: University of Chicago Press.

———. 1999a. Costs of memory: Ideas and predictions. *Journal of Theoretical Biology* 197:41–50.

———. 1999b. Ecological relevance of associative learning in fruit fly larvae. *Behavioral Ecology and Sociobiology* 45:195–200.

———. 2002. Behavioural and ecological consequences of limited attention. *Philosophical Transactions of the Royal Society of London Series B* 357:1539–1548.

——. 2004a. Evolutionary biology of animal cognition. *Annual Review of Ecology, Evolution, and Systematics* 35:347–374.
——. 2004b. Male fruit flies learn to avoid interspecific courtship. *Behavioral Ecology* 15:695–698.
——. 2005a. Experience improves courtship in male fruit flies. *Animal Behaviour* 69:1203–1209.
——. 2005b. Learning affects mate choice in female fruit flies. *Behavioral Ecology* 16:800–804.
——. 2006. Learning in the context of sexual behavior in insects. *Animal Biology* 56:125–141.
——. 2008a. Evolutionary biology of insect learning. *Annual Review of Entomology* 53:387–408.
——. 2008b. Learning decreases heterospecific courtship and mating in fruit flies. *Biology Letters* 4:645–647.
——. 2008c. Life history of learning: Performance curves of honeybees in the wild. *Ethology* 114:1195–1200.
——. 2008d. Life history of learning: Short and long term performance curves of honeybees in settings that minimize the role of learning. *Animal Behaviour* 75:1125–1130.
——. 2009. Dynamics of learning in the context of courtship in *Drosophila persimilis* and *D. pseudoobscura*. *Animal Behaviour* 77:253–259.
Dukas, R., and E. A. Bernays. 2000. Learning improves growth rate in grasshoppers. *Proceedings of the National Academy of Sciences of the United States of America* 97:2637–2640.
Dukas, R., and P. K. Visscher. 1994. Lifetime learning by foraging honey bees. *Animal Behaviour* 48:1007–1012.
Dunbar, R. I. M. 1992. Neocortex size as a constraint on group size in primates. *Journal of Human Evolution* 20:469–493.
Dunbar, R. I. M., and S. Shultz. 2007a. Evolution in the social brain. *Science* 317:1344–1347.
——. 2007b. Understanding primate brain evolution. Philosophical Transactions of the Royal Society of London Series B 362:649–658.
Dunning, D. C. 1968. Warning sounds of moths. *Zeitschrift für Tierpsychologie* 25:129–138.
Dunning, D. C., L. Acharya, C. B. Merriman, and L. D. Ferro. 1992. Interactions between bats and arctiid moths. *Canadian Journal of Zoology—Revue Canadienne de Zoologie* 55:1213–1224.
Dunning, D. C., and K. D. Roeder. 1965. Moth sounds and the insect catching behavior of bats. *Science* 147:173–174.
Durst, C., S. Eichmuller, and R. Menzel. 1994. Development and experience lead to increased volume of subcompartments of the honeybee mushroom body. *Behavioral and Neural Biology* 62:259–263.
Eales, L. A. 1985. Song learning in zebra finches: Some effects of model availability on what is learnt and when. *Animal Behaviour* 33:1293–1300.
Eckert, R. 1972. Bioelectric control of ciliary activity. *Science* 176:473–481.
Edmunds, M. 1974. Defence in Animals: A Survey of Anti-predator Defences. Burnt Mill, UK: Longman.
Eens, M., R. Pinxten, and R. F. Verheyen. 1991. Male song as a cue for mate choice in the European starling. *Behaviour* 116:210–238.
Efferson C., R. Lalive, P. J. Richerson, R. McElreath, and M. Lubell. 2008. Conformists and mavericks: The empirics of frequency-dependent cultural transmission. *Evolution and Human Behavior* 29:56–64.
Egashira, N., A. Tanoue, F. Higashihara, K. Mishima, Y. Fukue, Y. Takano, G. Tsujimoto, K. Iwasakia, and M. Fujiwara. 2004. V1a receptor knockout mice exhibit impairment of spatial memory in an eight-arm radial maze. *Neuroscience Letters* 356:195–198.

Ehmer, B., and W. Gronenberg. 2002. Segregation of visual input to the mushroom bodies in the honeybee (*Apis mellifera*). *Journal of Comparative Neurology* 451:362–373.

Ehmer, B., and R. R. Hoy. 2000. Mushroom bodies of vespid wasps. *Journal of Comparative Neurology* 416:93–100.

Ehmer, B., H. K. Reeve, and R. R. Hoy. 2001. Comparison of brain volumes between single and multiple foundresses in the paper wasp *Polistes dominulus*. *Brain, Behavior, and Evolution* 57:161–168.

Ehret, G. 1987. Categorical perception of sound signals: Facts and hypotheses from animal studies. In S. Harnad, ed., *Categorical Perception: The Groundwork of Cognition*, 301–331. Cambridge: Cambridge University Press.

Ehret, G., and B. Haack. 1981. Categorical perception of mouse pup ultrasounds by lactating females. *Naturwissenschaften* 68:208.

Eisenbach, M., and J. W. Lengeler. 2004. *Chemotaxis*. London: Imperial College Press.

Ejima, A., B. P. C. Smith, C. Lucas, J. D. Levine, and L. C. Griffith. 2005. Sequential learning of pheromonal cues modulates memory consolidation in trainer-specific associative courtship conditioning. *Current Biology* 15:194–206.

Eklöf, J., A. M. Svensson, and J. Rydell. 2002. Northern bats, *Eptesicus nilssonii*, use vision but not flutter-detection when searching for prey in clutter. *Oikos* 99:347–351.

Eliassen, S., C. Jørgensen, M. Mangel, and J. Giske. 2007. Exploration or exploitation: Life expectancy changes the value of learning in foraging strategies. *Oikos* 116:513–523.

Ellis, J. D., Jr., R. Hepburn, K. S. Delaplane, P. Neurmann, and P. J. Elzen. 2003. The effects of small hive beetles, *Arthina tumida* (Coleoptera: Nitidulidae), on nests and flight activity of Cape and European honey bees (*Apis mellifera*). *Apidologie* 34:399–408.

Emery, N. J., and N. S. Clayton. 2001. Effects of experience and social context on prospective caching strategies by scrub jays. *Nature* 414:443–446.

———. 2008. How to build a scrub-jay that reads minds. In S. Itakura and K. Fujita, eds., *Origins of the Social Mind*, 65–97. Tokyo: Springer.

Emery, N. J., J. M. Dally, and N. S. Clayton. 2004. Western scrub-jays (*Aphelocoma californica*) use cognitive strategies to protect their caches from thieving conspecifics. *Animal Cognition* 7:37–43.

Emery, N. J., A. M. Seed, A. M. P. von Bayern, and N. S. Clayton. 2007. Cognitive adaptions of social bonding in birds. *Philosophical Transactions of the Royal Society of London Series B* 362:489–505.

Emlen, S. T., and L. W. Oring. 1977. Ecology, sexual selection, and evolution of mating systems. *Science* 197:215–223.

Endler, J. A. 1992. Signals, signal conditions, and the direction of evolution. *American Naturalist* 139:S125–S153.

Endler, J. A., D. A. Westcott, J. R. Madden, and R. Robson. 2005. Animal visual systems and the evolution of color patterns; sensory processing illuminates signal evolution. *Evolution* 59:1795–1818.

Enggist-Dueblin, E., and U. Pfister. 2002. Cultural transmission of vocalizations in ravens, *Corvus corax*. *Animal Behaviour* 64:831–841.

Enquist, M., and S. Ghirlanda. 2005. *Neural Networks and Animal Behavior*. Princeton: Princeton University Press.

———. 2007. Evolution of imitation does not explain the origin of human cumulative culture. *Journal of Theoretical Biology* 246:129–135.

Erber, J., T. Masuhr, and R. Menzel. 1980. Localization of short-term memory in the brain of the bee, *Apis mellifera*. *Physiological Entomology* 5:343–358.

Ericsson, K. A., N. Charness, P. J. Feltovich, and R. R. Hoffman. 2006. *The Cambridge Handbook of Expertise and Expert Performance*. Cambridge: Cambridge University Press.

Eriksson K., M. Enquist, and S. Ghirlanda. 2007. Critical points in current theory of conformist social learning. *Journal of Evolutionary Psychology* 5:67–87.

Estes, J. A., M. T. Tinker, T. M. Williams, and D. F. Doak. 1998. Killer whale predation on sea otters linking oceanic and nearshore ecosystems. *Science* 282:473–476.

Etscorn, F. 1973. Effects of a preferred versus nonpreferred CS in the establishment of a taste aversion. *Physiological Psychology* 1:5–6.

Evans, C. S. 1997. Referential signals. In D. H. Owings, M. D. Beecher, and N. S. Thompson, eds., *Perspectives of Ethology*, 99–143. New York: Plenum Press.

Evans, C. S., and L. Evans. 2007. Representational signalling in birds. *Biology Letters* 3:8–11.

Evans, C. S., L. Evans, and P. Marler. 1993a. On the meaning of alarm calls: Functional reference in an avian vocal system. *Animal Behaviour* 46:23–38.

Evans, C. S., J. M. Macedonia, and P. Marler. 1993b. Effects of apparent size and speed on the response of chickens, *Gallus gallus*, to computer-generated simulations of aerial predators. *Animal Behaviour* 46:1–11.

Evans, P. D., S. L. Gilbert, N. Mekel-Bobrov, E. J. Vallender, J. R. Anderson, L. M. Vaez-Azizi, S. A. Tishkoff, R. R. Hudson, and B. T. Lahn. 2005. Microcephalin, a gene regulating brain size, continues to evolve adaptively in humans. *Science* 309:1717–1720.

Everitt, B. J., and T. W. Robbins. 2005. Neural systems of reinforcement for drug addiction: From actions to habits to compulsion. *Nature Neuroscience* 8:1481–1489.

Everts, H. G. J., A. J. H. DeRuiter, and J. M. Koolhaas. 1997. Differential lateral septal vasopressin in wild-type rats: Correlation with aggression. *Hormones and Behavior* 31:136–144.

Fabre, J.-H., A. Teixeira de Mattos, and B. Miall. 1918. *The Wonders of Instinct*. New York: Century Co.

Fagen, R. M. 1972. An optimal life history strategy in which reproductive effort decreases with age. *American Naturalist* 106:258–261.

Fahrbach, S. E. 2006. Structure of the mushroom bodies of the insect brain. *Annual Review of Entomology* 51:209–232.

Fahrbach, S. E., T. Giray, S. M. Farris, and G. E. Robinson. 1997. Expansion of the neuropil of the mushroom bodies in male honey bees is coincident with initiation of flight. *Neuroscience Letters* 236:135–138.

Fahrbach, S. E., T. Giray, and G. E. Robinson. 1995a. Volume changes in the mushroom bodies of adult honey bee queens. *Neurobiology of Learning and Memory* 63:181–191.

Fahrbach, S. E., D. Moore, E. A. Capaldi, S. M. Farris, and G. E. Robinson. 1998. Experience-expectant plasticity in the mushroom bodies of the honeybee. *Learning and Memory* 5:115–123.

Fahrbach, S. E., and G. E. Robinson. 1995. Behavioral development in the honey bee: Toward the study of learning under natural conditions. *Learning and Memory* 2:199–224.

Fahrbach, S. E., J. L. Strande, and G. E. Robinson. 1995b. Neurogenesis is absent in the brains of adult honey bees and does not explain behavioral neuroplasticity. *Neuroscience Letters* 197:145–148.

Falls, J. B. 1985. Song matching in western meadowlarks. *Canadian Journal of Zoology—Revue Canadienne de Zoologie* 63:2520–2524.

Falls, J. B., J. R. Krebs, and P. McGregor. 1982. Song matching in the great tit (*Parus major*): The effect of similarity and familiarity. *Animal Behavior* 30:997–1009.

Farooqui, T., K. Robinson, H. Vaessin, and B. H. Smith. 2003. Modulation of early olfactory processing by an octopaminergic reinforcement pathway in the honeybee. *Journal of Neuroscience* 23:5370–5380.

Farris, H. E., A. S. Rand, and M. J. Ryan. 2002. The effects of spatially separated call components on phonotaxis in túngara frogs: Evidence for auditory grouping. *Brain, Behavior and Evolution* 60:181–188.

———. 2005. The effect of time, space and spectrum on auditory grouping in túngara frogs. *Journal of Comparative Physiology A* 191:1173–1183.

Farris, S. M., G. E. Robinson, and S. E. Fahrbach. 2001. Experience- and age-related outgrowth of intrinsic neurons in the mushroom bodies of the worker honeybee. *Journal of Neuroscience* 21:6395–6404.

Farris, S. M., and N. S. Roberts. 2005. Coevolution of generalist feeding ecologies and gyrencephalic mushroom bodies in insects. *Proceedings of the National Academy of Sciences of the United States of America* 102:17394–17399.

Faure, P. A., and R. M. R. Barclay. 1994. Substrate-gleaning versus aerial hawking: Plasticity in the foraging and echolocation behavior of the long-eared bat, *Myotis evotis*. *Journal of Comparative Physiology A* 174:651–660.

Faure, P. A., J. H. Fullard, and R. M. R. Barclay. 1990. The response of tympanate moths to the echolocation calls of a substrate gleaning bat, *Myotis evotis*. *Journal of Comparative Physiology A* 166:843–849.

Faure, P. A., J. H. Fullard, and J. W. Dawson. 1993. The gleaning attacks of the northern long-eared bat, *Myotis septentrionalis,* are relatively inaudible to moths. *Journal of Experimental Biology* 178:173–189.

Faure, P. A., and R. R. Hoy. 2000. The sounds of silence: Cessation of singing and song pausing are ultrasound-induced acoustic startle behaviors in the katydid *Neoconocephalus ensiger* (Orthoptera; Tettigoniidae). *Journal of Comparative Physiology A* 186:129–142.

Federspiel, I. G. 2006. Social and cognitive aspects of cooperation in keas (*Nestor notabilis*). Thesis, University of Vienna.

Federspiel, I. G., and N. J. Emery. In preparation. Social learning in jackdaws (*Corvus monedula*).

Fee, M. S., A. A. Kozhevnikov, and R. H. R. Hahnloser. 2004. Neural mechanisms of vocal sequence generation in the songbird. *Annals of the New York Academy of Science* 1016:153–170.

Feldman M. W., K. Aoki, and J. Kumm. 1996. Individual versus social learning: Evolutionary analysis in a fluctuating environment. *Anthropological Science* 104:209–232.

Feldman M. W., and K. N. Laland. 1996. Gene-culture coevolutionary theory. *Trends in Ecology and Evolution* 11:453–457.

Fenton, M. B. 1990. The foraging behavior and ecology of animal-eating bats. *Canadian Journal of Zoology—Revue Canadienne De Zoologie* 68:411–422.

———. 1995. Natural history and biosonar signals. In A. N. Popper and R. R. Hoy, eds., *Hearing in Bats,* 37–86. Berlin: Springer.

Fenton, M. B., D. Audet, M. K. Obrist, and J. Rydell. 1995. Signal strength, timing, and self-deafening—the evolution of echolocation in bats. *Paleobiology* 21:229–242.

Fenton, M. B., and G. P. Bell. 1981. Recognition of species of insectivorous bats by their echolocation calls. *Journal of Mammalogy* 62:233–243.

Fenton, M. B., and J. H. Fullard. 1979. Influence of moth hearing on bat echolocation strategies. *Journal of Comparative Physiology* 132:77–86.

Fenton, M. B., D. S. Jacobs, E. J. Richardson, P. J. Taylor, and E. White. 2004. Individual signatures in the frequency-modulated sweep calls of African large-eared, free-tailed bats *Otomops martiensseni* (Chiroptera: Molossidae). *Journal of Zoology* 262:11–19.

Ferris, C. F., and Y. Delville. 1994. Vasopressin and serotonin interactions in the control of agonistic behavior. *Psychoneuroendocrinology* 19:593–601.

Ferry-Graham, L. A., D. I. Bolnick, and P. C. Wainwright. 2002. Using functional morphology to examine the ecology and evolution of specialization. *Integrative and Comparative Biology* 42:265–267.

Fiala, A., U. Müller, and R. Menzel. 1999. Reversible downregulation of protein kinase A during olfactory learning using antisense technique impairs long-term memory formation in the honeybee, *Apis mellifera*. *Journal of Neuroscience* 19:10125–10134.

Fichtel, C., and P. M. Kappeler. 2002. Anti-predator behavior of group-living Malagasy primates: Mixed evidence for a referential alarm call system. *Behavioural Ecology and Sociobiology* 51:262–275.

Ficken, R. W., J. W. Popp, and P. E. Matthiae. 1985. Avoidance of acoustic interference by ovenbirds. *Wilson Bulletin* 116:569–571.

Fiedler, J. 1979. Prey catching with and without echolocation in the Indian false vampire (*Megaderma lyra*). *Behavioral Ecology and Sociobiology* 6:155–160.

Fink, S., L. Excoffier, and G. Heckel. 2006. Mammalian monogamy is not controlled by a single gene. *Proceedings of the National Academy of Sciences of the United States of America* 103:10956–10960.

Fischer, J., and K. Hammerschmidt. 2001. Functional referents and acoustic similarity revisited: The case of Barbary macaque alarm calls. *Animal Cognition* 4:29–35.

Fisher, H., A. Aron, and L. L. Brown. 2005. Romantic love: An fMRI study of a neural mechanism for mate choice. *Journal of Comparative Neurology* 493:58–62.

Fisher, J., and R. A. Hinde. 1949. The opening of milk bottles by birds. *British Birds* 42:347–357.

Fisher, M. O., R. G. Nager, and P. Monaghan. 2006. Compensatory growth impairs adult cognitive performance. *Public Library of Science Biology* 4:1462–1466.

Fishman, Y. I., J. C. Arezzo, and M. Steinschneider. 2004. Auditory stream segregation in monkey auditory cortex: Effects of frequency separation, retention rate, and tone duration. *Journal of the Acoustic Society of America* 116:1656–1670.

Fitzgerald, R. W., and D. M. Madison. 1983. Social organization of a free-ranging population of pine voles, *Microtus pinetorum*. *Behavioral Ecology and Sociobiology* 13:183–187.

Fitzpatrick, M. J., Y. Ben-Shahar, H. M. Smid, L. E. M. Vet, G. E. Robinson, and M. B. Sokolowski. 2005. Candidate genes for behavioural ecology. *Trends in Ecology and Evolution* 20:96–104.

Fleishman, L. J. 1992. The influence of the sensory system and the environment on motion patterns in the visual displays of anoline lizards and other vertebrates. *American Naturalist* 139: S36–61.

Folkers, E. 1982. Visual learning and memory of *Drosophila melanogaster* wild type C---S and the mutants dunce1, amnesiac, turnip and rutabaga. *Journal of Insect Physiology* 28:535–539.

Foote, J. R., and C. A. Barber. 2007. High level of song sharing found in an eastern song sparrow (*Melospiza melodia*) population. *Auk* 124:53–62.

Forshaw, J. M. 1977. *Parrots of the World*. Neptune: TFH Publications.

Fragaszy D. M., and E. Visalberghi. 1996. Social learning in monkeys: Primate "primacy" reconsidered. In C. M. Heyes and B. G. Galef Jr., eds., *Social Learning in Animals: The Roots of Culture*, 65–81. London: Academic Press.

———. 2004. Socially biased learning in monkeys. *Learning and Behavior* 32:24–35.

Franks, N. R., and T. Richardson. 2006. Teaching in tandem-running ants. *Nature* 439:153.

Frederiksen, J. K., and C. N. Slobodchikoff. 2007. Referential specificity in the alarm calls of the black-tailed prairie dog. *Ethology, Ecology and Evolution* 19:87–99.

Freeberg, T. M. 1998. The cultural transmission of courtship patterns in cowbirds, *Molothrus ater*. *Animal Behaviour* 56:1063–1073.

——. 2004. Social transmission of courtship behavior and mating preferences in brown-headed cowbirds, *Molothrus ater*. *Learning and Behavior* 32:122–130.

Freeberg, T. M., M. J. West, A. P. King, S. D. Duncan, and D. R. Sengelaub. 2002. Cultures, genes, and neurons in the development of song and singing in brown-headed cowbirds (*Molothrus ater*). *Journal of Comparative Physiology A* 188:993–1002.

Fritz, J., A. Bisenberger, and K. Kotrschal. 2000. Stimulus enhancement in greylag geese: Socially mediated learning of an operant task. *Animal Behaviour* 59:1119–1125.

Fritz, J., and K. Kotrschal. 1999. Social learning in common ravens, *Corvus corax*. *Animal Behaviour* 57:785–793.

——. 2002. On avian imitation: Cognitive and ethological perspectives. In K. N. Dautenhahn and C. L. Nehaniv, eds., *Imitation in Animals and Artifacts*, 133–155. Cambridge: MIT Press.

Fukuda, M. T. H., A. L. Francolin-Silva, and S. S. Almeida. 2002. Early postnatal protein malnutrition affects learning and memory in the distal but not in the proximal cue version of the Morris water maze. *Behavioral Brain Research* 133:271–277.

Fullard, J. H. 1982. Echolocation assemblages and their effects on moth auditory systems. *Canadian Journal of Zoology—Revue Canadienne De Zoologie* 60:2572–2576.

——. 1984. Listening for bats: Pulse repetition rate as a cue for a defensive behavior in *Cycnia tenera* Hübner (Lepidoptera: Arctiidae). *Journal of Comparative Physiology A* 154:249–252.

——. 1988. The tuning of moth ears. *Experientia* 44:423–428.

——. 1992. The neuroethology of sound production in tiger moths (Lepidoptera, Arctiidae). I. Rhythmicity and central control. *Journal of Comparative Physiology A* 170:575–588.

——. 1994. Auditory changes in noctuid moths endemic to bat-free habitat. *Journal of Evolutionary Biology* 7:435–445.

——. 1998. The sensory coevolution of moths and bats. In R. R. Hoy, A. N. Popper, and R. R. Fay, eds., *Comparative Hearing: Insects*, 279–326. New York: Springer.

Fullard, J. H., and J. W. Dawson. 1997. The echolocation calls of the spotted bat, *Euderma maculatum*, are relatively inaudible to moths. *Journal of Experimental Biology* 200:129–137.

Fullard, J. H., J. W. Dawson, and D. S. Jacobs. 2003a. Auditory encoding during the last moment of a moth's life. *Journal of Experimental Biology* 206:281–294.

Fullard, J. H., J. W. Dawson, L. D. Otero, and A. Surlykke. 1997. Bat-deafness in day-flying moths (Lepidoptera, Notodontidae, Dioptinae). *Journal of Comparative Physiology A* 181:477–483.

Fullard, J. H., and M. B. Fenton. 1977. Acoustic and behavioural analyses of the sounds produced by some species of Nearctic Arctiidae (Lepidoptera). *Canadian Journal of Zoology—Revue Canadienne de Zoologie* 55:1213–1224.

Fullard, J. H., M. B. Fenton, J. A. and Simmons. 1979. Jamming bat echolocation: The clicks of arctiid moths. *Canadian Journal of Zoology—Revue Canadienne de Zoologie* 57:647–649.

Fullard, J. H., K. E. Muma, and J. W. Dawson. 2003b. Quantifying an anti-bat flight response by eared moths. *Canadian Journal of Zoology—Revue Canadienne de Zoologie* 81:395–399.

Fullard, J. H., and N. Napoleone. 2001. Diel flight periodicity and the evolution of auditory defences in the Macrolepidoptera. *Animal Behaviour* 62:349–368.

Fullard, J. H., J. M. Ratcliffe, and C. G. Christie. 2007a. Acoustic feature recognition in the dogbane tiger moth, *Cycnia tenera*. *Journal of Experimental Biology* 210:2481–2488.

Fullard, J. H., J. M. Ratcliffe, and C. Guignion. 2005. Sensory ecology of predator-prey interactions: Responses of the AN2 interneuron in the field cricket, *Teleogryllus oceanicus*, to the echolocation calls of sympatric bats. *Journal of Comparative Physiology A* 191:605–618.

Fullard, J. H., J. M. Ratcliffe, and A. R. Soutar. 2004. Extinction of the acoustic startle response in moths endemic to a bat-free habitat. *Journal of Evolutionary Biology* 17:856–861.

Fullard, J. H., J. M. Ratcliffe, and H. M. ter Hofstede. 2007b. Neural evolution in the bat-free habitat of Tahiti: Partial regression in an anti-predator auditory system. *Biology Letters* 3:26–28.

Fullard, J. H., J. A. Simmons, and P. A. Saillant. 1994. Jamming bat echolocation: The dogbane tiger moth *Cycnia tenera* times its clicks to the terminal attack calls of the big brown bat *Eptesicus fuscus*. *Journal of Experimental Biology* 194:285–298.

Furrer, R. D., and M. B. Manser. In press. The evolution of urgency-based and functionally referential alarm calls in ground-dwelling species. *American Naturalist.*

Futuyma, D. J. 1998. *Evolutionary Biology*. Sunderland, MA: Sinauer.

Gajdon, G. K., N. Fijn, and L. Huber. 2004. Testing social learning in a wild mountain parrot, the kea (*Nestor notabilis*). *Learning and Behavior* 32:62–71.

———. 2006. Limited spread of innovation in a wild parrot, the kea (*Nestor notabilis*). *Animal Cognition* 9:173–181.

Galas, D. J., and E. Branscomb. 1978. The enzymatic determination of DNA polymerase accuracy: Theory of T4 polymerase mechanisms. *Journal of Molecular Biology* 124:653–687.

Galas, D. J., T. B. L. Kirkwood, and R. F. Rosenberger. 1986. An introduction to the problem of accuracy. In T. B. L. Kirkwood, R. F. Rosenberger, and D. J. Galas, eds., *Accuracy of Molecular Processes,* 1–16. London: Chapman and Hall.

Galef, B. G., Jr. 1988. Imitation in animals: History, definition and interpretation of data from the psychological laboratory. In T. R. Zentall and B. G. Galef Jr., eds., *Social Learning: Psychological and Biological Perspectives,* 3–28. Hillsdale, NJ: Erlbaum.

———. 1991. Information centres of Norway rats: Sites for information exchange and information parasitism. *Animal Behaviour* 41:295–301.

———. 1995. Why behaviour patterns that animals learn socially are locally adaptive. *Animal Behaviour* 49:1325–1334.

———. 2006. Theoretical and empirical approaches to understanding when animals use socially acquired information and from whom they acquire it. In J. R. Lucas and L. Simmons, eds., *Essays in Animal Behaviour: Celebrating 50 Years of Animal Behaviour,* 161–182. San Diego: Academic Press.

Galef B. G., Jr., K. E. Dudley, and E. E. Whiskin. 2008. Social learning of food preferences in "dissatisfied" and "uncertain" Norway rats. *Animal Behaviour* 75:631–637.

Galef, B. G., Jr., and L.-A. Giraldeau. 2000. Evidence of social effects on mate choice in vertebrates. *Behavioural Processes* 51:167–175.

———. 2001. Social influences on foraging in vertebrates: Behavioural mechanisms and adaptive functions. *Animal Behaviour* 61:3–15.

Galef, B. G., Jr., and E. E. Whiskin. 2006. Increased reliance on socially acquired information while foraging in risky situations? *Animal Behaviour* 72:1169–1176.

———. 2008. "Conformity" in Norway rats? *Animal Behaviour* 75:2035–2039.

Galef, B. G., Jr., and D. J. White. 1998. Mate-choice copying in Japanese quail, *Coturnix coturnix japonica*. *Animal Behaviour* 55:545–552.

———. 2000. Evidence of social effects on mate choice in vertebrates. *Behavioral Processes* 51:167–175.

Ganeshina, O., S. Schafer, and D. Malun. 2000. Proliferation and programmed cell death of neuronal precursors in the mushroom bodies of the honeybee. *Journal of Comparative Neurology* 417:349–365.

Gannicott, A. M., and R. C. Tinsley. 1997. Egg hatching in the monogenean gill parasite *Discocotyle sagittata* from the rainbow trout (*Onchorhynchus mykiss*). *Parasitology* 114:569–579.

Garamszegi, L. Z., and M. Eens. 2004a. Brain space for a learned task: Strong intraspecific evidence for neural correlates of singing behavior in songbirds. *Brain Research Reviews* 44:187–193.

———. 2004b. The evolution of hippocampus volume and brain size in relation to food hoarding in birds. *Ecology Letters* 7:1216–1224.

Garamszegi, L. Z., A. P. Møller, and J. Erritzoe. 2002. Coevolving avian eye size and brain size in relation to prey capture and nocturnality. *Proceedings of the Royal Society of London Series B* 269:961–967.

Gauthier, M., V. C. Lozano, A. Zaoujal, and D. Richard. 1994. Effects of intracranial injections of scopolamine on olfactory conditioning in the honeybee. *Behavioral Brain Research* 63:145–149.

Gauthreaux, S. A., Jr. 1978. The ecological significance of behavioural dominance. *Perspectives in Ethology* 3:17–54.

Gerber, B., S. Scherer, K. Neuser, B. Michels, T. Hendel, R. F. Stocker, and M. Heisenberg. 2004. Visual learning in individually assayed *Drosophila* larvae. *Journal of Experimental Biology* 207:179–188.

Gerhardt, H. C., and F. Huber. 2002. *Acoustic Communication in Insects and Anurans*. Chicago: University of Chicago Press.

Gerhardt, H. C., S. D. Tanner, C. M. Corrigan, and H. C. Walton. 2000. Female preference functions based on call duration in the gray tree frog (*Hyla versicolor*). *Behavioral Ecology* 11:663–669.

Gershwin, M. E., R. S. Beach, and L. S. Hurley. 1985. *Nutrition and Immunity*. Orlando: Academic Press.

Getz, L. L., C. S. Carter, and L. Gavish. 1981. The mating system of the prairie vole, *Microtus ochrogaster*: Field and laboratory evidence for pair-bonding. *Behavioral Ecology and Sociobiology* 8:189–194.

Getz, L. L., and B. McGuire. 1993. A comparison of living singly and in male-female pairs in the prairie vole, *Microtus ochrogaster*. *Ethology* 94:265–278.

Getz, L. L., B. McGuire, T. Pizzuto, J. E. Hofmann, and B. Frase. 1993. Social-organization of the prairie vole (*Microtus ochrogaster*). *Journal of Mammalogy* 74:44–58.

Gherardi, F., and J. Tiedemann. 2004. Binary individual recognition in hermit crabs. *Behavioral Ecology and Sociobiology* 55:524–530.

Ghose, K., T. K. Horiuchi, P. S. Krishnaprasad, and C. F. Moss. 2006. Echolocating bats use a nearly time-optimal strategy to intercept prey. *Public Library of Science Biology* 4:865–873.

Gibson, K. R. 1986. Cognition, brain size and the extraction of embedded food resources. In J. G. Else and P. C. Lee, eds., *Primate Ontogeny, Cognition and Social Behaviour*, 93–103. Cambridge: Cambridge University Press.

———. 2002. Evolution of human intelligence: The roles of brain size and mental construction. *Brain, Behavior and Evolution* 59:10–20.

Gibson, R. M., J. W. Bradbury, and S. L. Vehrencamp. 1991. Mate choice in lekking sage grouse revisited the roles of vocal display female site fidelity and copying. *Behavioral Ecology* 2:165–180.

Gibson, R. M., and J. Höglund. 1992. Copying and sexual selection. *Trends in Ecology and Evolution* 7:229–232.

Gil, D., M. Naguib, K. Riebel, A. Rutstein, and M. Gahr. 2006. Early condition, song learning, and the volume of song brain nuclei in the zebra finch (*Taeniopygia guttata*). *Journal of Neurobiology* 66:1602–1612.

Giraldeau, L. A., T. J. Valone, and J. J. Templeton. 2002. Potential disadvantages of using socially acquired information. *Philosophical Transactions of the Royal Society of London Series B* 357:1559–1566.

Gittleman, J. L. 1994. Female brain size and parental care in carnivores. *Proceedings of the National Academy of Science of the United States of America* 91:5495–5497.

Giurfa, M. 2007. Behavioral and neural analysis of associative learning in the honeybee: A taste from the magic well. *Journal of Comparative Physiology A* 193:801–824.

Gjullin, C. M., C. P. Hegarty, and W. B. Bollen. 1941. The necessity of a low oxygen concentration for the hatching of *Aedes* mosquito eggs. *Journal of Cellular and Comparative Physiology* 17:193–202.

Glimcher, P. W., and A. Rustichini. 2004. Neuroeconomics: The consilience of brain and decision. *Science* 306:447–452.

Gnatzy, W., and G. Kämper. 1990. Digger wasp against crickets. II. An airborne signal produced by a running predator. *Journal of Comparative Physiology A* 167:551–556.

Goel, V., M. Makale, and J. Grafman. 2004. The hippocampal system mediates logical reasoning about familiar spatial environments. *Journal of Cognitive Neuroscience* 16:654–664.

Goldberg, J. L., J. W. A. Grant, and L. Lefebvre. 2001. Effects of the temporal predictability and spatial clumping of food on the intensity of competitive aggression in the Zenaida dove. *Behavioral Ecology* 12:490–495.

Gomez-Mestre, I., J. C. Touchon, and K. M. Warkentin. 2006. Amphibian embryo and parental defenses and a larval predator reduce egg mortality from water mold. *Ecology* 87:2570–2581.

Gomez-Mestre, I., and K. M. Warkentin. 2007. To hatch and hatch not: Similar selective trade-offs but different responses to egg predators in two closely related, syntopic treefrogs. *Oecologia* 153:197–206.

Gomez-Mestre, I., J. J. Wiens, and K. M. Warkentin. 2008. Evolution of adaptive plasticity: Risk-sensitive hatching in Neotropical leaf-breeding treefrogs (*Agalychnis:* Hylidae). *Ecological Monographs* 78, 205–224.

Goodson, J. L., and A. H. Bass. 2001. Social behavior functions and related anatomical characteristics of vasotocin/vasopressin systems in vertebrates. *Brain Research Reviews* 35:246–265.

Goss-Custard, J. D. 1996. *The Oystercatcher: From Individuals to Populations.* Oxford: Oxford University Press.

Gould, E., and P. Tanapat. 1999. Stress and neurogenesis. *Biological Psychiatry* 46:1472–1479.

Gould, E., P. Tanapat, T. Rydel, and N. Hastings. 2000. Regulation of hippocampal neurogenesis in adulthood. *Biological Psychiatry* 48:715–720.

Grafen, A. 1990. Biological signals as handicaps. *Journal of Theoretical Biology* 144:517–546.

Grant, J. W. A., and L. D. Green. 1996. Mate copying versus preference for actively courting males by female Japanese medaka (*Oryzias latipes*). *Behavioral Ecology* 7:165–167.

Greenberg, R. 1990. Feeding neophobia and ecological plasticity: A test of the hypothesis with captive sparrows. *Animal Behaviour* 39:375–379.

Greenfield, M. 1988. Interspecfic acoustic interaction among katydids *Neoconocephalus:* Inhibition-induced shifts in diel periodicity. *Animal Behaviour* 36:684–695.

———. 2005. Mecahnisms and evolution of communal sexual displays in arthropods and anurans. *Advances in the Study of Behavior* 35:1–62.

Greenspan, R. J. 2006. *An Introduction to Nervous Systems.* Cold Spring Harbor, NY: Cold Spring Harbor Laboratory Press.

Grieco, F., A. J. van Noordwijk, and M. E. Visser. 2002. Evidence for the effect of learning on timing of reproduction in blue tits. *Science* 296:136–138.

Griem, J. N., and K. L. M. Martin. 2000. Wave action: The environmental trigger for hatching in the California grunion *Leuresthes tenuis* (Teleostei: Atherinopsidae). *Marine Biology* 137:177–181.

Griesser, M. 2008. Referential calls signal predator behavior in a group-living bird species. *Current Biology* 18:69–73.

Griffin, A. S. 2004. Social learning about predators: A review and prospectus. *Learning and Behavior* 32:31–140.

Griffin, A. S., and C. S. Evans. 2003. Social learning of antipredator behaviour in a marsupial. *Animal Behaviour* 66:485–492.

Griffin, A. S., J. M. Pemberton, P. N. M. Brotherton, G. McIlrath, D. Gaynor, R. Kansky, J. O'Riain, and T. H. Clutton-Brock. 2003. A genetic analysis of breeding success in the cooperative meerkat (*Suricata suricatta*). *Behavioral Ecology* 14:472–480.

Griffin, D. R. 1958. *Listening in the Dark*. New Haven: Yale University Press.

———. 1971. The importance of atmospheric attenuation for the echolocation of bats (Chiroptera). *Animal Behaviour* 19:55–61.

Griffin, D. R., F. A. Webster, and C. R. Michael. 1960. The echolocation of flying insects by bats. *Animal Behaviour* 8:141–154.

Griffith, S. C., I. P. F. Owens, and K. A. Thuman. 2002. Extra pair paternity in birds: A review of interspecific variation and adaptive function. *Molecular Ecology* 11:2195–2212.

Griffiths, S. W. 2003. Learned recognition of conspecifics by fishes. In C. Brown, K. N. Laland, and J. Krause, eds., Learning in fishes: Why they are smarter than you think. Special issue, *Fish and Fisheries* 4:256–268.

Gronenberg, W. 2001. Subdivisions of hymenopteran mushroom body calyces by their afferent supply. *Journal of Comparative Neurology* 435:474–489.

Gronenberg, W., S. Heeren, and B. Holldöbler. 1996. Age-dependent and task-related morphological changes in the brain and mushroom bodies of the ant *Camponotus floridanus*. *Journal of Experimental Biology* 199:2011–2019.

Guidugli, K., A. Nascimento, G. Amdam, A. Barchuk, S. Omholt, Z. Simões, and K. Hartfelder. 2005. Vitellogenin regulates hormonal dynamics in the worker caste of a eusocial insect. *FEBS Letters* 579:4961–4965.

Guilford, T., and M. S. Dawkins. 1991. Receiver psychology and the evolution of animal signals. *Animal Behaviour* 42:1–14.

———. 1995. What are conventional signals? *Animal Behaviour* 49:1689–1695.

Gundersen, H. J. G., P. Bagger, T. F. Bendtsen, S. M. Evans, L. Korbo, N. Marcussen, A. Møller, K. Nielsen, J. R. Nyengaard, and B. Pakkenberg. 1988. The new stereological tools: Disector, fractionator, nucleator and point sampled intercepts and their use in pathological research and diagnosis. *Acta Pathologica, Microbiologica et Immunologica Scandinavica* 96:857–881.

Gwinner, E. 1964. Untersuchungen über das Ausdrucks- und Sozialverhalten des Kolkraben (*Corvus corax corax* L.). *Zeitschrift für Tierpsychologie* 21:657–748.

Gyger, M., P. Marler, and R. Pickert. 1987. Semantics of an avian alarm call system: The male domestic fowl, *Gallus domesticus*. *Behavior* 102:15–40.

Gyllström, M., and L.-A. Hansson. 2004. Dormancy in freshwater zooplankton: Induction, termination and the importance of benthic-pelagic coupling. *Aquatic Science* 66:274–295.

Hammer, M. 1993. An identified neuron mediates the unconditioned stimulus in associative olfactory learning in honeybees. *Nature* 366:59–63.

Hammer, M., and R. Menzel. 1995. Learning and memory in the honeybee. *Journal of Neuroscience* 15:1617–1630.

Hammock, E. A. D., and L. J. Young. 2004. Functional microsatellite polymorphism associated with divergent social structure in vole species. *Molecular Biology and Evolution* 21:1057–1063.

———. 2005. Microsatellite instability generates diversity in brain and sociobehavioral traits. *Science* 308:1630–1634.

Hampton, R. R., D. F. Sherry, S. J. Shettleworth, M. Khurgel, and G. Ivy. 1995. Hippocampal volume and food-storing behavior are related in parids. *Brain, Behavior and Evolution* 45:54–61.

Hampton, R. R., and Shettleworth, S. J. 1996. Hippocampal lesions impair memory for location but not color in passerine birds. *Behavioral Neuroscience* 110:831–835.

Harker, K. T., and I. Q. Whishaw. 2004. A reaffirmation of the retrosplenial contribution to rodent navigation: Reviewing the influences of lesion, strain, and task. *Neuroscience and Biobehavioral Reviews* 28:485–496.

Harnad, S. 1987. *Categorical Perception: The Groundwork of Cognition*. New York: Cambridge University Press.

Harris, M. A., and R. E. Lemon. 1972. Songs of song sparrows (*Melospiza melodia*): Individual variation and dialects. *Canadian Journal of Zoology—Revue Canadienne de Zoologie* 50:301–309.

———. 1974. Songs of song sparrows: Reactions of males to songs of different localities. *Condor* 76:33–44.

Harvey, S., P. G. Phillips, A. Rees, and T. R. Hall. 1984. Stress and adrenal function. *Journal of Experimental Zoology* 232:633–645.

Hasselquist, D., S. Bensch, and T. von Schantz. 1996. Correlation between male song repertoire, extra-pair paternity and offspring survival in the great reed warbler. *Nature* 381:229–232.

Hatchwell, B. J., D. J. Ross, M. K. Fowlie, and A. McGowan. 2001. Kin discrimination in cooperatively breeding long-tailed tits. *Proceedings of the Royal Society of London Series B* 268:885–890.

Hayes, U. L., and K. C. Chambers. 2005. High doses of vasopressin delay the onset of extinction and strengthen acquisition of LiCl-induced conditioned taste avoidance. *Physiology and Behavior* 84:625–633.

Healy, S. D., and T. Guilford. 1990. Olfactory bulb size and nocturnality in birds. *Evolution* 44:339–346.

Healy, S. D., E. Gwinner, and J. R. Krebs. 1996. Hippocampal volume in migratory and non-migratory warblers: Effect of age and experience. *Behavioral Brain Research* 81:61–68.

Healy, S. D., and J. R. Krebs. 1992. Food storing and the hippocampus in corvids: Amount and volume are correlated. *Proceedings of the Royal Society of London Series B* 248:241–245.

———. 1996. Food storing and the hippocampus in Paridae. *Brain, Behavior and Evolution* 47:195–199.

Healy, S. D., and C. Rowe. 2007. A critique of comparative studies of brain size. *Proceedings of the Royal Society of London Series B* 274:453–464.

Hebets, E. A. 2003. Subadult experience influences adult mate choice in an arthropod: Exposed female wolf spiders prefer males of a familiar phenotype. *Proceedings of the National Academy of Sciences of the United States of America* 100:13390–13395.

Hebets, E. A., and D. R. Papaj. 2005. Complex signal function: Developing a framework of testable hypotheses. *Behavioral Ecology and Sociobiology* 57:197–214.

Heinrich, B. 1999. *Mind of the Raven*. New York: Harper Collins.

Heinrich, B., and J. Pepper. 1998. Influence of competitors on caching behaviour in common ravens, *Corvus corax*. *Animal Behaviour* 56:1083–1090.

Heisenberg, M. 2003. Mushroom body memoir: From maps to models. *Nature Reviews Neuroscience* 4:266–275.

Heisenberg, M., M. Heusipp, and C. Wanke. 1995. Structural plasticity in the *Drosophila* brain. *Journal of Neuroscience* 15:1951–1960.

Helton, W. S. 2008. Expertise acquisition as sustained learning in humans and other animals: Commonalities across species. *Animal Cognition* 11:99–107.

Henderson, I. G., P. J. B. Hart, and T. Burke. 2000. Strict monogamy in a semi-colonial passerine: The jackdaw *Corvus monedula*. *Journal of Avian Biology* 31:177–182.

Henrich, J., and R. Boyd. 1998. The evolution of conformist transmission and between-group differences. *Evolution and Human Behavior* 19:215–242.

Herculano-Houzel, S., C. E. Collins, P. Y. Wong, and J. H. Kaas. 2007. Cellular scaling rules for primate brains. *Proceedings of the National Academy of Sciences of the United States of America* 104:3562–3567.

Herculano-Houzel, S., B. Mota, and R. Lent. 2006. Cellular scaling rules for rodent brains. *Proceedings of the National Academy of Sciences of the United States of America* 103:12138–12143.

Heyes, C. M. 1994. Social learning in animals: Categories and mechanisms. *Biological Reviews of the Cambridge Philosophical Society* 69:207–231.

———. 1999. Imitation, cognition, and culture: On the psychological mechanism and transmission function of imitation learning. *Advances in Ethology* 34:9.

Heyes, C. M., and B. G. Galef Jr. 1996. *Social Learning in Animals: The Roots of Culture*. San Diego: Academic Press.

Hiebert, S. M., P. K. Stoddard, and P. Arcese. 1989. Repertoire size, territory acquisition and reproductive success in the song sparrow. *Animal Behaviour* 37:266–273.

Higginson, A. D., and C. J. Barnard. 2004. Accumulating wing damage affects foraging decisions in honey bees (*Apis mellifera* L.). *Ecological Entomology* 29:52–59.

Hill, C. E., S. E. Campbell, and M. D. Beecher. In preparation. Does female choice select for male song repertoires? Female choice of extra-pair mates and male song characteristics in the song sparrow.

Hill, C. E., S. E. Campbell, J. C. Nordby, J. M. Burt, and M. D. Beecher. 1999. Song sharing in two populations of song sparrows. *Behavioral Ecology and Sociobiology* 46:341–349.

Hill, P. S. M. 2008. *Vibrational Communication in Animals*. Cambridge: Harvard University Press.

Hill, S., and M. J. Ryan. 2006. The role of female quality in the mate choice copying behaviour of sailfin mollies. *Biology Letters* 2:203–205.

Hinde, R. A., and J. Fisher. 1951. Further observations on the opening of milk bottles by birds. *British Birds* 44:393–396.

Hitchcock, C. L., and D. F. Sherry. 1995. Cache pilfering and its prevention in pairs of black-capped chickadees. *Journal of Avian Biology* 26:187–192.

Hoelzer, G. A. 1989. The good parent process of sexual selection. *Animal Behaviour* 38:1067–1078.

Höglund, J., R. V. Alatalo, R. M. Gibson, and A. Lundberg. 1995. Mate-choice copying in black grouse. *Animal Behaviour* 49:1627–1633.

Hogue, M.-E., J. P. Beaugrand, and P. C. Lague. 1996. Coherent use of information by hens observing their former dominant defeating or being defeated by a stranger. *Behavioural Processes* 38:241–252.

Hoke, K. L., M. J. Ryan, and W. Wilczynski. 2007. Integration of sensory and motor processing underlying social behaviour in tungara frogs. *Proceedings of the Royal Society of London Series B* 274:641–649.

Holderied, M. W., C. Korine, M. B. Fenton, S. Parsons, S. Robson, and G. Jones. 2005. Echolocation call intensity in the aerial hawking bat *Eptesicus bottae* (Vespertilionidae) studied using stereo videogrammetry. *Journal of Experimental Biology* 208:1321–1327.

Holderied, M. W., and O. von Helversen. 2003. Echolocation range and wingbeat period match in aerial-hawking bats. *Proceedings of the Royal Society of London Series B* 270:2293–2299.

Hollén, L. I., T. Clutton-Brock, and M. B. Manser. 2008. Ontogenetic changes in alarm call production and usage in meerkats (*Suricata suricatta*): Adaptations or constraints? *Behavioral Ecology and Sociobiology* 62:639–654.

Honey Bee Genome Sequencing Consortium. 2006. Insights into social insects from the genome of the honeybee *Apis mellifera*. *Nature* 443:931–949.

Hoogland, R. D., D. Morris, and N. Tinbergen. 1957. The spines of sticklebacks (*Gasterosteus* and *Pygosteus*) as a means of defence against predators (*Perca* and *Esox*). *Behaviour* 10:205–237.

Horel, J. A. 1991. Use of cold to reversibly suppress local brain function in behaving animals. In P. M. Conn, ed., *Methods in Neurosciences: Lesions and Tranplantation* 7:97–110. New York: Academic Press.

Horn, A., and J. B. Falls. 1988. Structure of western meadowlark (*Sturnella neglecta*) song repertoires. *Canadian Journal of Zoology—Revue Canadienne de Zoologie* 66:284–288.

Horning, C. L., M. D. Beecher, P. K. Stoddard, and S. E. Campbell. 1993. Song perception in the song sparrow: Importance of different parts of the song in song type classification. *Ethology* 94:46–58.

Horrocks, J. A., and W. Hunte. 1986. Sentinel behaviour in vervet monkeys: Who sees whom first? *Animal Behaviour* 34:1566–1567.

Hoshooley, J. S., and D. F. Sherry. 2007. Greater hippocampal neuronal recruitment in food-storing than in non-food-storing birds. *Developmental Neurobiology* 67:406–414.

Houck, L. D., and L. C. Drickamer. 1996. *Foundations of Animal Behavior: Classic Papers with Commentaries*. Chicago: University of Chicago Press.

Houde, E. D. 2002. Mortality. In L. A. Fuiman and R. G. Werner, eds., *Fishery Science: The Unique Contributions of Early Life Stages*, 64–87. Oxford: Blackwell Science.

Hoy, R. R. 1992. The evolution of hearing in insects as an adaptation of predation from bats. In D. B. Webster, R. R. Fay, and A. N. Popper, eds., *The Evolutionary Biology of Hearing*, 115–129. New York: Springer.

———. 1998. Acute as a bug's ear: An informal discussion of hearing in insects. In R. R. Hoy, A. N. Popper, and R. R. Fay, eds., *Comparative Hearing: Insects*, 1–17. New York: Springer.

Hoy, R. R., T. G. Nolen, and G. C. Casaday. 1985. Dendritic sprouting and compensatory synaptogenesis in an identified interneuron following auditory deprivation in a cricket. *Proceedings of the National Academy of Sciences of the United States of America* 82:7772–7776.

Hoy, R. R., A. N. Popper, and R. R. Fay. 1998. *Comparative Hearing: Insects*. New York: Springer.

Hristov, N. I., and W. E. Conner. 2005a. Effectiveness of tiger moth (Lepidoptera, Arctiidae) chemical defenses against an insectivorous bat (*Eptesicus fuscus*). *Chemoecology* 15:105–113.

———. 2005b. Sound strategy: Acoustic aposematism in the bat–tiger moth arms race. *Naturwissenschaften* 92:164–169.

Huber, F. 1960. Untersuchungen über die Funktion des Zentralnervensystems und insbesondere des Gehirns bei der Fortbewegung und der Lauterzeugung der Grillen. *Zeitschrift für Vergleichende Physiologie* 44:60–132.

Huber, L., S. Rechberger, and M. Taborsky. 2001. Social learning affects object exploration and manipulation in keas, *Nestor notabilis*. *Animal Behaviour* 62:945–954.

Huey, R. B., P. E. Hertz, and B. Sinervo. 2003. Behavioral drive versus behavioral inertia in evolution: A null model approach. *American Naturalist* 161:357–366.

Hughes, M., R. C. Anderson, W. A. Searcy, L. M. Bottensek, and S. Nowicki. 2007. Song type sharing and territory tenure in eastern song sparrows: Implications for the evolution of song repertoires. *Animal Behaviour* 73:701–710.

Hughes, M., S. Nowicki, W. A. Searcy, and S. Peters. 1998. Song-type sharing in song sparrows: Implications for repertoire function and song learning. *Behavioral Ecology and Sociobiology* 42:437–446.

Hulse, S. H. 1988. The social function of intellect. In R. W. Byrne and A. Whiten, eds., *Machiavellian Intelligence: Social Expertise and the Evolution of Intellect in Monkeys, Apes, and Humans*, 13–26. Oxford: Clarendon Press.

———. 2002. Auditory sense analysis in animal communication. *Advances in the Study of Behavior* 31:163–200.

Humphrey, N. K. 1976. The social function of intellect. In P. P. G. Bateson and R. A. Hinde, eds., *Growing Points in Ethology*, 303–317. Cambridge: Cambridge University Press.

Hunt, G. J., G. V. Amdam, D. Schlipalius, C. Emore, N. Sardesai, C. E. Williams, O. Rueppell, E. Guzmán-Novoa, M. Arechavaleta-Velasco, S. Chandra, M. K. Fondrk, M. Beye, and R. E. Page. 2007. Behavioral genomics of honeybee foraging and nest defense. *Naturwissenschaften* 94:247–267.

Hunter, M. L., and J. R. Krebs. 1979. Geographical variation in the song of the great tit (*Parus major*) in relation to ecological factors. *Journal of Animal Ecology* 48:759–785.

Ihara, Y. 2002. A model for evolution of male parental care and female multiple mating. *American Naturalist* 160:235–244.

Immelmann, K. 1969. Song development in the zebra finch and other estrildid finches. In R. A. Hinde, ed., *Bird Vocalizations*, 61–74. Cambridge: Cambridge University Press.

Inglis, I. R. 1983. Towards a cognitive theory of exploratory behaviour. In J. Archer and L. Birke, eds., *Exploration in Animals and Humans*, 72–116. London: Van Nostrand Reinhold.

Insel, T. R., S. Preston, and J. T. Winslow. 1995. Mating in the monogamous male: Behavioral consequences. *Physiology and Behavior* 57:615–627.

Insel, T. R., Z. X. Wang, and C. F. Ferris. 1994. Patterns of brain vasopressin receptor distribution associated with social organization in microtine rodents. *Journal of Neuroscience* 14:5381–5392.

Ireland, D. H., A. J. Wirsing, and D. L. Murray. 2007. Phenotypically plastic responses of green frog embryos to conflicting predation risk. *Oecologia* 152:162–168.

Irschick, D., L. Dyer, and T. W. Sherry. 2005. Phylogenetic methodologies for studying specialization. *Oikos* 110:404–408.

Isler, K., and C. P. van Schaik. 2006. Metabolic costs of brain size evolution. *Biology Letters* 2:557–560.

Ismail, N., S. Christine, G. E. Robinson, and S. E. Fahrbach. 2008. Pilocarpine improves recognition of nestmates in young honey bees. *Neuroscience Letters* 439:178–181.

Ismail, N., G. E. Robinson, and S. E. Fahrbach. 2006. Stimulation of muscarinic receptors mimics experience-dependent plasticity in the honey bee brain. *Proceedings of the National Academy of Sciences of the United States of America* 103:207–211.

Iwaniuk, A. N., K. M. Dean, and J. E. Nelson. 2004. A mosaic pattern characterizes the evolution of the avian brain. *Proceedings of the Royal Society of London Series B* 271:S148–S151.

Iwaniuk, A. N., and D. A. Nelson. 2003. Developmental differences are correlated with relative brain size in birds: A comparative analysis. *Canadian Journal of Zoology—Revue Canadienne de Zoologie* 81:1913–1928.

Iwaniuk, A. N., and J. E. Nelson. 2002. Can endocranial volume be used as an estimate of brain size in birds? *Canadian Journal of Zoology—Revue Canadienne de Zoologie* 80:16–23.

Jackson, J. R. 1963. The nesting of keas. *Notornis* 10:334–337.

Jacobs, D. S., J. M. Ratcliffe, and J. H. Fullard. 2008. Beware of bats, beware of birds: The auditory responses of eared moths to bat and bird predation. *Behavioral Ecology* 19:1333-1342.

Jacobs, L. F. 1996. Sexual selection and the brain. *Trends in Ecology and Evolution* 11:A82–A86.

James, W. 1890. *The Principles of Psychology*. New York: Dover.

Janetos, A. C. 1980. Strategies of female mate choice: A theoretical-analysis. *Behavioral Ecology and Sociobiology* 7:107–112.

Janik, V. M., and P. J. B. Slater. 1997. Vocal learning in mammals. *Advances in the Study of Behavior* 26:59–99.

Jansson, C., J. Ekman, and A. von Brömssen. 1981. Winter mortality and food supply in tits *Parus* species. *Oikos* 37:313–322.

Jenkins, P. F. 1978. Cultural transmission of song patterns and dialect development in a free-living bird population. *Animal Behaviour* 26:50–78.

Jennings, H. S. 1906. *Behavior of the Lower Organisms*. Bloomington: Indiana University Press.

Jennions, M. D., and M. Petrie. 1997. Variation in mate choice and mating preferences: A review of causes and consequences. *Biological Reviews of the Cambridge Philosophical Society* 72:283–327.

Jensen, C. 1979. Learning performance in mice genetically selected for brain weight: Problems of generality. In M. E. Hahn, C. Jenson, and B. C. Dudek, eds., *Development and Evolution of Brain Size: Behavioral Implications*, 205–220. New York: Academic Press.

Jerison, H. J. 1973. *Evolution of the Brain and Intelligence*. New York: Academic Press.

Johnson, F., and S. W. Bottjer. 1992. Growth and regression of thalamic efferents in the song-control system of male zebra finches. *Journal of Comparative Neurology* 326:442–450.

Johnson, J. B., D. Saenz, C. K. Adams, and R. N. Conner. 2003. The influence of predator threat on the timing of a life-history switch point: Predator-induced hatching in the southern leopard frog (*Rana sphenocephala*). *Canadian Journal of Zoology—Revue Canadienne de Zoologie* 81:1608–1613.

Johnston, V. H., and J. P. Ryder. 1987. Divorce in larids: A review. *Colonial Waterbirds* 10:16–26.

Jolly, A. 1966. Lemur social behaviour and primate intelligence. *Science* 153:501–507.

Jones, B. C., L. M. DeBruine, A. C. Little, R. P. Burriss, and D. R. Feinberg. 2007. Social transmission of face preferences among humans. *Proceedings of the Royal Society of London Series B* 274:899–903.

Jones, G., and J. Rydell. 2003. Attack and defense: Interactions between echolocating bats and their insect prey. In T. H. Kunz and M. B. Fenton, eds., *Bat Ecology*, 301–345. Chicago: University of Chicago Press.

Jones, G., and E. C. Teeling. 2006. The evolution of echolocation in bats. *Trends in Ecology and Evolution* 21:149–156.

Jones, G., P. I. Webb, J. A. Sedgeley, and C. F. J. O'Donnell. 2003. Mysterious *Mystacina:* How the New Zealand short-tailed bat (*Mystacina tuberculata*) locates insect prey. *Journal of Experimental Biology* 206:4209–4216.

Jones, K. E., A. Purvis, A. MacLarnon, O. R. P. Bininda-Emonds, and N. B. Simmons. 2002. A phylogenetic supertree of the bats (Mammalia: Chiroptera). *Biological Reviews* 77:223–259.

Jose, D. G., O. Stutman, and R. Good. 1973. Long term effects on immune function of early nutritional deprivation. *Nature* 241:57–58.

Kahlert, J. 2003. The constraint on habitat use in wing-moulting greylag geese *Anser anser* caused by anti-predator displacements. *Ibis* 145:E45–E52.

———. 2006. Factors affecting escape behaviour in moulting greylag geese *Anser anser*. *Journal of Ornithology* 147:569–577.

Kalka, M. B., A. R. Smith, and E. K. V. Kalko. 2008. Bats limit arthropods and herbivory in a tropical forest. *Science* 320:71.

Kalko, E. K. V. 1995. Insect pursuit, prey capture and echolocation in pipistrelle bats (Microchiroptera). *Animal Behaviour* 50:861–880.

Kameda, T., and D. Nakanishi. 2002. Cost-benefit analysis of social/cultural learning in a nonstationary uncertain environment: An evolutionary simulation and an experiment with human subjects. *Evolution and Human Behavior* 23:373–393.

Kamil, A. C. 1988. A synthetic approach to the study of animal intelligence. In D. W. Leger, ed., *Nebraska Symposium on Motivation, 1987: Comparative Perspectives in Modern Psychology*, 257–308. Lincoln: University of Nebraska Press.

———. 1998. On the proper definition of cognitive ethology. In R. P. Balda, I. M. Pepperberg, and A. C. Kamil, eds., *Animal Cognition in Nature*, 1–28. San Diego: Academic Press.

Kamil, A. C., and T. D. Sargent. 1981. Foraging Behavior: Ecological, Ethological, and Psychological Approaches. New York: Garland.

Kandel, E. R., J. H. Schwartz, and T. M. Jessell. 1995. *Essentials of Neural Science and Behavior*. Norwalk, CT: Appleton and Lange.

Kats, L. B., and L. M. Dill. 1998. The scent of death: Chemosensory assessment of predation risk by prey animals. *Ecoscience* 5:361–394.

Keller, R. 1975. Das Spielverhalten der Keas (*Nestor notabilis* Gould) des Züricher Zoos. *Zeitschrift für Tierpsychologie* 38:393–408.

Kempenaers, B., G. R. Verheyen, and A. A. Dhondt. 1997. Extrapair paternity in the blue tit (*Parus caeruleus*): Female choice, male characteristics, and offspring quality. *Behavioral Ecology* 8:481–492.

Kempermann, G. 2002. Why new neurons? Possible functions for adult hippocampal neurogenesis. *Journal of Neuroscience* 22:635–638.

Kendal, R. L., I. Coolen, and K. N. Laland. 2004. The role of conformity in foraging when personal and social information conflict. *Behavioral Ecology* 15:269–277.

Kendal, R. L., I. Coolen, Y. Van Bergen, and K. N. Laland. 2005. Trade-offs in the adaptive use of social and asocial learning. *Advances in the Study of Behavior* 35:333–379.

Kenyon, F. C. 1896. The brain of the bee: A preliminary contribution to the morphology of the nervous system of the arthropoda. *Journal of Comparative Neurology* 6:133–210.

Kick, S. A., and J. A. Simmons. 1984. Automatic gain control in the bat's sonar receiver and the neuroethology of echolocation. *Journal of Neuroscience* 4:2725–2737.

Kiesecker, J. M., and A. R. Blaustein. 1997. Influences of egg laying behavior on pathogenic infection of amphibian eggs. *Conservation Biology* 11:214–220.

Kirkpatrick, B., C. S. Carter, S. W. Newman, and T. R. Insel. 1994. Axon-sparing lesions of the medial nucleus of the amygdala decrease affiliative behaviors in the prairie vole (*Microtus ochrogaster*): Behavioral and anatomical specificity. *Behavioral Neuroscience* 108:501–513.

Kirkpatrick, M., and L. A. Dugatkin. 1994. Sexual selection and the evolutionary effects of copying mate choice. *Behavioral Ecology and Sociobiology* 34:443–449.

Kirkpatrick, M., A. S. Rand, and M. J. Ryan. 2006. Mate choice rules in animals. *Animal Behaviour* 71:1215–1225.

Kirkpatrick, M., and M. J. Ryan. 1991. The evolution of mating preferences and the paradox of the lek. *Nature* 350:33–38.

Kirkwood, T. B. L., and S. N. Austad. 2000. Why do we age? *Nature* 408:233–238.

Kirkwood, T. B. L., R. F. Rosenberger, and D. J. Galas. 1986. *Accuracy in Molecular Processes*. London: Chapman and Hall.

Kirn, J. R., R. P. Clower, D. E. Kroodsma, and T. J. DeVoogd. 1989. Song-related brain regions in the red-winged blackbird are affected by sex and season but not repertoire size. *Journal of Neurobiology* 20:139–163.

Kitaysky, A. S., E. V. Kitaiskaia, J. F. Piatt, and J. C. Wingfield. 2003. Benefits and costs of increased levels of corticosterone in seabird chicks. *Hormones and Behavior* 43:140–149.

———. 2006. A mechanistic link between chick diet and decline in seabirds? *Proceedings of the Royal Society of London Series B* 273:445–450.

Kitaysky, A. S., J. F. Piatt, and J. C. Wingfield. 1999. The adrenocortical stress-response of black-legged kittiwake chicks in relation to dietary restrictions. *Journal of Comparative Physiology B* 169:303–310.

Kitaysky, A. S., J. C. Wingfield, and J. F. Piatt. 2001. Corticosterone facilitates begging and affects resource allocation in the black-legged kittiwake. *Behavioral Ecology* 12:619–625.

Kleiman, D. G. 1977. Monogamy in mammals. *Quarterly Review of Biology* 52:39–69.

Klopfer, P. H. 1959. Social interactions in discrimination learning with special reference to feeding behaviour in birds. *Behaviour* 14:282–299.

———. 1962. *Behavioral aspects of ecology*. London: Prentice-Hall.

Kluender, K. R., R. L. Diehl, and P. R. Killeen. 1987. Japanese quail can learn phonetic categories. *Science* 237:1195–1197.

Kohler, R. E. 1994. Lords of the Fly: *Drosophila* Genetics and the Experimental Life. Chicago: University of Chicago Press.

Kokko, H. 1999. Cuckoldry and the stability of biparental care. *Ecology Letters* 2:247–255.

Kokko, H., and L. J. Morrell. 2005. Mate guarding, male attractiveness, and paternity under social monogamy. *Behavioral Ecology* 16:724–731.

Komers, P. E., and P. N. M. Brotherton. 1997. Female space use is the best predictor of monogamy in mammals. *Proceedings of the Royal Society of London Series B* 264:1261–1270.

Komischke, B., J. C. Sandoz, D. Malun, and M. Giurfa. 2005. Partial unilateral lesions of the mushroom bodies affect olfactory learning in honeybees *Apis mellifera* L. *European Journal of Neuroscience* 21:477–485.

Konarzewski, M., J. Kowalczyk, T. Swierubska, and B. Lewonczuk. 1996. Effect of short-term feed restriction, realimentation and overfeeding on growth of song thrush (*Turdus philomelos*) nestlings. *Functional Eclogy* 10:97–105.

Konishi, M., and E. Akutagawa. 1985. Neuronal growth, atrophy and death in a sexually dimorphic song nucelus in the zebra finch brain. *Nature* 315:145–147.

Koops, M. A. 2004. Reliability and the value of information. *Animal Behaviour* 67:103–111.

Koops, M. A., and M. V. Abrahams. 1998. Life history and the fitness consequences of imperfect information. *Evolutionary Ecology* 12:601–613.

Koshland, D. 1980. Bacterial Chemotaxis as a Model Behavioral System. New York: Raven Press.

Kozhevnikov, A. A., and M. S. Fee. 2007. Singing-related activity of identified HVC neurons in the zebra finch. *Journal of Neurophysiology* 97:4271–4283.

Kramer, H. G., and R. E. Lemon. 1983. Dynamics of territorial singing between neighboring song sparrows (*Melospiza melodia*). *Behaviour* 85:198–223.

Krebs, J. R. 1990. Food-storing birds: Adaptive specialization in brain and behaviour? *Philosophical Transactions of the Royal Society Series B* 329:153–160.

Krebs, J. R., R. Ashcroft, and K. Van Orsdol. 1981. Song matching in the great tit *Parus major* L. *Animal Behaviour* 29:918–923.

Krebs, J. R., N. S. Clayton, S. D. Healy, D. A. Cristol, S. N. Patel, and A. R. Jolliffe. 1996. The ecology of the avian brain: Food-storing memory and the hippocampus. *Ibis* 138:34–46.

Krebs, J. R., and N. B. Davies, eds. 1978. *Behavioural Ecology: An Evolutionary Approach*. Oxford: Blackwell.

Krebs, J. R., and R. Dawkins. 1984. Animal signals: Mind-reading and manipulation. In J. R. Krebs and N. B. Davies, eds., *Behavioural Ecology: An Evolutionary Approach*, 380–402. 2d ed. Oxford: Blackwell Scientific.

Krebs, J. R., D. F. Sherry, S. D. Healy, V. H. Perry, and A. L. Vaccarino. 1989. Hippocampal specialization of food-storing birds. *Proceedings of the National Academy of Sciences of the United States of America* 86:1388–1392.

Kroodsma, D. E. 1974. Song learning, dialects, and dispersal in the Bewick's wren. *Zeitschrift für Tierpsychologie* 35:352–380.

———. 1982. Song repertoires: Problems in their definition and use. In D. E. Kroodsma and E. H. Miller, *Acoustic Communication in Birds*, 125–146. New York: Academic Press.

———. 1983. The ecology of avian vocal learning. *BioScience* 33:165–171.

———. 1988. Contrasting styles of song development and their consequences among passerine birds. In R. C. Bolles and M. D. Beecher, eds., *Evolution and Learning*, 157–184. Hillsdale, NJ: Erlbaum.

———. 1996. Ecology of passerine song development. In D. E. Kroodsma and E. H. Miller, eds., *Ecology and Evolution of Acoustic Communication in Birds*, 3–19. Ithaca: Cornell University Press.

Kroodsma, D. E., W. C. Liu, E. Goodwin, and P. A. Bedell. 1999a. The ecology of song improvisation as illustrated by North American sedge wrens. *Auk* 116:373–386.

Kroodsma, D. E., and R. Pickert. 1984. Repertoire size, auditory templates, and selective vocal learning in songbirds. *Animal Behaviour* 32:395–399.

Kroodsma, D. E., J. Sanchez, D. W. Stemple, E. Goodwin, M. L. da Silva, and J. M. E. Vielliard. 1999b. Sedentary life style of Neotropical sedge wrens promotes song imitation. *Animal Behaviour* 57:855–863.

Kroodsma, D. E., and J. Verner. 1978. Complex singing behaviors among *Cistothorus* wrens. *Auk* 95:703–716.

Kruijt, J. P., I. Bossema, and G. J. Lammers. 1982. Effects of early experience and male activity on mate choice in mallard females (*Anas platyrhynchos*). *Behaviour* 80:32–43.

Kruuk, H. 2003. Niko's Nature: The Life of Niko Tinbergen and His Science of Animal Behaviour. Oxford: Oxford University Press.

Kuang, S., S. A. Doran, R. J. A. Wilson, G. G. Goss, and J. I. Goldberg. 2002. Serotonergic sensory-motor neurons mediate a behavioral response to hypoxia in pond snail embryos. *Journal of Neurobiology* 52:73–83.

Kubat, S. 1992. Die Rolle von Neuigkeit, Andersartigkeit und sozialer Struktur für die Exploration von Objekten beim Kea (*Nestor notabilis*). Thesis, University of Vienna.

Kudo, H., and R. I. M. Dunbar. 2001. Neocortex size and social network size in primates. *Animal Behaviour* 62:711–722.

Kuhl, P. K. 1981. Discrimination of speech by nonhuman animals: Basic auditory sensitivities conducive to the perception of speech sound categories. *Journal of the Acoustical Society of America* 70:340–349.

Kunc, H. P., V. Amrhein, and M. Naguib. 2006. Vocal interactions in nightingales, *Luscinia megarhynchos*: More aggressive males have higher pairing success. *Animal Behaviour* 72:25–30.

Kunz, T. H., and M. B. Fenton. 2003. *Bat Ecology*. Chicago: University of Chicago Press.

Kusch, R. C., and D. P. Chivers. 2004. The effects of crayfish predation on phenotypic and life-history variation in fathead minnows. *Canadian Journal of Zoology—Revue Canadienne de Zoologie* 82:917–921.

Lachlan, R. F., and M. R. Servedio. 2004. Song learning accelerates allopatric speciation. *Evolution* 58:2049–2063.

Lagerhans, R. B., and T. J. DeWitt. 2002. Plasticity constrained: Over-generalized induction cues cause maladaptive phenotypes. *Evolutionary Ecology Research* 4:857–870.

Laland, K. N. 2004. Social learning strategies. *Learning and Behavior* 32:4–14.

Laland, K. N., and J. R. Kendal. 2003. What the models say about social learning. In D. M. Fragaszy and S. Perry, eds., *The Biology of Traditions: Models and Evidence*, 33–55. Cambridge: Cambridge University Press.

Laland K. N., J. R. Kendal, and R. L. Kendal. In press. Animal culture: Problems and solutions. In K. N. Laland and B. G. Galef Jr., eds., *The Question of Animal Culture*. Cambridge: Harvard University Press.

Laland, K. N., and K. Williams. 1998. Social transmission of maladaptive information in the guppy. *Behavioral Ecology* 9:493–499.

Lampe, H. M., and G.-P. Saetre. 1995. Female pied flycatchers prefer males with larger song repertoires. *Proceedings of the Royal Society of London Series B* 262:163–167.

Lande, R. 1981. Models of speciation by sexual selection on polygenic traits. *Proceedings of the National Academy of Science of the United States of America* 78:3721–3725.

Latham, K. E., and J. J. Just. 1989. Oxygen availability provides a signal for hatching in the rainbow trout (*Salmo gairdneri*) embryo. *Canadian Journal of Fisheries and Aquatic Sciences* 46:55–58.

Laurila, A., S. Pakkasmaa, P.-A. Crochet, and J. Merilä. 2002. Predator-induced plasticity in early life history and morphology in two anuran amphibians. *Oecologia* 132:524–530.

Laverty, T. M. 1994. Costs to foraging bumble bees of switching plant species. *Canadian Journal of Zoology—Revue Canadienne de Zoologie* 72:43–47.

Laverty, T. M., and R. C. Plowright. 1988. Flower handling by bumblebees: A comparison of specialists and generalists. *Animal Behaviour* 36:733–740.

Lawrence, B. D., and J. A. Simmons. 1982. Measurements of atmospheric attenuation at ultrasonic frequencies and the significance for echolocation by bats. *Journal of the Acoustical Society of America* 71:585–590.

Leavesley, A. J., and R. D. Magrath. 2005. Communicating about danger: Urgency alarm calling in a bird. *Animal Behaviour* 70:365–373.

Lee, P. C. 1991. Adaptations to environmental change: An evolutionary perspective. In H. O. Box, ed., *Primate Responses to Environmental Change*, 39–56. London: Chapman and Hall.

Lefebvre, L. 1983. Equilibrium distribution of feral pigeons at multiple food sources. *Behavioural Ecology and Sociobiology* 12:11–17.

———. 2000. Feeding innovations and their cultural transmission in bird populations. In C. M. Heyes and L. Huber, eds., *The Evolution of Cognition*, 311–328. Cambridge: MIT Press.

Lefebvre, L., and J. J. Bolhuis. 2003. Positive and negative correlates of feeding innovations in birds: Evidence for limited modularity. In S. M. Reader and K. N. Laland, eds., *Animal Innovation*, 39–61. Oxford: Oxford University Press.

Lefebvre, L., and L.-A. Giraldeau. 1996. Is social learning an adaptive specialisation? In C. M. Heyes and B. G. Galef Jr., eds., *Social Learning in Animals: The Roots of Culture*, 107–128. San Diego: Academic Press.

Lefebvre, L., N. Nicolakakis, and D. Boire. 2002. Tools and brains in birds. *Behaviour* 139:939–973.

Lefebvre, L., and B. Palameta. 1988. Mechanism, ecology, and population diffusion of socially learned, food-finding behaviour in feral pigeons. In T. R. Zentall and B. G. Galef Jr., eds., *Social learning: Psychological and Biological Perspectives*, 141–164. Hillsdale, NJ: Erlbaum.

Lefebvre, L., B. Palameta, and K. K. Hatch. 1996. Is group-living associated with social learning? A comparative test of a gregarious and a territorial columbid. *Behaviour* 133:241–261.

Lefebvre, L., S. M. Reader, and D. Boire. 2006. The evolution of encephalization. In J. H. Kaas, ed., *The Evolution of Nervous Systems,* 121–142. Oxford: Elsevier.

Lefebvre, L., S. M. Reader, and D. Sol. 2004. Brains, innovations and evolution in birds and primates. *Brain, Behavior and Evolution* 63:233–246.

Leggio, M. G., L. Mandolesi, F. Federico, F. Spirito, B. Ricci, F. Gelfo, and L. Petrosini. 2005. Environmental enrichment promotes improved spatial abilities and enhanced dendritic growth in the rat. *Behavioral Brain Research* 163:78–90.

Lehrer, M. 1997. Honeybee's visual orientation at the feeding site. In M. Lehrer, ed., *Orientation and Communication in Arthropods,* 115–144. Basel: Birkhauser.

Lemaire, V., Koehl, M., Le Moal, M., Abrous, D. N. 2000. Prenatal stress produces learning deficits associated with inhibition of neurogenesis in the hippocampus. *Proceedings of the National Academy of Sciences of the United States of America* 97:11032–11037.

Lemon, R. E., S. Perreault, and D. M. Weary. 1994. Dual strategies of song development in American redstarts, *Setophaga ruticilla. Animal Behaviour* 47:317–329.

Lengagne, T., and P. J. Slater. 2002. The effects of rain on acoustic communication: Tawny owls have good reason for calling less in wet weather. *Proceedings of the Royal Society of London Series B* 269:2121–2125.

Leonardo, A., and M. S. Fee. 2005. Ensemble coding of vocal control in birdsong. *Journal of Neuroscience* 25:652–661.

Le Roux, A., M. I. Cherry, and M. B. Manser. 2008. The audience effect in a facultatively social mammal, the yellow mongoose, *Cynictis penicillata. Animal Behaviour* 75:943–949.

Leuner, B., E. Gould, and T. J. Shors. 2006. Is there a link between adult neurogenesis and learning? *Hippocampus* 16:216–224.

Lewis, E. R., and P. M. Narins. 1985. Do frogs communicate with seismic signals? *Science* 215:1641–1643.

Lewis, J. W. 2006. Cortical networks related to human use of tools. *Neuroscientist* 12:211–231.

Li, D. Q. 2002. Hatching responses of subsocial spitting spiders to predation risk. *Proceedings of the Royal Society of London Series B* 269:2155–2161.

Li, D. Q., and R. R. Jackson. 2005. Influence of diet-related chemical cues from predators on the hatching of egg-carrying spiders. *Journal of Chemical Ecology* 31:333–342.

Liker, A., and T. Székely. 2005. Mortality costs of sexual selection and parental care in natural populations of birds. *Evolution* 59:890–897.

Lim, M. M., Z. X. Wang, D. E. Olazabal, X. H. Ren, E. F. Terwilliger, and L. J. Young. 2004. Enhanced partner preference in a promiscuous species by manipulating the expression of a single gene. *Nature* 429:754–757.

Lim, M. M., and L. J. Young. 2004. Vasopressin-dependent neural circuits underlying pair bond formation in the monogamous prairie vole. *Neuroscience* 125:35–45.

Lindenfors, P., C. L. Nunn, and R. A. Barton. 2007. Primate brain architecture and selection in relation to sex. *BMC Biology* 5:20.

Lindstrom, L., R. V. Alatalo, A. Lyytinen, and J. Mappes. 2001. Predator experience on cryptic prey affects the survival of conspicuous aposematic prey. *Proceedings of the Royal Society of London Series B* 268:357–361.

Lint, K. C. 1958. High haunts and strange habits of the kea. *Zoonooz* 31:3–6.

Liu, D., J. Diorio, B. Tannenbaum, C. Caldji, D. Francis, A. Freedman, S. Sharma, D. Pearson, P. M. Plotsky, and M. J. Meaney. 1997. Maternal care, hippocampal glucocorticoid receptors, and hypothalamic-pituitary-adrenal responses to stress. *Science* 277:1659–1662.

Liu, G., H. Seiler, A. Wen, T. Zars, K. Ito, R. Wolf, M. Heisenberg, and L. Liu. 2006. Distinct memory traces for two visual features in the *Drosophila* brain. *Nature* 439:551–556.

Liu, W.-C., and D. E. Kroodsma. 2006. Song learning by chipping sparrows: When, where, and from whom. *Condor* 108:509–517.

Liu, Y., J. T. Curtis, and Z. X. Wang. 2001. Vasopressin in the lateral septum regulates pair bond formation in male prairie voles (*Microtus ochrogaster*). *Behavioral Neuroscience* 115:910–919.

Livdahl, T. P., and J. S. Edgerly. 1987. Egg hatching inhibition: Field evidence for population regulation in a treehole mosquito. *Ecological Entomology* 12:395–399.

Lively, C. M. 1986a. Competition, comparative life histories, and maintenance of shell dimorphism in a barnacle. *Ecology* 67:858–864.

———. 1986b. Predator-induced shell dimorphism in the acorn barnacle *Chthamalus anisopoma*. *Evolution* 40:232–242.

Livezey, B. C. 1986. A phylogenetic analysis of recent anseriform genera using morphological characters. *Auk* 103:737–754.

Lombard, E. 1911. Le signe de l'elevation de la voix. *Annales des Maladies de l'Oreille et du Larynx* 37:101–119.

Lorenz, K. Z. 1970. *Studies in Animal and Human Behavior*. Cambridge: Harvard Univeristy Press.

Losos, J. B., T. W. Schoener, and D. A. Spiller. 2003. Effect of immersion in seawater on egg survival in the lizard *Anolis sagrei*. *Oecologia* 137:360–362.

———. 2004. Predator-induced behaviour shifts and natural selection in field experimental lizard populations. *Nature* 432:505–508.

Louissant, A., Jr., S. Rao, C. Leventhal, and S. A. Goldman. 2002. Coordinated interaction of neurogenesis and angiogenesis in the adult songbird brain. *Neuron* 34:945–960.

Lucas, J. R., A. Brodin, S. de Kort, and N. S. Clayton. 2004. Does hippocampal volume correlate with the degree of caching specialization? *Proceedings of the Royal Society of London Series B* 271:2423–2429.

Lutz, C. C., N. Ismail, A. Brockmann, S. E. Fahrbach, and G. E. Robinson. 2007. Regulation of gene expression in the mushroom bodies of the honey bee by foraging experience and pilocarpine. Paper presented at the Society for Neuroscience annual meeting.

Lynch, K. S., A. S. Rand, M. J. Ryan, and W. Wilczynski. 2005. Reproductive state influences female plasticity in mate choice. *Animal Behaviour* 69:689–699.

Lynch, M., and S. J. Arnold. 1988. The measurement of selection on size and growth. In B. Ebenman and L. Person, eds., *Size Structured Populations*, 47–59. Berlin: Springer-Verlag.

Lynn, S. E., C. W. Breuner, and J. C. Wingfield, J. C. 2003. Short-term fasting affects locomotor activity, corticosterone, and corticosterone binding globulin in a migratory songbird. *Hormones and Behavior* 43:150–157.

MacDonald, I. F., B. Kempster, L. Zanette, and S. A. MacDougall-Shackleton. 2006. Early nutritional stress impairs development of a song-control brain region in both male and female juvenile song sparrows (*Melospiza melodia*) at the onset of song learning. *Proceedings of the Royal Society of London Series B* 273:2559–2564.

Macedonia, J. M. 1990. What is communicated in the antipredator calls of lemurs: Evidence from playback experiments with ringtailed and ruffed lemurs. *Ethology* 86:177–190.

Macedonia, J. M., and C. S. Evans. 1993. Variation among mammalian alarm call systems and the problem of meaning in animal signals. *Ethology* 93:177–197.

Machado, C. A., R. M. Kliman, J. A. Markert, and J. Hey. 2002. Inferring the history of speciation from multilocus DNA sequence data: The case of *Drosophila pseudoobscura* and close relatives. *Molecular Biology and Evolution* 19:472–488.

Macmillan, N. A., and C. D. Creelman. 2005. *Detection Theory: A User's Guide*. 2d ed. Mahwah, NJ: Lawrence Erlbaum Associates.

Macphail, E. M. 2002. The role of avian hippocampus in spatial memory. *Psicológica* 23:93–108.

Macphail, E. M., and J. J. Bolhuis. 2001. The evolution of intelligence: Adaptive specialization versus general process. *Biological Reviews of the Cambridge Philosophical Society* 76:341–364.

Madden, J. 2001. Sex, bowers and brains. *Proceedings of the Royal Society of London Series B* 268:833–838.

Madge, S., and P. J. K. McGowan. 2002. *Pheasants, Partridges and Grouse*. Princeton: Princeton University Press.

Maguire, E. A. 2001. The retrosplenial contribution to human navigation: A review of lesion and neuroimaging findings. *Scandinavian Journal of Psychology* 42:225–238.

Maguire, E. A., D. G. Gadian, I. S. Johnsrude, C. D. Good, J. Ashburner, R. S. J. Frackowiak, and C. D. Frith. 2000. Navigation-related structural change in the hippocampi of taxi drivers. *Proceedings of the National Academy of Sciences of the United States of America* 97:4398–4403.

Maguire, E. A., H. J. Spiers, C. D. Good, T. H. Richard, S. J. Frackowiak, and N. Burgess. 2003. Navigation expertise and the human hippocampus: A structural brain imaging analysis. *Hippocampus* 13:250–259.

Maguire, E. A., K. Woollett, and H. J. Spiers. 2006. London taxi drivers and bus drivers: A structural MRI and neuropsychological analysis. *Hippocampus* 16:1091–1101.

Magurran, A. E. 1989. Acquired recognition of predator odour in the European minnow (*Phoexinus phoexinus*). *Ethology* 82:216–223.

Magurran, A. E., and I. W. Ramnarine. 2004. Learned mate recognition and reproductive isolation in guppies. *Animal Behaviour* 67:1077–1082.

Malun, D., M. Giurfa, C. G. Galizia, N. Plath, R. Brandt, B. Gerber, and B. Eisermann. 2002a. Hydroxyurea-induced partial mushroom body ablation does not affect acquisition and retention of olfactory differential conditioning in honeybees. *Journal of Neurobiology* 53:343–360.

Malun, D., N. Plath, M. Giurfa, A. D. Moseleit, and U. Muller. 2002b. Hydroxyurea-induced partial mushroom body ablation in the honeybee *Apis mellifera:* Volumetric analysis and quantitative protein determination. *Journal of Neurobiology* 50:31–44.

Manser, M. B. 1998. The evolution of auditory communication in suricates, *Suricata suricatta*. Thesis, University of Cambridge.

———. 1999. Response of foraging group members to sentinel calls in suricates, *Suricata suricatta*. *Proceedings of the Royal Society of London Series B* 266:1013–1019.

———. 2001. The acoustic structure of suricates' alarm calls varies with predator type and the level of response urgency. *Proceedings of the Royal Society of London Series B* 268:2315–2324.

———. In preparation. Functionally referential alarm calls in meerkats are denotative rather than imperative.

Manser, M. B., and M. B. Bell. 2004. Spatial representation of shelter locations in meerkats, *Suricata suricatta*. *Animal Behaviour* 68:151–157.

Manser, M. B., M. B. Bell, and L. B. Fletcher. 2001. The information that receivers extract from alarm calls in suricates. *Proceedings of the Royal Society of London Series B* 268:2485–2491.

Manser, M. B., M. Dewas, and L. Hollén. In preparation. Non-specific alarm call in meerkats is functional referential regarding movement of other animals.

Manser, M. B., and L. B. Fletcher. 2004. Vocalize to localize—a test on functionally referential alarm calls. *Interactions Studies* 5:325–342.

Manser, M. B., R. M. Seyfarth, and D. L. Cheney. 2002. Suricate alarm calls signal predator class and urgency. *Trends in Cognitive Sciences* 6:55–57.

Marchetti, C., and P. J. Drent. 2000. Individual differences in the use of social information in foraging by captive great tits. *Animal Behaviour* 60:131–140.

Marden, J. H., and P. Chai. 1991. Aerial predation and butterfly design: How palatability, mimicry, and the need for evasive flight constrain mass allocation. *American Naturalist* 138:15–36.

Marfori, M. A., P. G. Parker, T. G. Gregg, J. G. Vandenbergh, and N. G. Solomon. 1997. Using DNA fingerprinting to estimate relatedness within social groups of pine voles. *Journal of Mammalogy* 78:715–724.

Margulies, C., T. Tully, and J. Dubnau. 2005. Deconstructing memory in *Drosophila*. *Current Biology* 15:R700–R713.

Marino, L. 2005. Big brains do matter in new environments. Proceedings of the National Academy of Sciences of the United States of America 102:5306–5307.

Markow, T. A. 1988. Reproductive behavior of *Drosophila melanogaster* and *D. nigrospiracula* in the field and in the laboratory. *Journal of Comparative Psychology* 102:169–173.

Marler, P. 1955. Characteristics of some animal calls. *Nature* 176:6–8.

———. 1961. Logical analysis of animal communication. *Journal of Theoretical Biology* 1:295–317.

———. 1967. Animal communication signals. *Science* 157:769–774.

Marler, P., A. Dufty, and R. Pickert. 1986. Vocal communication in the domestic chicken. 2. Is a sender sensitive to the presence and nature of a receiver? *Animal Behaviour* 34:194–198.

Marler, P., C. S. Evans, and M. D. Hauser. 1992. Animal signals: Motivational, referential, or both? In H. Papousek, U. Jurgens, and M. Papousek, eds., *Nonverbal Communication: Comparative and Developmental Approaches*, 66–86. Cambridge: Cambridge University Press.

Marler, P., and S. Peters. 1987. A sensitive period for song acquisition in the song sparrow, *Melospiza melodia*, a case of age-limited learning. *Ethology* 76:89–100.

———. 1988. The role of song phonology and syntax in vocal learning preferences in the song sparrow, *Melospiza melodia*. *Ethology* 77:125–149.

Martin, K. L. M. 1999. Ready and waiting: Delayed hatching and extended incubation of anamniotic vertebrate terrestrial eggs. *American Zoologist* 39:279–288.

Martin, T. E. 2004. Avian life-history evolution has an eminent past: Does it have a bright future? *Auk* 121:289–301.

Marzluff, J. M., and R. P. Balda. 1992. The Pinyon Jay: Behavioral Ecology of a Colonial and Cooperative Bird. London: Poyser.

Maynard Smith, J. 1982. *Evolution and the Theory of Games*. Cambridge: Cambridge University Press.

Maynard Smith, J., and D. Harper. 2003. *Animal Signals*. Oxford: Oxford University Press.

Maynard Smith, J., and G. A. Parker. 1976. Logic of asymmetric contests. *Animal Behaviour* 24:159–175.

Mayr, E. 1946. Experiments on sexual isolation in *Drosophila*. VII. The nature of the isolating mechanisms between *Drosophila pseudoobscura* and *Drosophila persimilis*. *Proceedings of the Royal Society of London Series B* 32:128–137.

———. 1965. The nature of colonising birds. In H. G. Baker and G. L. Stebbins, eds., *The Genetics of Colonizing Species*, 29–43. New York: Academic Press.

McArthur, P. D. 1986. Similarity of playback songs to self song as a determinant of response strength in song sparrows (*Melospiza melodia*). *Animal Behavior* 34:199–207.

McCasland, J. S. 1987. Neuronal control of bird song production. *Journal of Neuroscience* 7:23–39.

McElreath, R., M. Lubell, P. J. Richerson, T. M. Waring, W. Baum, E. Edsten, C. Efferson, and B. Paciotti. 2005. Applying evolutionary models to the laboratory study of social learning. *Evolution and Human Behavior* 26:483–508.

McGowan, K. J., and G. E. Woolfenden. 1989. A sentinel system in the Florida scrub-jay. *Animal Behaviour* 37:1000–1006.

McGregor, P. K. 1991. The singer and the song: On the receiving end of bird song. *Biological Reviews of the Cambridge Philosophical Society* 66:57–82.

McGregor, P. K., and T. Dabelsteen. 1996. Communication networks. In D. E. Kroodsma and E. H. Miller, eds., *Ecology and Evolution of Acoustic Communication in Birds,* 409–425. Ithaca: Cornell University Press.

McGregor, P. K., and J. R. Krebs. 1982. Song types in a population of great tits (*Parus major*): Their distribution, abundance and acquisition by individuals. *Behaviour* 79:126–152.

———. 1989. Song learning in adult great tits (*Parus major*): Effects of neighbours. *Behaviour* 108:139–159.

McGregor, P. K., and D. B. A. Thompson. 1988. Constancy and change in local dialects of the corn bunting. *Ornis Scandinavica* 19:153–159.

McGuire, S. E., P. T. Le, and R. L. Davis. 2001. The role of *Drosophila* mushroom body signaling in olfactory memory. *Science* 293:1330–1333.

Meaney, M. J., J. Diorio, D. Francis, J. Widdowson, P. LaPlante, C. Caldji, S. Sharma, J. R. Seckl, and P. M. Plotsky. 1996. Early environmental regulation of forebrain glucocorticoid receptor gene expression: Implications for adrenocortical responses. *Developmental Neuroscience* 18:49–72.

Meech, R. W., and G. O. Mackie. 2007. Evolution of excitability in lower metazoans. In G. North and R. J. Greenspan, eds., *Invertebrate Neurobiology,* 581–615. Cold Spring Harbor, NY: Cold Spring Harbor Laboratory Press.

Mennill, D. J., P. T. Boag, and L. M. Ratcliffe. 2003. The reproductive choices of eavesdropping female black-capped chickadees, *Poecile atricapillus*. *Naturwissenschaften* 90:577–582.

Mennill, D. J., and L. M. Ratcliffe. 2004. Overlapping and matching in the song contests of black-capped chickadees. *Animal Behaviour* 67:441–450.

Mennill, D. J., L. M. Ratcliffe, and P. T. Boag. 2002. Female eavesdropping on male song contests in songbirds. *Science* 296:873.

Menzel, R. 1985. Learning in honey bees in an ecological and behavioral context. In B. Hölldobler and M. Lindauer, eds., *Experimental Behavioral Ecology and Sociobiology,* 55–74. New York: Gustav Fischer Verlag.

Mercer, A. R., P. G. Mobbs, A. P. Davenport, and P. D. Evans. 1983. Biogenic amines in the brain of the honeybee, *Apis mellifera*. *Cell and Tissue Research* 234:655–677.

Merilla, J., and D. A. Wiggins. 1995. Offspring number and quality in the blue tit: A quantitative genetic approach. *Journal of Zoology* 237:615–623.

Mery, F., and T. J. Kawecki. 2003. A fitness cost of learning ability in *Drosophila melanogaster*. *Proceedings of the Royal Society of London Series B* 270:2465–2469.

———. 2004. An operating cost of learning in *Drosophila melanogaster*. *Animal Behaviour* 68:589–598.

———. 2005. A cost of long-term memory in *Drosophila*. *Science* 308:1148.

Mesoudi, A., and M. J. O'Brien. 2008. The cultural transmission of Great Basin projectile-point technology: An experimental simulation. *American Antiquity* 73:3–28.

Metcalfe, N. B., and P. Monaghan. 2001. Compensation for a bad start: Grow now, pay later? *Trends in Ecology and Evolution* 16:254–260.

Mettke-Hofmann, C. 2000. Niche expansion and exploratory behavior in islands: Are they linked? *International Zoo Yearbook* 37:244–256

Mettke-Hofmann, C., and E. Gwinner. 2003. Long-term memory for a life on the move. *Proceedings of the National Academy of Sciences of the United States of America* 100:5863–5866.

Mettke-Hofmann, C., H. Winkler, and B. Leisler. 2002. The significance of ecological factors for exploration and neophobia in parrots. *Ethology* 108:249–272.

Michel, R. P., and L. M. Cruz-Orive. 1988. Application of the Cavalieri principle and vertical sections method to lung: Estimation of volume and pleural surface area. *Journal of Microscopy* 150:117–136.

Midford, P. E., J. P. Hailman, and G. E. Woolfenden. 2000. Social learning of a novel foraging patch in families of free-living Florida scrub-jays. *Animal Behaviour* 59:1199–1207.

Miller, G. F., and L. Penke. 2007. The evolution of human intelligence and the coefficient of additive genetic variance in human brain size. *Intelligence* 35:97–114.

Miller, L. A., and A. Surlykke. 2001. How some insects detect and avoid being eaten by bats: Tactics and countertactics of prey and predator. *BioScience* 51:570–581.

Miller, P. L. 1992. The effect of oxygen lack on egg hatching in an Indian dragonfly, *Potamarcha congener*. *Physiological Entomology* 17:68–72.

Mineka, S., Davidson, M., M. Cook, and R. Keir. 1984. Observational conditioning of snake fear in rhesus-monkeys. *Journal of Abnormal Psychology* 93:355–372.

Mirescu, C., and E. Gould. 2006. Stress and adult neurogenesis. *Hippocampus* 16:233–238.

Mirescu, C., J. D. Peters, and E. Gould. 2004. Early life experience alters response of adult neurogenesis to stress. *Nature Neuroscience* 7:841–846.

Mirescu, C., J. D. Peters, L. Noiman, and E. Gould. 2006. Sleep deprivation inhibits adult neurogenesis in the hippocampus by elevating glucocorticoids. *Proceedings of the National Academy of Sciences of the United States of America* 103:19170–19175.

Mobbs, P. G. 1982. The brain of the honeybee *Apis mellifera* L.: The connections and spatial organization of the mushroom bodies. *Philosophical Transactions of the Royal Society of London Series B* 298:309–354.

———. 1984. Neural networks in the mushroom bodies of the honeybee. *Journal of Insect Physiology* 30:43–58.

Molles, L. E., and S. L. Vehrencamp. 2001. Songbird cheaters pay a retaliation cost: Evidence for auditory conventional signals. *Proceedings of the Royal Society of London Series B* 268:2013–2019.

Moore, B. R. 1992. Avian movement imitation and a new form of mimicry: Tracing the evolution of a complex form of learning. *Behaviour* 122:231–263.

Moore, R. D., B. Newton, and A. Sih. 1996. Delayed hatching as a response of streamside salamander eggs to chemical cues from predatory sunfish. *Oikos* 77:331–335.

Moreira, P. L., and M. Barata. 2005. Egg mortality and early embryo hatching caused by fungal infection of Iberian rock lizard (*Lacerta monticola*) clutches. *Herpetological Journal* 15:265–272.

Morgan, L. C. 1890. *Animal Life and Intelligence*. Boston: Ginn.

Morgan, S. G. 1995. The timing of larval release. In L. McEdward, ed., *Ecology of Marine Invertebrate Larvae*, 157–191. Boca Raton, FL: CRC Press.

Morse, D. H. 1980. Behavioral mechanisms in ecology. Cambridge: Harvard University Press.

Morton, E. S. 1975. Ecological sources of selection on avian sounds. *American Naturalist* 109:17–34.

———. 1977. Occurrence and significance of motivation structural rules in some bird and mammal sounds. *American Naturalist* 111:855–869.

Moscarini, G., M. Ottaviani, and L. Smith. 1998. Social learning in a changing world. *Economic Theory* 11:657–665.

Moss, C. F., and A. Surlykke. 2001. Auditory scene analysis by echolocation in bats. *Journal of the Acoustical Society of America* 110:2207–2226.

Mountjoy, D. J., and R. E. Lemon. 1995. Extended song learning in wild European starlings. *Animal Behaviour* 49:357–366.

Murphy, S., I. Beall, A. Renner, and T. Smock. 1997. Orthodromic activation of peptidergic cells in the medial amygdala. *Peptides* 18:1175–1177.

Myers, R. E. 1976. Comparative neurology of vocalization and speech: Proof of a dichotomy. *Annals of the New York Academy of Sciences* 280:745–757.

Myrvik, Q. N. 1994. Immunology and nutrition. In M. E. Shils, J. A. Olson, and M. Shike, eds., *Modern Nutrition in Health and Disease,* 623–662. Philadelphia: Lea and Febiger.

Nagarajan, R., S. E. G. Lea, and J. D. Goss-Custard. 2002. Mussel valve discrimination and strategies used in valve discrimination by the oystercatcher, *Haematopus ostralegus. Journal of Animal Ecology* 16:339–345.

Naguib, M. 1999. Effects of song overlapping and alternating on nocturnally singing nightingales. *Animal Behaviour* 58:1061–1067.

Naguib, M., V. Amrhein, and H. P. Kunc. 2004. Effects of territorial intrusions on eavesdropping neighbors: Communication networks in nightingales. *Behavioral Ecology* 15:1011–1015.

Naguib, M., C. Fichtel, and D. Todt. 1999. Nightingales respond more strongly to vocal leaders of simulated dyadic interactions. *Proceedings of the Royal Society of London Series B* 266:537–542.

Naguib, M., and D. Todt. 1997. Effects of dyadic vocal interactions on other conspecific receivers in nightingales. *Animal Behaviour* 54:1535–1543.

Negus, N. C., and P. J. Berger. 1977. Experimental triggering of reproduction in a natural-population of *Microtus montanus. Science* 196:1230–1231.

Negus, N. C., P. J. Berger, and L. G. Forslund. 1977. Reproductive strategy of *Microtus montanus. Journal of Mammalogy* 58:347–353.

Nelson, D. A. 1992. Song overproduction and selective attrition lead to song sharing in the field sparrow (*Spizella pusilla*). *Behavioral Ecology and Sociobiology* 30:415–424.

———. 1999. Ecological influences on vocal development in the white-crowned sparrow. *Animal Behaviour* 58:21–36.

Nelson, D. A., and P. Marler. 1989. Categorical perception of a natural stimulus continuum: Birdsong. *Science* 244:976–978.

———. 1994. Selection-based learning in bird song development. Proceedings of the National Academy of Sciences of the United States of America 91:10498–10501.

Nelson, R. J., and B. C. Trainor. 2007. Neural mechanisms of aggression. *Nature Reviews Neuroscience* 8:536–546.

Neuweiler, G. 1989. Foraging ecology and audition in echolocating bats. *Trends in Ecology and Evolution* 4:160–166.

———. 1990. Auditory adaptations for prey capture in echolocating bats. *Physiological Review* 70:615–641.

Neuweiler, G., W. Metzner, U. Heilmann, R. Rübsamen, M. Eckrich, and H. H. Costai. 1987. Foraging behaviour and echolocation in the rufous horseshoe bat (*Rhinolophus rouxi*) of Sri Lanka. *Behavioral Ecology and Sociobiology* 20:53–67.

Nicolakakis, N., and L. Lefebvre. 2001. Forebrain size and innovation rate in European birds: Feeding, nesting and confounding variables. *Behaviour* 137:1415–1429.

Nicolakakis, N., D. Sol, and L. Lefebvre. 2003. Behavioural flexibility predicts species richness in birds, but not extinction risk. *Animal Behaviour* 65:445–452.

Nielsen, B. M. B., and S. L. Vehrencamp. 1995. Responses of song sparrows to song-type matching via interactive playback. *Behavioral Ecology and Sociobiology* 37:109–117.

Nishida, R. 2002. Sequestration of defensive substances from plants by Lepidoptera. *Annual Review of Entomology* 47:57–92.

Nol, E., and J. N. M. Smith. 1987. Effects of age and breeding experience on seasonal reproductive success in the song sparrow. *Journal of Animal Ecology* 56:301–313.

Noor, M. A. 1995. Speciation driven by natural selection in *Drosophila*. *Nature* 375:674–675.

Norberg, U. M., and M. B. Fenton. 1988. Carnivorous bats? *Biological Journal of the Linnaen Society* 33:383–394.

Norberg, U. M., and J. M. V. Rayner. 1987. Ecological morphology and flight in bats (Mammalia; Chiroptera): Wing adaptations, flight performance, foraging strategy and echolocation. *Philosophical Transactions of the Royal Society of London Series B* 316:335–427.

Nordby, J. C., S. E. Cambell, and M. D. Beecher. 1999. Ecological correlates of song learning in song sparrows. *Behavioral Ecology* 10:287–297.

———. 2001. Late song learning in song sparrows. *Animal Behaviour* 61:835–846.

———. 2002. Adult song sparrows do not alter their song repertoires. *Ethology* 108:39–50.

———. 2007. Selective attrition and individual song repertoire development in song sparrows. *Animal Behaviour* 74:1413–1418.

Nordby, J. C., S. E. Campbell, J. M. Burt, and M. D. Beecher. 2000. Social influences during song development in the song sparrow: A laboratory experiment simulating field conditions. *Animal Behaviour* 59:1187–1197.

Nordeen, E. J., A. Grace, M. J. Burek, and K. W. Nordeen. 1992. Sex-dependent loss of projection neurons involved in avian song learning. *Journal of Neurobiology* 23:671–679.

Nordeen, E. J., and K. W. Nordeen. 1988. Sex and regional differences in the incorporation of neurons born during song learning in zebra finches. *Journal of Neuroscience* 8:2869–2874.

North, G., and R. J. Greenspan. 2007. *Invertebrate Neurobiology*. Cold Spring Harbor, NY: Cold Spring Harbor Laboratory Press.

Norton-Griffiths, M. N. 1969. The organisation, control and development of parental feeding in the oystercatcher (*Haematopus ostralegus*). *Behaviour* 34:55–114.

Nottebohm, F. 2002a. Neuronal replacement in adult brain. *Brain Research Bulletin* 57:737–749.

———. 2002b. Why are some neurons replaced in adult brain? *Journal of Neuroscience* 22:624–628.

———. 2004. The road we travelled: Discovery, choreography, and significance of brain replaceable neurons. *Annals of the New York Acedemy of Sciences* 1016:628–658.

Nottebohm, F., S. Kasparian, and C. Pandazis. 1981. Brain space for a learned task. *Brain Research* 213:99–109.

Nottebohm, F., M. E. Nottebohm, and L. Crane. 1986. Developmental and seasonal changes in canary song and their relation to changes in the anatomy of song-control nuclei. *Behavioral and Neural Biology* 46:445–471.

Nottebohm, F., T. M. Stokes, and C. M. Leonard. 1976. Central control of song in the canary, *Serinus canarius*. *Journal of Comparative Neurology* 165:457–486.

Nowicki, S., S. Peters, and J. Podos. 1998. Song learning, early nutrition and sexual selection in songbirds. *American Zoologist* 38:179–190.

Nowicki, S., J. Podos, and F. Valdes. 1994. Temporal patterning of within-song type and between-song type variation in song repertoires. *Behavioral Ecology and Sociobiology* 34:329–335.

Nowicki, S., and W. A. Searcy. 2005a. Adaptive priorities in brain development: Theoretical comment on Pavosudov et al. (2005). *Behavioral Neuroscience* 119:1415–1418.

———. 2005b. Song and mate chice in birds: How the development of behavior helps us understand function. *Auk* 122:1–14.

Nowicki, S., W. A. Searcy, M. Hughes, and J. Podos. 2001. Evolution of bird song: Male and female response to song innovation in swamp sparrows. *Animal Behaviour* 62:1189–1195.

Nowicki, S., W. A. Searcy, and S. Peters. 2002a. Brain development, song learning and mate choice in birds: A review and experimental test of the "nutritional stress hypothesis." *Journal of Comparative Physiology A* 188:1003–1014.

———. 2002b. Quality of song learning affects female response to male bird song. *Proceedings of the Royal Society of London Series B* 269:1949–1954.

Oberweger, K., and F. Goller. 2001. The metabolic cost of birdsong production. *Journal of Experimental Biology* 204:3379–3388.

Obrist, M. K., and J. J. Wenstrup. 1998. Hearing and hunting in red bats (*Lasiurus borealis,* Vespertilionidae): Audiogram and ear properties. *Journal of Experimental Biology* 201:143–154.

O'Connor, R. J. 1984. *The Growth and Development of Birds*. New York: Wiley and Sons.

O'Donnell, S., N. A. Donlan, and T. A. Jones. 2004. Mushroom body structural change is associated with division of labor in eusocial wasp workers (*Polybia aequatorialis,* Hymenoptera: Vespidae). *Neuroscience Letters* 356:159–162.

———. 2007. Developmental and dominance-associated differences in mushroom body structure in the paper wasp *Mischocyttarus mastigophorus*. *Developmental Neurobiology* 67:39–46.

O'Loghlen, A. L., and M. D. Beecher. 1997. Sexual preferences for mate song types in female song sparrows. *Animal Behaviour* 53:835–841.

———. 1999. Mate, neighbour and stranger songs: A female song sparrow perspective. *Animal Behaviour* 58:13–20.

O'Loghlen, A. L., and S. I. Rothstein. 1995. Culturally correct song dialects are correlated with male age and female song preferences in wild populations of brown-headed cowbirds. *Behavioral Ecology and Sociobiology* 36:251–259.

Ophir, A. G., P. Campbell, K. Hanna, and S. M. Phelps. 2008a. Field tests of cis-regulatory variation at the prairie vole *avpr1a* locus: Association with V1aR abundance but not sexual or social fidelity. *Hormones and Behavior* 54:694–702.

Ophir, A. G., and J. DelBarco-Trillo. 2007. Anogenital distance predicts female choice and male potency in prairie voles. *Physiology and Behavior* 92:533–540.

Ophir, A. G., and B. G. Galef Jr. 2004. Sexual experience can affect use of public information in mate choice. *Animal Behaviour* 68:1221–1227.

Ophir, A.G., S. M. Phelps, A. B. Sorin, and J. O. Wolff. 2007. Morphological, genetic and behavioral comparisons of two prairie vole populations in the field and laboratory. *Journal of Mammalogy* 88:989–999.

Ophir, A. G., Sorin, A. B., Phelps, S. M., and Wolff, J. O. 2008b. Social but not genetic monogamy is associated with greater breeding success in prairie voles. *Animal Behaviour* 75:1143–1154.

Ophir, A. G., J. O. Wolff, and S. M. Phelps. 2008c. Variation in neural V1aR predicts sexual fidelity and space use among male prairie voles in semi-natural settings. *Proceedings of the National Academy of Sciences of the United States of America* 105:1249–1254.

Ortiz-Barrientos, D., B. A. Counterman, and M. A. F. Noor. 2004. The genetics of speciation by reinforcement. *Public Library of Science Biology* 2:2256–2263.

Otter, K., P. K. McGregor, A. M. R. Terry, F. R. L. Burford, T. M. Peake, and T. Dabelsteen. 1999. Do female great tits (*Parus major*) assess males by eavesdropping? A field study using interactive song playback. *Proceedings of the Royal Society of London Series B* 266:1305–1309.

Owens, I. P. F., and P. M. Bennett. 1997. Variation in mating system among birds: Ecological basis revealed by hierarchical comparative analysis of mate desertion. *Proceedings of the Royal Society of London Series B* 264:1103–1110.

——. 2000. Ecological basis of extinction risk in birds: Habitat loss versus human persecution and introduced predators. *Proceedings of the National Academy of Sciences of the United States of America* 97:12144–12148.

Page, R. A., and M. J. Ryan. 2005. Flexibility in assessment of prey cues: Frog-eating bats and frog calls. *Proceedings of the Royal Society of London Series B* 272:841–847.

——. 2006. Social transmission of novel foraging behavior in bats: Frog calls and their referents. *Current Biology* 16:1201–1205.

Page, R. E., G. E. Robinson, D. S. Britton, and M. K. Fondrk. 1992. Genotypic variability for rates of behavioral development in worker honeybees (*Apis mellifera* L.). *Behavioral Ecology* 3:173–180.

Palleroni, A., M. Hauser, and P. Marler. 2005. Do responses of galliform birds vary adaptively with predator size? *Animal Cognition* 8:200–210.

Pankiw, T. 2004. Brood pheromone regulates foraging activity of honey bees (Hymenoptera: Apidae). *Journal of Economic Entomology* 97:748–751.

Parker, S. T., and K. R. Gibson. 1977. Object manipulation, tool use and sensorimotor intelligence as feeding adaptations in cebus monkeys and great apes. *Journal of Human Evolution* 6:623–641.

Partan, S. R., and P. Marler. 1999. Communication goes multimodal. *Science* 283:1272–1273.

——. 2005. Issues in the classification of multimodal communication signals. *American Naturalist* 166:231–245.

Passingham, R. E. 1975. The brain and intelligence. *Brain, Behavior and Evolution* 11:1–15.

Passmore, N. I., and S. R. Telford. 1981. The effect of chorus organization on mate localization in the painted reed frog (*Hyperolius marmoratus*). *Behavioral Ecology and Sociobiology* 9:291–293.

Patel, S. N., N. S. Clayton, and J. R. Krebs. 1997. Spatial learning induces neurogenesis in the avian brain. *Behavioral Brain Research* 89:115–128.

Pavey, C. R., and C. J. Burwell. 2004. Foraging ecology of the horseshoe bat, *Rhinolophus megaphyllus* (Rhinolophidae), in eastern Australia. *Wildlife Research* 31:403–413.

Pavey, C. R., C. J. Burwell, and D. J. Milne. 2006. The relationship between echolocation-call frequency and moth predation of a tropical bat fauna. *Canadian Journal of Zoology—Revue Canadienne de Zoologie* 84:425–433.

Paxinos, G., and C. Watson. 2006. *The Rat Brain Atlas in Stereotaxic Coordinates*. New York: Academic Press.

Payne, R. B. 1981. Song learning and social interaction in indigo buntings. *Animal Behaviour* 29:688–697.

——. 1982. Ecological consequences of song matching: Breeding success and intraspecific song mimicry in indigo buntings. *Ecology* 63:401–411.

——. 1983. The social context of song mimicry: Song-matching dialects in indigo buntings (*Passerina cyanea*). *Animal Behaviour* 31:788–805.

——. 1985. Behavioral continuity and change in local song populations of village indigobirds *Vidua chalybeata*. *Zeitschrift für Tierpsychologie* 70:1–44.

Paz-y-Miño, G., A. B. Bond, A. C. Kamil, and R. P. Balda. 2004. Pinyon jays use transitive inference to predict social dominance. *Nature* 430:778–781.

Peake, T. M. 2005. Eavesdropping in communication networks. In P. K. McGregor, ed., *Animal Communication Networks*, 13–37. Cambridge: Cambridge University Press.

Peake, T. M., G. Matessi, P. K. McGregor, and T. Dabelsteen. 2005. Song type matching, song type switching and eavesdropping in male great tits. *Animal Behaviour* 69:1063–1068.

Peake, T. M., and P. K. McGregor. 2004. Information and aggression in fishes. *Learning and Behavior* 32:114–121.

Peake, T. M., A. M. Terry, P. K. McGregor, and T. Dabelsteen. 2001. Male great tits eavesdrop on simulated male-to-male vocal interactions. *Proceedings of the Royal Society of London Series B* 268:1183–1187.

Penke, Z., K. Felszeghy, B. Fernette, D. Sage, C. Nyaka, and A. Burlet. 2001. Postnatal maternal deprivation produces long-lasting modifications of the stress response, feeding and stress-related behavior in the rat. *European Journal of Neuroscience* 14:747–755.

Pepperberg, I. M. 1985. Social modeling theory: A possible framework for understanding avian vocal learning. *Auk* 102:854–864.

———. 2007. Grey parrots do not always "parrot": The roles of imitation and phonological awareness in the creation of new labels from existing vocalizations. *Language Sciences* 29:1–13.

Persons, M. H., L. J. Fleishman, M. A. Frye, and M. E. Stimphil. 1999. Sensory response patterns and the evolution of visual signal design in anoline lizards. *Journal of Comparative Physiology A* 184:585–607.

Pesendorfer, M. B. 2007. Individual and "observational" learning of object affordances in keas (*Nestor notabilis*). Thesis, University of Vienna.

Peters, S., W. A. Searcy, M. D. Beecher, and S. Nowicki. 2000. Geographic variation in the organization of song sparrow repertoires. *Auk* 117:936–942.

Petranka, J. W., J. J. Just, and E. C. Crawford. 1982. Hatching of amphibian embryos: The physiological trigger. *Science* 217:257–259.

Petrie, M., and B. Kempenaers. 1998. Extra-pair paternity in birds: Explaining variation between species and populations. *Trends in Ecology and Evolution* 13:52–58.

Pettifor, R. A., C. M. Perrins, and R. H. McCleery. 2001. The individual optimization of fitness: Variation in reproductive output, including clutch size, mean nestling mass and offspring recruitment, in manipulated broods of great tits *Parus major*. *Journal of Animal Ecology* 70:62–79.

Pfaff, J. A., L. Zanette, S. A. MacDougall-Shackleton, and E. A. MacDougall-Shackleton. 2007. Song repertoire size varies with HVC volume and is indicative of male quality in song sparrows (*Melospiza melodia*). *Proceedings of the Royal Society of London Series B* 274:2035–2040.

Pfaus, J. G., and M. M. Heeb. 1997. Implications of immediate-early gene induction in the brain following sexual stimulation of female and male rodents. *Brain Research Bulletin* 44:397–407.

Phelps, S. M., A. S. Rand, and M. J. Ryan. 2006. A cognitive framework for mate choice and species recognition. *American Naturalist* 167:28–42.

Phelps, S. M., and L. J. Young. 2003. Extraordinary diversity in vasopressin (V1a) receptor distributions among wild prairie voles (*Microtus ochrogaster*): Patterns of variation and covariation. *Journal of Comparative Neurology* 466:564–576.

Pierce, B. A. 2002. *Genetics: A Conceptual Approach*. New York: Freeman.

Pies-Schulz-Hofen, R. 2004. *Die Tierpflegerausbildung*. Stuttgart: Parey.

Pigliucci, M. 2001. *Phenotypic Plasticity: Beyond Nature and Nurture*. Baltimore: John Hopkins University Press.

Pitkow, L. J., C. A. Sharer, X. L. Ren, T. R. Insel, E. F. Terwilliger, and L. J. Young. 2001. Facilitation of affiliation and pair-bond formation by vasopressin receptor gene transfer into the ventral forebrain of a monogamous vole. *Journal of Neuroscience* 21:7392–7396.

Pitnick, S., K. E. Jones, and G. S. Wilkinson. 2006. Mating system and brain size in bats. *Proceedings of the Royal Society of London Series B* 273:719–724.

Platt, M. L. 2002. Neural correlates of decisions. *Current Opinion in Neurobiology* 12:141–148.

Plotkin, H. C., and F. J. Odling-Smee. 1979. Learning, change and evolution: An enquiry into the telenomy of learning. *Advances in the Study of Behavior* 10:1–41.

Plowright, C. M. S., and F. Landry. 2000. A direct effect of competition on food choice by pigeons. *Behavioural Processes* 50:59–64.

Podos, J. 1997. A performance constraint on the evolution of trilled vocalizations in a songbird family (Passeriformes: Emberizidae). *Evolution* 51:537–551.

Podos, J., S. Peters, T. Rudnicky, P. Marler, and S. Nowicki. 1992. The organization of song repertoires in song sparrows: Themes and variations. *Ethology* 90:89–106.

Poizat, G., E. Rosecchi, and A. J. Crivelli. 1999. Empirical evidence of a trade-off between reproductive effort and expectation of future reproduction in female three-spined sticklebacks. *Proceedings of the Royal Society of London Series B* 266:1543–1543.

Pollen, A. A., A. P. Dobberfuhl, J. Scace, M. M. Igulu, S. C. P. Renn, C. A. Shumway, and H. A. Hofmann. 2007. Environmental complexity and social organization sculpt the brain in Lake Tanganyikan cichlid fish. *Brain, Behavior and Evolution* 70:21–39.

Popper, A. N., and R. R. Fay. 1995. *Hearing in Bats*. Berlin: Springer.

Porter, S. 1947. The breeding of the kea (*Nestor notabilis*). *Avicultural Magazine* 53:50–55.

Poulin, P., and Q. J. Pittman. 1993. Arginine vasopressin-induced sensitization in brain: Facilitated inositol phosphate production without changes in receptor number. *Journal of Neuroendocrinology* 5:23–31.

Pravosudov, V. V. 2007. The relationship between environment, food caching, spatial memory, and the hippocampus in chickadees. In K. Otter, ed., *The Ecology and Behavior of Chickadees and Titmice*, 25–41. Oxford: Oxford University Press.

———. 2008. Mountain chickadees discriminate between potential cache pilferers and non-pilferers. *Proceedings of the Royal Society of London Series B* 275:55–61.

Pravosudov, V. V., and N. S. Clayton. 2002. A test of the adaptive specialization hypothesis: Population differences in caching, memory and the hippocampus in black-capped chickadees (*Poecile atricapilla*). *Behavioral Neuroscience* 116:515–522.

Pravosudov, V. V., and S. de Kort. 2006. Is the western scrub-jay (*Aphelocoma californica*) really an underdog among food-caching corvids when it comes to hippocampal volume and food caching propensity? *Brain, Behavior and Evolution* 67:1–9.

Pravosudov, V. V., and A. S. Kitaysky. 2006. Effects of nutritional restrictions during post-hatching development on adrenocortical function in western scrub-jays (*Aphelocoma californica*). *General and Comparative Endocrinology* 145:25–31.

Pravosudov, V. V., A. S. Kitaysky, and A. Omanska. 2006. The relationship between migratory behavior, memory and the hippocampus: An intraspecific comparison. *Proceedings of the Royal Society of London Series B* 273:2641–2649.

Pravosudov, V. V., P. Lavenex, and A. Omanska. 2005. Nutritional deficits during early development affect hippocampal structure and spatial memory later in life. *Behavioral Neuroscience* 119:1368–1374.

Pravosudov, V. V., and A. Omanska. 2005. Dominance-related changes in spatial memory are associated with changes in hippocampal cell proliferation rates in mountain chickadees. *Journal of Neurobiology* 62:31–41.

Pravosudov, V. V., K. Sanford, and T. P. Hahn. 2007. On the evolution of brain size in relation to migratory behaviour in birds. *Animal Behaviour* 73:535–539.

Premack, D. 1972. Concordant preferences as a precondition for affective but not for symbolic communication (or how to do experimental anthropology). *Cognition* 1:251–264.

Price, T. 2008. *Speciation in Birds*. Greenwood Village, CO: Roberts.

Price, T., A. Qvarnstrom, and D. Irwin. 2003. The role of phenotypic plasticity in driving genetic evolution. *Proceedings of the Royal Society of London Series B* 270:1433–1440.

Proctor, H. C. 1991. Courtship in the water mite *Neumania papillator:* Males capitalize on female adaptations for predation. *Animal Behaviour* 42:589–598.

Prokopy, R. J., A. L. Averill, S. S. Cooley, and C. A. Roitberg. 1982. Associative learning in egglaying site selection by apple maggot flies. *Science* 218:76–77.

Pugesek, B. H. 1981. Increased reproductive effort with age in the California gull (*Larus californicus*). *Science* 212:822–823.

Pulliam, H. R. 1981. Learning to forage optimally. In A. C. Kamil and T. D. Sargent, eds., *Foraging Behavior: Ecological, Ethological, and Psychological Approaches,* 379–388. New York: Garland Press.

Pyburn, W. F. 1970. Breeding behavior of the leaf-frogs *Phyllomedusa callidryas* and *Phyllomedusa dacnicolor* in Mexico. *Copeia* 1970:209–218.

Quinn, W. G. 2005. Nematodes learn: Now what? *Nature Neuroscience* 8:1639–1640.

Quinn, W. G., W. A. Harris, and S. Benzer. 1974. Conditioned behavior in Drosophila melanogaster. Proceedings of the National Academy of Sciences of the United States of America 71:708–712.

Raby, C. R., D. M. Alexis, A. Dickinson, and N. S. Clayton. 2007. Planning for the future in western scrub-jays. *Nature* 445:919–921.

Raff, R. A. 1996. The Shape of Life: Genes, Development, and the Evolution of Animal Form. Chicago: University of Chicago Press.

Rainey, H. J., K. Zuberbühler, and P. J. B. Slater. 2004. The response of black-casqued hornbills to predator vocalizations and primate alarm calls. *Behaviour* 141:1263–1277.

Rao, C. V., D. M. Wolf, and A. P. Arkin. 2002. Control, exploitation and tolerance of intracellular noise. *Nature* 420:231–237.

Raser, J. M., and E. K. O'Shea. 2005. Noise in gene expression: Origins, consequences, and control. *Science* 309:2010–2013.

Ratcliffe, J. M. 2009. Neuroecology and diet selection in phyllostomid bats. *Behavioral Processes* 80:247–251.

Ratcliffe, J. M., and J. W. Dawson. 2003. Behavioural flexibility: The little brown bat, *Myotis lucifugus,* and the northern long-eared bat, *M. septentrionalis,* both glean and hawk prey. *Animal Behaviour* 66:847–856.

Ratcliffe, J. M., M. B. Fenton, and S. J. Shettleworth. 2006. Behavioral flexibility positively correlated with relative brain volume in predatory bats. *Brain, Behavior and Evolution* 67:165–176.

Ratcliffe, J. M., and J. H. Fullard. 2005. The adaptive function of tiger moth clicks against echolocating bats: An experimental and synthetic approach. *Journal of Experimental Biology* 208:4689–4698.

Ratcliffe, J. M., J. H. Fullard, B. J. Arthur, and R. R. Hoy. 2009. Tiger moths and the threat of bats: Decision-making based on the activity of a single sensory neuron. *Biology Letters* 5:368–371.

Ratcliffe, J. M., C. Guignion, K. E. Muma, A. R. Soutar, and J. H. Fullard. 2008. Anti-bat flight activity in sound producing versus silent moths. *Canadian Journal of Zoology—Revue Canadienne de Zoologie* 86:582–587.

Ratcliffe, J. M., and M. L. Nydam. 2008. Multimodal warning signals for a multiple predator world. *Nature* 455:96–99.

Ratcliffe, J. M., H. Raghuram, G. Marimuthu, J. H. Fullard, and M. B. Fenton. 2005. Hunting in unfamiliar space: Echolocation in the Indian false vampire bat, *Megaderma lyra,* when gleaning prey. *Behavioral Ecology and Sociobiology* 58:157–164.

Rattiste, K. 2004. Reproductive success in presenescent common gulls (*Larus canus*): The importance of the last year of life. *Proceedings of the Royal Society of London Series B* 271:2059–2064.

Reader, S. M. 2000. Social learning and innovation: Individual differences, diffusion dynamics and evolutionary issues. Thesis, University of Cambridge.

———. 2004. Don't call me clever. *New Scientist* 183:34.

Reader, S. M., and K. N. Laland. 2002. Social intelligence, innovation, and enhanced brain size in primates. *Proceedings of the National Academy of Science of the United States of America* 99:4436–4441.

———. 2003. Animal innovation: An Introduction. In S. M. Reader and K. N. Laland, eds., *Animal Innovation,* 3–38. Oxford: Oxford University Press.

———. 2004. The cognitive face of avian life histories. *Wilson Bulletin* 116:119–196.

Reader, S. M., and K. MacDonald. 2003. Environmental variability and primate behavioural flexibility. In S. M. Reader and K. N. Laland, eds., *Animal Innovation,* 83–116. Oxford: Oxford University Press.

Real, L. 1990. Search theory and mate choice. 1. Models of single-sex discrimination. *American Naturalist* 136:376–405.

Reddy, E., and M. B. Fenton. 2003. Exploiting vulnerable prey: Moths and red bats (*Lasiurus borealis;* Vespertilionidae). *Canadian Journal of Zoology—Revue Canadienne de Zoologie* 81:1553–1560.

Reeves, B. J., and M. D. Beecher. In preparation. Song sharing and neighbor recognition in two western populations of song sparrows.

Reid, J. M., P. Arcese, A. L. E. V. Cassidy, S. M. Hiebert, J. N. M. Smith, P. K. Stoddard, A. B. Marr, and L. F. Keller. 2004. Song repertoire size predicts initial mating success in male song sparrows, *Melospiza melodia. Animal Behaviour* 68:1055–1063.

Reid, W. V. 1988. Age-specific patterns of reproduction in the glaucous-winged gull: Increased effort with age? *Ecology* 69:1454–1465.

Reiner, A., D. J. Perkel, C. V. Mello, and E. D. Jarvis. 2004. Songbirds and the revised avian brain nomenclature. *Annals of the New York Academy of Sciences* 1016:77–108.

Remage-Healey, L., D. P. Nowacek, and A. H. Bass. 2006. Dolphin foraging sounds suppress calling and elevate stress hormone levels in a prey species, the Gulf toadfish. *Journal of Experimental Biology* 209:4444–4451.

Reznick, D. N., M. J. Bryant, D. Roff, C. K. Ghalambor, and D. E. Ghalambor. 2004. Effect of extrinsic mortality on the evolution of senescence in guppies. *Nature* 431:1095–1099.

Reznick, D. N., F. A. Shaw, F. H. Rodd, and R. G. Shaw. 1997. Evaluation of the rate of evolution in natural populations of guppies (*Poecilia reticulata*). *Science* 275:1934–1937.

Richards, D. G., and R. H. Wiley. 1980. Reverberations and amplitude fluctuations in the propagation of sound in a forest: Implications for animal communication. *American Naturalist* 115:381–399.

Richards, R. J. 1987. Darwin and the Emergence of Evolutionary Theories of Mind and Behavior. Chicago: University of Chicago Press.

Richner, H., P. Schneiter, and H. Stirnimann. 1989. Life-history consequences of growth rate depression: An experimental study on carrion crows (*Corvus corone corone* L.). *Functional Ecology* 3:617–624.

Ricklefs, R. E. 1983. Avian postnatal development. In D. S. Farner, J. R. King, and K. C. Parkes, eds., *Avian Biology,* 1–83. New York: Academic Press.

——. 2004. The cognitive face of avian life histories. *Condor* 116:119–196.

Ricklefs, R. E., and A. Scheuerlein. 2001. Comparison of aging-related mortality among birds and mammals. *Experimental Gerontology* 36:845–857.

Ricklefs, R. E., and M. Wikelski. 2002. The physiology/life history nexus. *Trends in Ecology and Evolution* 17:462–468.

Riebel, K. 2003. The "mute" sex revisted: Vocal production and perception learning in female songbirds. *Advances in the Study of Behavior* 33:48–96.

Ritchie, M. G. 1996. The shape of female mating preferences. Proceedings of the National Academy of Sciences of the United States of America 93:14628–14631.

Ritter, L. V. 1984. Growth of nestling scrub jays in California. *Journal of Field Ornithology* 55: 48–53.

Robinson, B. W., and R. Dukas. 1999. The influence of phenotypic modifications on evolution: The Baldwin effect and modern perspectives. *Oikos* 85:582–589.

Robinson, G. E. 1992. Regulation of division of labor in insect societies. *Annual Review of Entomology* 37:637–665.

Robinson, G. E., R. E. Page, C. Strambi, and A. Strambi. 1989. Hormonal and genetic control of behavioral integration in honey bee colonies. *Science* 246:109–112.

Robinson, S. R. 1980. Antipredator behavior and predator recognition in Beldings ground-squirrels. *Animal Behaviour* 28:840–852.

Rodd, F. H., K. A. Hughes, G. F. Grether, and C. T. Baril. 2002. A possible non-sexual origin of mate preference: Are male guppies mimicking fruit? *Proceedings of the Royal Society of London Series B* 269:475–481.

Roeder, K. D. 1966. Acoustic sensitivity of the noctuid tympanic organ and its range for the cries of bats. *Journal of Insect Physiology* 12:843–59.

——. 1967. Nerve Cells and Insect Behavior. Cambridge: Harvard University Press.

——. 1974. Acoustic sensory responses and possible bat evasion tactics of certain moths. In M. B. D. Burt, ed., *Proceedings of the Canadian Society of Zoologists' Annual Meeting*, 71–78. Ottawa: National Research Council of Canada.

Roeder, K. D., and M. B. Fenton. 1973. Acoustic responsiveness of *Scoliopteryx libatrix* L. (Lepidoptera: Noctuidae), a moth that shares its hibernacula with some insectivorous bats. *Canadian Journal of Zoology—Revue Canadienne de Zoologie* 51:681–685.

Roeder, K. D., and A. E. Treat. 1962. The acoustic detection of bats by moths. In H. Strouhal and M. Beier, eds., *Proceedings of the 11th International Congress of Entomology*, 7–11. Vienna: Naturhistorisches Museum.

Röell, A. 1978. Social behaviour of the jackdaw, *Corvus monedula*, in relation to its niche. *Behaviour* 64:1–124.

Roff, D. A. 2002. *Life History Evolution*. Sunderland, MA: Sinauer.

Rogers, A. R. 1988. Does biology constrain culture? *American Anthropologist* 90:819–831.

Rogge, J. R., and K. M. Warkentin. 2008. External gills and adaptive embryo behavior facilitate synchronous development and hatching plasticity under respiratory constraint. *Journal of Experimental Biology* 211:3627–3635.

Romanes, G. 1883. *Mental Evolution in Animals*. London: Kegan Paul, Trench.

Roper, A., and R. Zann. 2006. The onset of song learning and song tutor selection in fledgling zebra finches. *Ethology* 112:458–470.

Roper, T. J., and S. Redston. 1987. Conspicuousness of distasteful prey affects the strength and durability of one-trial avoidance-learning. *Animal Behaviour* 35:739–747.

Rose, M. R. 1991. *Evolutionary Biology of Aging*. Oxford: Oxford University Press.

Rosengaus, R. B., C. Jordan, M. L. Lefebvre, and J. F. A. Traniello. 1999. Pathogen alarm behavior in a termite: A new form of communication in social insects. *Naturwissenschaften* 86:544–548.

Rosenthal, G. 2007. Spatiotemporal dimensions of visual signals in animal communication. *Annual Review of Ecology, Evolution and Systematics* 38:155–178.

Rosenthal, R., and K. L. Fode. 1963. The effect of experimenter bias on the performance of the albino rat. *Behavioral Science* 8:183–189.

Røskraft, E. 1983. Male promiscuity and female adultery by the rook *Corvus frugilegus*. *Ornis Scandinavica* 14:175–179.

Rothschild, M., T. Reichstein, J. von Euw, R. Aplin, and R. R. M. Harman. 1970. Toxic Lepidoptera. *Toxicon* 8:293–299.

Royle, N. J., J. Lindstrom, and N. B. Metcalfe. 2005. A poor start in life negatively affects dominance status in adulthood independent of body size in green swordtails *Xiphophorus helleri*. *Proceedings of the Royal Society of London Series B* 272:1917–1922.

Rumrill, S. S. 1990. Natural mortality of marine invertebrate larvae. *Ophelia* 32:163–198.

Rüppell, O., S. Christine, C. Mulcrone, and L. Groves. 2007. Aging without functional senescence in honey bee workers. *Current Biology* 17:R274–R275.

Rüppell, O., T. Pankiw, and R. J. Page. 2004. Pleiotropy, epistasis and new QTL: The genetic architecture of honey bee foraging behavior. *Journal of Heredity* 95:481–491.

Russell-Hunter, W. D., M. L. Apley, and R. D. Hunter. 1972. Early life history of *Melampus* and the significance of semilunar synchrony. *Biological Bulletin* 143:623–656.

Russo, D., G. Jones, and R. Arlettaz. 2007. Echolocation and passive listening by foraging mouse-eared bats *Myotis myotis* and *M. blythii*. *Journal of Experimental Biology* 210:166–176.

Ruxton, G. D., T. N. Sherratt, and M. P. Speed. 2004. *Avoiding Attack: The Evolutionary Ecology of Crypsis, Warning Signals and Mimicry*. Oxford: Oxford University Press.

Ryan, B. C., and J. G. Vandenbergh. 2002. Intrauterine position effects. *Neuroscience and Biobehavioral Reviews* 26:665–678.

Ryan, M. J. 1998. Receiver biases, sexual selection and the evolution of sex differences. *Science* 281:1999–2003.

Ryan, M. J., K. L. Akre, and M. Kirkpatrick. 2007. Mate choice. *Current Biology* 17:313–316.

Ryan, M. J., and E. A. Brenowitz. 1985. The role of body size phylogeny and ambient noise in the evolution of bird song. *American Naturalist* 126:87–100.

Ryan, M. J., and A. S. Rand. 1993. Species recognition and sexual selection as a unitary problem in animal communication. *Evolution* 47:647–657.

———. 2003. Sexual selection in female perceptual space: How female tungara frogs perceive and respond to complex population variation in acoustic mating signals. *Evolution* 57:2608–2618.

Ryan, M. J., W. Rand, P. L. Hurd, S. M. Phelps, and A. S. Rand. 2003. Generalization in response to mate recognition signals. *American Naturalist* 161:380–394.

Rybak, J., and R. Menzel. 1993. Anatomy of the mushroom bodies in the honey bee brain: The neuronal connections of the alpha-lobe. *Journal of Comparative Neurology* 334:444–465.

———. 1998. Integrative properties of the PE1 neuron, a unique mushroom body output neuron. *Learning and Memory* 5:133–145.

Rydell, J., and R. Arlettaz. 1994. Low-frequency echolocation enables the bat *Tadarida teniotis* to feed on tympanate insects. *Proceedings of the Royal Society of London Series B* 257:175–178.

Sacher, G. A. 1978. Longevity and aging in vertebrate evolution. *BioScience* 28:497–501.

Saenz, D., J. B. Johnson, C. K. Adams, and G. H. Dayton. 2003. Accelerated hatching of southern leopard frog (*Rana sphenocephala*) eggs in response to the presence of a crayfish (*Procambarus nigrocinctus*) predator. *Copeia* 2003:646–649.

Safi, K., M. A. Seid, and D. K. N. Dechmann. 2005. Bigger is not always better: When brains get smaller. *Biology Letters* 1:283–286.

Saigusa, M. 2002. Hatching controlled by the circatidal clock, and the role of the medulla terminalis in the optic peduncle of the eyestalk, in an estuarine crab, *Sesarma haematocheir*. *Journal of Experimental Biology* 205:3487–3504.

Saimi, Y., and C. Kung. 1987. Behavioral genetics of paramecium. *Annual Review of Genetics* 21: 47–65.

Saino, N., C. Suffritti, R. Martinelli, D. Rubolini, and A. P. Møller. 2003. Immune response covaries with corticosterone plasma levels under experimentally stressful conditions in nestling barn swallows (*Hirundo rustica*). *Behavioral Ecology* 14:318–325.

Samuelson, P. A. 1947. *The Foundations of Economic Analysis*. New York: Atheneum.

Sandell, M., and O. Liberg. 1992. Roamers and stayers: A model on male mating tactics and mating systems. *American Naturalist* 139:177–189.

Sanfey, A. G., G. Loewenstein, S. M. McClure, and J. D. Cohen. 2006. Neuroeconomics: Cross-currents in research on decision-making. *Trends in Cognitive Sciences* 10:108–116.

Schehka, S., K. H. Esser, and E. Zimmermann. 2007. Acoustical expression of arousal in conflict situations in tree shrews (*Tupaia belangeri*). *Journal of Comparative Physiology A* 193:845–852.

Scheiber, I. B. R., B. M. Weiss, D. Frigerio, and K. Kortschal. 2005. Active and passive social support in families of greylag geese (*Anser anser*). *Behaviour* 142:1535–1557.

Scheid, C., F. Range, and T. Bugnyar. 2007. When, what, and whom to watch? Quantifying attention in ravens (*Corvus corax*) and jackdaws (*Corvus monedula*). *Journal of Comparative Psychology* 121:380–386.

Scheifele, P. M., S. Andrew, R. A. Cooper, M. Darre, F. E. Musiek, and L. Max. 2005. Indication of a Lombard vocal response in the St. Lawrence River beluga. *Journal of the Acoustical Society of America* 117:1486–1492.

Schew, W. A., and R. E. Ricklefs. 1998. Developmental plasticity. In J. M. Starck and R. E. Ricklefs, eds., *Avian Growth and Development: Evolution within the Altricial-Precocial Spectrum*, 288–304. Oxford: Oxford University Press.

Schippers, M.-P., R. Dukas, R. W. Smith, J. Wang, K. Smolen, and G. B. McClelland. 2006. Lifetime performance in foraging honeybees: Behaviour and physiology. *Journal of Experimental Biology* 209:3828–3836.

Schlag, K. H. 1998. Why imitate, and if so, how? A bounded rational approach to multi-armed bandits. *Journal of Economic Theory* 78:130–156.

Schlupp, I., C. Marler, and M. J. Ryan. 1994. Benefit to male sailfin mollies of mating with heterospecific females. *Science* 263:373–3374.

Schlupp, I., and M. J. Ryan. 1997. Male sailfin mollies (*Poecilia latipinna*) copy the mate choice of other males. *Behavioral Ecology* 8:104–107.

Schluter, D., and J. N. M. Smith. 1986. Natural selection on beak and body size in the song sparrow. *Evolution* 40:221–231.

Schmidt, K. A. 2006. Non-additivity among multiple cues of predation risk: A behaviorally-driven trophic cascade between owls and songbirds. *Oikos* 113:82–90.

Schmidt, S. 1988. Evidence for a spectral basis of texture perception in bat sonar. *Nature* 331: 617–619.

Schmidt, S., S. Hanke, and J. Pillat. 2000. The role of echolocation in the hunting of terrestrial prey: Evidence for an underestimated strategy in the gleaning bat, *Megaderma lyra*. *Journal of Comparative Physiology A* 186:975–988.

Schnitzler, H.-U., and O. W. Henson. 1980. Performance of airborne animal sonar systems. In R. G. Busnel and J. F. Fish, eds., *Animal Sonar Systems*, 109–181. New York: Plenum Press.

Schnitzler, H.-U., and E. K. V. Kalko. 2001. Echolocation by insect-eating bats. *BioScience* 51: 557–569.

Schnitzler, H.-U., C. F. Moss, and A. Denzinger. 2003. From spatial orientation to food acquisition in echolocating bats. *Trends in Ecology and Evolution* 18:386–394.

Schoeman, M. C., and D. S. Jacobs. 2003. Support for the allotonic frequency hypothesis in an insectivorous bat community. *Oecologia* 134:154–162.

Schroeder, D. J., and R. H. Wiley. 1983. Communication with repertoires of shared song themes in tufted titmice. *Auk* 100:414–424.

Schröter, U., and R. Menzel. 2003. A new ascending sensory tract to the calyces of the honeybee mushroom body, the subesophageal-calycal tract. *Journal of Comparative Neurology* 465:168–178.

Schuck-Paim, C., A. Pompilio, and A. Kacelnik. 2004. State-dependent decisions cause apparent violations of rationality in animal choice. *Public Library of Science Biology* 2:2305–2315.

Schultz, W. 2006. Behavioral theories and the neurophysiology of reward. *Annual Review of Psychology* 57:87–115.

Schumm, A., D. Krull, and G. Neuweiler. 1991. Echolocation in the notch-eared bat, *Myotis emarginatus*. *Behavioral Ecology and Sociobiology* 28:255–261.

Schwab, C., T. Bugnyar, C. Schloegl, and K. Kotrschal. 2008. Enhanced social learning between siblings in common ravens, *Corvus corax*. *Animal Behaviour* 75:501–508.

Schwartz, J. J. 1987. The function of call alteration in anuran amphibians: A test of three hypotheses. *Evolution* 41:461–471.

Schwartz, J. J., and H. C. Gerhardt. 1989. Spatially mediated release from auditory masking in an anuran amphibian. *Journal of Comparative Physiology A* 166:37–41.

Schwartz, J. J., K. Huth, and T. Hutchin. 2004. How long do females really listen? Assessment time for female mate choice in the grey treefrog, *Hyla versicolor*. *Animal Behaviour* 68:533–540.

Scott, B. B., and C. Lois. 2007. Developmental origin and identity of song system neurons born during vocal learning in songbirds. *Journal of Comparative Neurology* 502:202–214.

Searcy, W. A. 1984. Song repertoire size and female preferences in song sparrows. *Behavioral Ecology and Sociobiology* 14:281–286.

Searcy, W. A., R. C. Anderson, and S. Nowicki. 2006. Bird song as a signal of aggressive intent. *Behavioral Ecology and Sociobiology* 60:234–241.

Searcy, W. A., and M. Andersson. 1986. Sexual selection and the evolution of song. *Annual Review of Ecology and Systematics* 17:507–533.

Searcy, W. A., and P. Marler. 1981. A test for responsiveness to song structure and programming in female sparrows. *Science* 213:926–928.

Searcy, W. A., and S. Nowicki. 2005. *The Evolution of Animal Communication*. Princeton: Princeton University Press.

Searcy, W. A., S. Nowicki, M. Hughes, and S. Peters. 2002. Geographic song discrimination in relation to dispersal distances in song sparrows. *American Naturalist* 159:221–230.

Searcy, W. A., S. Peters, and S. Nowicki. 2004. Effects of early nutrition on growth rate and adult size in song sparrows *Melospiza melodia*. *Journal of Avian Biology* 35:269–279.

Searcy, W. A., and K. Yasukawa. 1996. Song and female choice. In D. E. Kroodsma and E. H. Miller, eds., *Ecology and Evolution of Acoustic Communication in Birds*, 454–473. Ithaca: Cornell University Press.

Seed, A. M., N. S. Clayton, and N. J. Emery. 2007. Postconflict third party affiliation in rooks, *Corvus frugilegus*. *Current Biology* 17:152–158.

Seehausen, O., J. M. van Alphen, and F. Witte. 1997. Cichlid fish diversity threatened by eutrophication that curbs sexual selection. *Science* 277:1808–1811.

Seeley, T. D. 1982. Adaptive significance of the age polyethism schedule in honeybee colonies. *Behavioral Ecology and Sociobiology* 11:287–293.

———. 1996. The Wisdom of the Hive: The Social Physiology of Honey Bee Colonies. Cambridge: Harvard University Press.

Seid, M. A., K. M. Harris, and J. F. Traniello. 2005. Age-related changes in the number and structure of synapses in the lip region of the mushroom bodies in the ant *Pheidole dentata*. *Journal of Comparative Neurology* 488:269–277.

Servedio, M. R. 2000. The effects of predator learning, forgetting, and recognition errors on the evolution of warning coloration. *Evolution* 54:751–763.

Servedio, M. R., and M. Kirkpatrick. 1996. The evolution of mate choice copying by indirect selection. *American Naturalist* 148:848–867.

Seyfarth, R. M., and D. L. Cheney. 2003a. Meaning and emotion in animal vocalizations. *Annals of the New York Academy of Sciences* 1000:32–55.

———. 2003b. Signalers and receivers in animal communication. *Annual Review of Psychology* 54:145–173.

Seyfarth, R. M., D. L. Cheney, and P. Marler. 1980. Monkey responses to three different alarm calls: Evidence of predator classification and semantic communication. *Science* 210:801–803.

Seymour, R. S., J. D. Roberts, N. J. Mitchell, and A. J. Blaylock. 2000. Influence of environmental oxygen on development and hatching of aquatic eggs of the Australian frog, *Crinia georgiana*. *Physiological and Biochemical Zoology* 73:501–507.

Shafir, S. 1994. Intransitivity of preferences in honey bees: Support for "comparative" evaluation of foraging options. *Animal Behaviour* 48:55–67.

Shannon, C. E. 1948. A mathematical theory of communication. *Bell System Technical Journal* 27:379–423, 623–656.

Shelton, G. A. B. 1982. Electrical Conduction and Behaviour in "Simple" Invertebrates. Oxford: Clarendon Press.

Sherry, D. F., and B. G. Galef Jr. 1984. Cultural transmission without imitation: Milk bottle opening by birds. *Animal Behaviour* 32:937–938.

Sherry, D. F., L. F. Jacobs, and S. J. C. Gaulin. 1992. Spatial memory and adaptive specialization of the hippocampus. *Trends in Neurosciences* 15:298–303.

Sherry, D. F., and D. L. Schacter. 1987. The evolution of multiple memory systems. *Psychological Review* 94:439–454.

Sherry, D. F., and A. L. Vaccarino. 1989. Hippocampus and memory for food caches in black-capped chickadees. *Behavioral Neuroscience* 103:308–318.

Sherry, D. F., A. L. Vaccarino, K. Buckenham, and R. S. Hertz. 1989. The hippocampal complex of food-storing birds. *Brain, Behavior and Evolution* 34:308–317.

Shettleworth, S. J. 1990. Spatial memory in food-storing birds. *Philosophical Transactions of the Royal Society of London Series B* 329:143–151.

———. 1995. Memory in food-storing birds: From the field to the Skinner box. In E. Alleva, A. Fasolo, H.-P. Lipp, and L. Nadel, eds., *Behavioral Brain Research in Naturalistic and Semi-naturalistic Settings*, 158–179. The Hague: Kluwer Academic Publishers.

———. 1998. *Cognition, Evolution, and Behavior*. Oxford: Oxford University Press.

———. 2003. Memory and hippocampal specialization in food-storing birds: Challenges for research on comparative cognition. *Brain, Behavior and Evolution* 62:108–116.

———. 2009. *Cognition, Evolution, and Behavior.* 2d ed. Oxford: Oxford University Press.

Shiflett, M. W., K. L. Gould, T. V. Smulders, and T. J. DeVoogd. 2002. Septum volume and food-storing behavior are related in parids. *Journal of Neurobiology* 51:215–222.

Shinkai, M., J. Yokofujita, S. Oda, K. Murakami, H. Igarashi, and M. Kuroda. 2005. Dual axonal terminations from the retrosplenial and visual association cortices in the laterodorsal thalamic nucleus of the rat. *Anatomy and Embryology* 210:317–326.

Short, L. L. 1970. A new anseriform genus and species from the Nebraska Pliocene. *Auk* 87: 537–543.

Shultz, S., R. Bradbury, K. Evans, R. D. Gregory, and T. M. Blackburn. 2005. Brain size and resource specialisation predict long-term population trends in British birds. *Proceedings of the Royal Society of London Series B* 272:2305–2311.

Shultz, S., and R. I. M. Dunbar. 2007. The evolution of the social brain: Anthropoid primates contrast with other vertebrates. *Proceedings of the Royal Society of London Series B* 274:2429–2436.

Shump, K. A., Jr., and A. V. Shump. 1982. *Lasiurus borealis. Mammalian Species* 183:1–6.

Si, A., P. Helliwell, and R. Maleszka. 2004. Effects of NMDA receptor antagonists on olfactory learning and memory in the honeybee (*Apis mellifera*). *Pharmacology, Biochemistry and Behavior* 77:191–197.

Si, A., S. W. Zhang, and R. Maleszka. 2005. Effects of caffeine on olfactory and visual learning in the honey bee (*Apis mellifera*). *Pharmacology, Biochemistry and Behavior* 82:664–672.

Siegel, R. W., and J. C. Hall. 1979. Conditioned courtship in *Drosophila* and its mediation by association of chemical cues. *Proceedings of the National Academy of Sciences of the United States of America* 76:3430–3434.

Siemers, B. M. 2001. Finding prey by associative learning in gleaning bats: Experiments with a Natterer's bat *Myotis nattereri*. *Acta Chiropterologica* 3:211–215.

Siemers, B. M., and T. Ivanova. 2004. Ground gleaning in horseshoe bats: Comparative evidence from *Rhinolophus blasii, R. euryale* and *R. mehelyi*. *Behavioral Ecology and Sociobiology* 56:464–471.

Siemers, B. M., and R. A. Page. In press. Behavioral studies of bats in captivity: Methodology, training, and experimental design. In T. H. Kunz and S. Parsons, eds., *Ecological and Behavioral Methods for the Study of Bats.* Washington: Smithsonian Institution Press.

Siemers, B. M., and H.-U. Schnitzler. 2000. Natterer's bat (*Myotis nattereri* Kuhl, 1818) hawks for prey close to vegetation using echolocation signals of very broad bandwidth. *Behavioral Ecology and Sociobiology* 47:400–412.

———. 2004. Echolocation signals reflect niche differentiation in five sympatric congeneric bat species. *Nature* 429:657–661.

Siemers, B. M., P. Stilz, and H.-U. Schnitzler. 2001. The acoustic advantage of hunting at low heights above water: Behavioural experiments on the European "trawling" bats *Myotis capaccinii, M. dasycneme* and *M. daubentonii*. *Journal of Experimental Biology* 204:3843–3854.

Sigg, D., C. M. Thompson, and A. R. Mercer. 1997. Activity-dependent changes to the brain and behavior of the honey bee, *Apis mellifera* (L.). *Journal of Neuroscience* 17:7148–7156.

Sih, A., L. B. Kats, and E. F. Maurer. 2003. Behavioural correlations across situations and the evolution of antipredator behaviour in a sunfish-salamander system. *Animal Behaviour* 65:29–44.

Sih, A., and R. D. Moore. 1993. Delayed hatching of salamander eggs in response to enhanced larval predation risk. *American Naturalist* 142:947–960.

Simmons, J. A., M. B. Fenton, and M. J. O'Farrell. 1979. Echolocation and pursuit of prey by bats. *Science* 203:16–21.

Simmons, J. A., and R. A. Stein. 1980. Acoustic imaging in bat sonar: Echolocation signals and the evolution of echolocation. *Journal of Comparative Physiology A* 135:61–84.

Simmons, N. B., and J. B. Geisler. 1998. Phylogenetic relationships of *Icaronycteris, Archaeonycteris, Hassianycteris,* and *Palaeochiropteryx* to extant bat lineages, with comments on the evolution of echolocation and foraging strategies in Microchiroptera. *Bulletin of the American Museum of Natural History* 235:1–182.

Simmons, N. B., K. L. Seymour, J. Habersetzer, and G. F. Gunnell. 2008. Primitive early Eocene bat from Wyoming and the evolution of flight and echolocation. *Nature* 451:818–822.

Simpson, S. J., and D. Raubenheimer. 2000. The hungry locust. *Advances in the Study of Behavior* 29:1–44.

Sinnot, J. M., W. C. Stebbins, and D. B. Moody. 1975. Regulation of voice amplitude by the monkey. *Journal of the Acoustical Society of America* 58:412–414.

Sirot, E. 2001. Mate-choice copying by females: The advantages of a prudent strategy. *Journal of Evolutionary Biology* 14:418–423.

Skagen, S. K. 1988. Asynchronous hatching and food limitation: A test of Lack's hypothesis. *Auk* 105:78–88.

Skals, N., P. Anderson, M. Kanneworff, C. Lofstedt, and A. Surlykke. 2005. Her odours make him deaf: Crossmodal modulation of olfaction and hearing in a male moth. *Journal of Experimental Biology* 208:595–601.

Skinner, J. D., and Chimimba, T. C. 2006. *The Mammals of the Southern African Sub-region*. Cambridge: Cambridge University Press.

Slabbekoorn, H., and M. Peet. 2003. Birds sing at a higher pitch in urban noise. *Nature* 424:267.

Slater, P. J. B. 1989. Bird song learning: Causes and consequences. *Ethology, Ecology and Evolution* 1:19–46.

Slater, P. J. B., and S. A. Ince. 1982. Song development in chaffinches *Fringilla coelebs*: What is learned and when? *Ibis* 124:21–26.

Smid, H. M., G. Wang, T. Bukovinszky, J. L. M. Steidle, M. A. K. Bleeker, J. J. A. van Loon, and L. E. M. Vet. 2007. Species-specific acquisition and consolidation of long-term memory in parasitic wasps. *Proceedings of the Royal Society of London Series B* 274:1539–1546.

Smith, G. T., Brenowitz, E. A., Beecher, M. D., and Wingfield, J. C. 1997. Seasonal changes in testosterone, neural attributes of song control nuclei, and song structure in wild songbirds. *Journal of Neuroscience* 17:6001–6010.

Smith, H. G. 1995. Experimental demonstration of a trade-off between mate attraction and paternal care. *Proceedings of the Royal Society of London Series B* 260:45–51.

Smith, H. R., and K. C. H. Pang. 2005. Orexin-sapolin lesions of the medial septum impair spatial memory. *Neuroscience* 132:261–271.

Smith, W. J. 1981. Referents of animal communication. *Animal Behaviour* 29:1273–1275.

———. 1991. Singing is based on two markedly different kinds of signaling. *Journal of Theoretical Biology* 152:241–253.

Snoeijs, T., R. Pinxten, and M. Eens. 2005. Experimental removal of the male parent negatively affects growth and immunocompetence in nestling great tits. *Oecologia* 145:165–173.

Sohrabji, F., E. J. Nordeen, and K. W. Nordeen. 1990. Selective impairment of song learning following lesions of a forebrain nucleus in the juvenile zebra finch. *Behavioral and Neural Biology* 53:51–63.

Sol, D. 2003. Behavioural flexibility: A neglected issue in the ecological and evolutionary literature? In S. M. Reader and K. N. Laland, eds., *Animal Innovation,* 63–82. Oxford: Oxford University Press.

Sol, D., S. Bacher, S. M. Reader, and L. Lefebvre. 2008. Brain size predicts the success of mammal species introduced into novel environments. *American Naturalist* 172:S63–S71.

Sol, D., R. P. Duncan, T. M. Blackburn, P. Cassey, and L. Lefebvre. 2005a. Big brains, enhanced cognition, and response of birds to novel environments. *Proceedings of the National Academy of Sciences of the United States of America* 102:5460–5465.

Sol, D., and L. Lefebvre. 2000. Behavioural flexibility predicts invasion success in birds introduced to New Zealand. *Oikos* 90:599–605.

Sol, D., L. Lefebvre, and J. D. Rodriguez-Teijeiro. 2005b. Brain size, innovative propensity and migratory behaviour in temperate Palearctic birds. *Proceedings of the Royal Society of London Series B* 272:1471–2954.

Sol, D., and T. D. Price. 2008. Brain size and body size diversification in birds. *American Naturalist* 172:170–177.

Sol, D., D. G. Stirling, and L. Lefebvre. 2005c. Behavioral drive or behavioral inhibition in evolution: Subspecific diversification in Holarctic passerines. *Evolution* 59:2669–2677.

Sol, D., T. Szekely, A. Liker, and L. Lefebvre. 2007. Big-brained birds survive better in nature. *Proceedings of the Royal Society of London Series B* 274:763–769.

Sol, D., S. Timmermans, and L. Lefebvre. 2002. Behavioural flexibility and invasion success in birds. *Animal Behaviour* 63:495–502.

Soler, J. J., L. de Neve, T. Perez-Contreras, M. Soler, and G. Sorci. 2003. Trade-off between immunocompetence and growth in magpies: An experimental study. *Proceedings of the Royal Society of London Series B* 270:241–248.

Solomon, N. G., and J. J. Jacquot. 2002. Characteristics of resident and wandering prairie voles, *Microtus ochrogaster. Canadian Journal of Zoology—Revue Canadienne de Zoologie* 80:951–955.

Soutar, A. R., and J. H. Fullard. 2004. Nocturnal anti-predator adaptations in eared and earless Nearctic Lepidoptera. *Behavioral Ecology* 15:1016–1022.

Spangler, H. G. 1988. Moth hearing, defense, and communication. *Annual Review of Entomology* 33:59–81.

Speakman, J. R., and P. A. Racey. 1991. No cost of echolocation for bats in flight. *Nature* 350:421–423.

Speed, M. P. 2000. Warning signals, receiver psychology and predator memory. *Animal Behaviour* 60:269–278.

Spencer, K. A., K. L. Buchanan, A. R. Goldsmith, and C. K. Catchpole. 2003. Song as an honest signal of developmental stress in the zebra finch (*Taeniopygia guttata*). *Hormones and Behavior* 44:132–139.

———. 2004. Developmental stress, social rank and song complexity in the European starling (*Sturnus vulgaris*). *Proceedings of the Royal Society of London Series B* 271:S121–S123.

Spencer, K. A., K. L. Buchanan, S. Leitner, A. R. Goldsmith, and C. K. Catchpole. 2005a. Parasites affect song complexity and neural development in a songbird. *Proceedings of the Royal Society of London Series B* 272:2037–2043.

Spencer, K. A., and S. Verhulst. 2007. Delayed behavioral effects of postnatal exposure to corticosterone in the zebra finch (*Taeniopygia guttata*). *Hormones and Behavior* 51:273–280.

Spencer, K. A., J. H. Wimpenny, K. L. Buchanan, P. G. Lovell, A. R. Goldsmith, and C. K. Catchpole.

2005b. Developmental stress affects the attractiveness of male song and female choice in the zebra finch (*Taeniopygia guttata*). *Behavioral Ecology and Sociobiology* 58:423–428.

Spencer, R. J., M. B. Thompson, and P. B. Banks. 2001. Hatch or wait? A dilemma in reptilian incubation. *Oikos* 93:401–406.

Spitzer, M. D., and L. A. M. Galea. 2007. Testosterone and dihydrotestosterone, but not estradiol, enhance survival of new hippocampal neurons in adult male rats. *Developmental Neurobiology* 67:1321–1333.

Srygley, R. B., and P. Chai. 1990. Flight morphology of Neotropical butterflies: Palatability and distribution of mass to the thorax and abdomen. *Oecologia* 84:491–499.

Stankowitch, T., and D. T. Blumstein. 2005. Fear in animals: A meta-analysis and review of risk assessment. *Proceedings of the Royal Society of London Series B* 272:2627–2634.

Stearns, S. 1992. *The Evolution of Life Histories*. Oxford: Oxford University Press.

Stephens, D. W., J. S. Brown, and R. C. Ydenberg. 2007. *Foraging: Behavior and Ecology*. Chicago: University of Chicago Press.

Steppan, S. J., R. M. Adkins, and J. Anderson. 2004. Phylogeny and divergence-date estimates of rapid radiations in muroid rodents based on multiple nuclear genes. *Systematic Biology* 53:533–553.

Stoddard, P. K., M. D. Beecher, S. E. Campbell, and C. L. Horning. 1992a. Song-type matching in the song sparrow. *Canadian Journal of Zoology—Revue Canadienne de Zoologie* 70:1440–1444.

Stoddard, P. K., M. D. Beecher, C. L. Horning, and S. E. Campbell. 1991. Recognition of individual neighbors by song in the song sparrow, a species with song repertoires. *Behavioral Ecology and Sociobiology* 29:211–215.

Stoddard, P. K., M. D. Beecher, C. L. Horning, and M. S. Willis. 1990. Strong neighbor-stranger discrimination in song sparrows. *Condor* 92:1051–1056.

Stoddard, P. K., M. D. Beecher, P. Loesche, and S. E. Campbell. 1992b. Memory does not constrain individual recognition in a bird with song repertoires. *Behaviour* 122:274–287.

Stoddard, P. K., M. D. Beecher, and M. S. Willis. 1988. Response of territorial male song sparrows to song types and variations. *Behavioral Ecology and Sociobiology* 22:124–130.

Stopfer, M., S. Bhagavan, B. H. Smith, and G. Laurent. 1997. Impaired odour discrimination on desynchronization of odour-encoding neural assemblies. *Nature* 390:70–74.

Strausfeld, N. J. 1980. The Golgi method, its application to the insect nervous system and the phenomenon of stochastic impregnation. In N. J. Strausfeld and T. A. Miller, eds., *Neuroanatomical Techniques: Insect Nervous System*, 131–190. Heidelberg: Springer.

Strausfeld, N. J., L. Hansen, Y. Li, R. S. Gomez, and K. Ito. 1998. Evolution, discovery, and interpretations of arthropod mushroom bodies. *Learning and Memory* 5:11–37.

Stubbs, C. S., and F. A. Drummond. 2001. *Bombus impatiens* (Hymenoptera: Apidae): An alternative to *Apis mellifera* (Hymenoptera: Apidae) for lowbush blueberry pollination. *Journal of Economic Entomology* 94:609–616.

Sullivan, K. A. 1988a. Age-specific profitability and prey choice. *Animal Behaviour* 36:613–615.

———. 1988b. Ontogeny of time budgets in yellow-eyed juncos: Adaptation to ecological constraints. *Ecology* 69:118–124.

———. 1989. Predation and starvation: Age specific mortality in juvenile juncos (*Junco phaeonotus*). *Journal of Animal Ecology* 58:275–286.

Sullivan, M. S. 1994. Mate choice as an information gathering process under time constraint: Implications for behavior and signal design. *Animal Behaviour* 47:141–151.

Surlykke, A. 1984. Hearing in notodontid moths—a tympanic organ with a single auditory neuron. *Journal of Experimental Biology* 113:323–335.

———. 1988. Interaction between echolocating bats and their prey. In P. E. Nachtigall and P. W. B. Moore, eds., *Animal Sonar: Processes and Performance*, 551–566. New York: Plenum Press.

Surlykke, A., V. Futtrup, and J. Tougaard. 2003. Prey-capture success revealed by echolocation signals in pipistrelle bats (*Pipistrellus pygmaeus*). *Journal of Experimental Biology* 206:93–104.

Surlykke, A., and E. K. V. Kalko. 2008. Echolocating bats cry out loud to detect their prey. *Public Library of Science One* 3:e2036.

Surlykke, A., and C. F. Moss. 2000. Echolocation behavior of big brown bats, *Eptesicus fuscus*, in the field and the laboratory. *Journal of the Acoustical Society of America* 108:2419–2429.

Svensson, A. M., and J. Rydell. 1998. Mercury vapour lamps interfere with the bat defence of tympanate moths (*Operophtera* spp.; Geometridae). *Animal Behaviour* 55:223–226.

Svensson, G. P., C. Lofstedt, and N. Skals N. 2004. The odour makes the difference: Male moths attracted by sex pheromones ignore the threat by predatory bats. *Oikos* 104:91–97.

Swaney, W., J. R. Kendal, H. Capon, C. Brown, and K. N. Laland. 2001. Familiarity facilitates social learning of foraging behaviour in the guppy. *Animal Behaviour* 62:591–598.

Székely, T., C. K. Catchpole, A. DeVoogd, Z. Marchl, and T. J. DeVoogd. 1996. Evolutionary changes in a song control area of the brain (HVC) are associated with evolutionary changes in song repertoire among European warblers (Sylviidae). *Proceedings of the Royal Society of London Series B* 263:607–610.

Székely, T., A. Kosztolányi, C. Küpper, and G. H. Thomas. 2007. Sexual conflict over parental care: A case study of shorebirds. *Journal of Ornithology* 148:S211–S217.

Taylor, P. 1991. Optimal life histories with age dependent tradeoff curves. *Journal of Theoretical Biology* 148:33–48.

Tebbich, S., and R. Bshary. 2004. Cognitive abilities related to tool use in the woodpecker finch, *Cactospiza pallida*. *Animal Behaviour* 67:689–697.

Tebbich, S., M. Taborsky, and H. Winkler. 1996. Social manipulation causes cooperation in keas. *Animal Behaviour* 52:1–10.

Teeling, E. C., O. Madsen, R. A. Van Den Bussche, W. W. de Jong, M. J. Stanhope, and M. S. Springer. 2002. Microbat paraphyly and the convergent evolution of a key innovation in Old World rhinolophoid microbats. *Proceedings of the National Academy of Sciences of the United States of America* 99:1431–1436.

Teeling, E. C., M. S. Springer, O. Madsen, P. Bates, S. J. O'Brien, and W. J. Murphy. 2005. A molecular phylogeny for bats illuminates biogeography and the fossil record. *Science* 307:580–584.

Tempel, B. L., N. Bonini, D. R. Dawson, and W. G. Quinn. 1983. Reward learning in normal and mutant *Drosophila*. *Proceedings of the National Academy of Sciences of the United States of America* 80:1482–1486.

Templeton, C. N., E. Greene, and K. Davis. 2005. Allometry of alarm calls: Black-capped chickadees encode information about predator size. *Science* 308:1934–1937.

Templeton, J. T., A. C. Kamil, and R. P. Balda. 1999. Sociality and social learning in two species of corvids: The pinyon jay (*Gymnorhinus cyanocephalus*) and the Clark's nutcracker (*Nucifraga columbiana*). *Journal of Comparative Psychology* 113:450–455.

Temrin, H., and B. S. Tullberg. 1995. A phylogenetic analysis of the evolution of avian mating systems in relation to altricial and precocial young. *Behavioral Ecology* 6:296–307.

ten Cate, C., and D. Vos. 1999. Sexual imprinting and evolutionary processes in birds. *Advances in the Study of Behavior* 28:1–31.

ter Hofstede, H. M., and J. H. Fullard. 2008. The neuroethology of song cessation in response to gleaning bat calls in two species of katydids, *Neoconocephalus ensiger* and *Amblycorypha oblongifolia*. *Journal of Experimental Biology* 211:2431–2441.

ter Hofstede, H. M., J. M. Ratcliffe, and J. H. Fullard. 2008a. The effectiveness of katydid (*Neoconocephalus ensiger*) song cessation as antipredator defence against the gleaning bat *Myotis septentrionalis*. *Behavioral Ecology and Sociobiology* 63:217–226.

———. 2008b. Nocturnal activity positively correlated with auditory sensitivity in eared moths. *Biology Letters* 4:262–265.

Terkel, J. 1996. Cultural transmission of feeding behavior in the black rat (*Rattus rattus*). In C. M. Heyes and B. G. Galef Jr., eds., *Social Learning in Animals: The Roots of Culture*, 17–47. San Diego: Academic Press.

Thorpe, W. H. 1956. *Learning and Instinct in Animals*. London: Methuen.

Timmermans, S., L. Lefebvre, D. Boire, and P. Basu. 2001. Relative size of the hyperstriatum ventrale is the best predictor of feeding innovation rate in birds. *Brain, Behavior and Evolution* 56:196–203.

Timson, J. 1975. Hydroxyurea. *Mutation Research* 32:115–132.

Tinbergen, N. 1951. *The Study of Instinct*. Oxford: Oxford University Press.

———. 1963. On aims and methods of ethology. *Zeitschrift für Tierpsychologie* 20:410–433.

Titus, A. D., B. S. Shankaranarayana Rao, H. N. Harsha, K. Ramkumar, B. N. Srikumar, S. B. Singh, S. Chattarji, et al. 2007. Hypobaric hypoxia-induced dendritic atrophy of hippocampal neurons is associated with cognitive impairment in adult rats. *Neuroscience* 145:265–278.

Tollrian, R., and C. D. Harvell. 1999. *The Ecology and Evolution of Inducible Defenses*. Princeton: Princeton University Press.

Tomasello, M. 1990. Cultural transmission in the tool use and communicatory signalling of chimpanzees? In S. Parker and K. Gibson, eds., *"Language" and Intelligence in Monkeys and Apes: Comparative Developmental Perspectives*, 271–311. Cambridge: Cambridge University Press.

———. 1996. Do apes ape? In C. M. Heyes and B. G. Galef Jr., eds., *Social Learning in Animals: The Roots of Culture*, 319–346. San Diego: Academic Press.

Tomkins, J. L., J. Radwan, J. S. Kotiaho, and T. Tregenza. 2004. Genic capture and resolving the lek paradox. *Trends in Ecology and Evolution* 19:323–328.

Tonkinson, S. 1994. The Lombard effect in choral singing. *Journal of Voice* 8:24–29.

Touchon, J. C., I. Gomez-Mestre, and K. M. Warkentin. 2006. Hatching plasticity in two temperate anurans: Responses to a pathogen and predation cues. *Canadian Journal of Zoology—Revue Canadienne de Zoologie* 84:556–563.

Trivers, R. L. 1971. The evolution of reciprocal altruism. *Quarterly Review of Biology* 46:35–57.

———. 1972. Parental investment and sexual selection. In B. Campbell, ed., *Sexual Selection and the Descent of Man*, 136–179. Chicago: Aldine Press.

Tsipoura, N., and E. S. Morton. 1988. Song-type distribution in a population of Kentucky warblers. *Wilson Bulletin* 100:9–16.

Turner, C. A., M. H. Lewis, and M. A. King. 2003. Environmental enrichment: Effects on stereotyped behavior and dendritic morphology. *Developmental Psychobiology* 43:20–27.

Tversky, A. 1969. Intransitivity of preferences. *Psychological Review* 76:31–48.

Tversky, A., and I. Simonson. 1993. Context-dependent preferences. *Management Sciences* 39:1179–1189.

Valadares, C. T., and S. D. Almeida. 2005. Early protein malnutrition changes learning and memory in spaced but not in condensed trials in the Morris water maze. *Nutritional Neuroscience* 8:39–47.

Vallee, M., S. Maccari, F. Dellu, H. Simon, M. L. Moal, and W. Mayo. 1999. Long-term effects of prenatal stress and postnatal handling on age-related glucocorticoid secretion and cognitive performance: A longitudinal study in the rat. *European Journal of Neuroscience* 11:2906–2916.

Vallet, E., I. Beme, and M. Kreutzer. 1998. Two-note syllables in canary songs elicit high levels of sexual display. *Animal Behaviour* 55:291–297.

Vallet, E., and M. Kreutzer. 1995. Female canaries are sexually responsive to special song phrases. *Animal Behaviour* 49:1603–1610.

Valone, T. J. 2007. From eavesdropping on performance to copying the behavior of others: A review of public infomration use. *Behavioral Ecology and Sociobiology* 62:1–14.

van Bergen, Y., I. Coolen, and K. N. Laland. 2004. Nine-spined sticklebacks exploit the most reliable source when public and private information conflict. *Proceedings of the Royal Society of London Series B* 271:957–962.

van Groen, T., I. Kadish, and J. M. Wyss. 2002. The role of the laterodorsal nucleus of the thalamus in spatial learning and memory in the rat. *Behavioural Brain Research* 136:329–337.

van Groen, T., and J. M. Wyss. 2003. Connections of the retrosplenial granular b cortex in the rat. *Journal of Comparative Neurology* 463:249–263.

Vasquez, R. A. 1994. Assessment of predation risk via illumination level: Facultative central place foraging in the cricetid rodent *Phyllotis darwini*. *Behavioral Ecology and Sociobiology* 34:375–381.

Vegoz, V., E. Roussel, J. C. Sandoz, and M. Giurfa. 2007. Aversive learning in honeybees revealed by the olfactory conditioning of the sting extension reflex. *Public Library of Science One* 2:e288.

Verner, J. 1975. Complex song repertoire of male long-billed marsh wrens in eastern Washington. *Living Bird* 14: 263–300.

Verzijden, M. N., and C. ten Cate. 2007. Early learning influences species assortative mating preferences in Lake Victoria cichlid fish. *Biology Letters* 3:134–136.

Villa, J. 1979. Two fungi lethal to frog eggs in Central America. *Copeia* 1979:650–655.

———. 1980. "Frogflies" from Central and South America with notes on other organisms of the amphibian egg microhabitat. *Brenesia* 17:49–68.

Volman, S. F., T. C. Grubb Jr., and K. C. Schuett. 1997. Relative hippocampal volume in relation to food-storing behavior in four species of woodpeckers. *Brain, Behavior and Evolution* 49:110–120.

von Bayern, A. M. P., S. R. de Kort, N. S. Clayton, and N. J. Emery. 2007. The role of food- and object-sharing in the development of social bonds in juvenile jackdaws (*Corvus monedula*). *Behaviour* 144:711–733.

von Bayern, A. M. P., and N. J. Emery. In press. Bonding, mentalising and rationality. In S. Watanabe, ed., *Rational Animals, Irrational Humans.* Tokyo: Keio University Press.

Vonesh, J. R. 2000. Dipteran predation on the eggs of four *Hyperolius* frog species in western Uganda. *Copeia* 2000:560–566.

———. 2005. Egg predation and predator-induced hatching plasticity in the African reed frog, *Hyperolius spinigularis*. *Oikos* 110:241–252.

von Frisch, K. 1967. *The Dance Language and Orientation of Bees*. Cambridge: Harvard University Press.

Voronezhskaya, E. E., M. Y. Khabarova, and L. P. Nezlin. 2004. Apical sensory neurones mediate developmental retardation induced by conspecific environmental stimuli in freshwater pulmonate snails. *Development* 131:3671–3680.

Waddell, S., J. D. Armstrong, T. Kitamoto, K. Kaiser, and W. G. Quinn. 2000. The amnesiac gene product is expressed in two neurons in the *Drosophila* brain that are critical for memory. *Cell* 103:805–813.

Waddington, K. D. 1989. Implications of variation in worker body size for the honey bee recruitment system. *Journal of Insect Behavior* 2:91–103.

Wagner, W. E., M. R. Smeds, and D. D. Wiegmann. 2001. Experience affects female responses to male song in the variable field cricket *Gryllus lineaticeps* (Orthoptera, Gryllidae). *Ethology* 107:769–776.

Wakano, J. Y., and Y. Ihara. 2005. Evolution of male parental care and female multiple mating: Game-theoretical and two-locus diploid models. *American Naturalist* 166:E32–E44.

Wang, Z. X., C. F. Ferris, and G. J. Devries. 1994. Role of septal vasopressin innervation in paternal behavior in prairie voles (*Microtus ochrogaster*). *Proceedings of the National Academy of Sciences of the United States of America* 91:400–404.

Wang, Z. X., Y. Liu, L. J. Young, and T. R. Insel. 2000. Hypothalamic vasopressin gene expression increases in both males and females postpartum in a biparental rodent. *Journal of Neuroendocrinology* 12:111–120.

Wang, Z. X., and M. A. Novak. 1992. Influence of the social-environment on parental behavior and pup development of meadow voles (*Microtus pennsylvanicus*) and prairie voles (*Microtus ochrogaster*). *Journal of Comparative Psychology* 106:163–171.

Wang, Z. X., L. J. Young, G. J. De Vries, and T. R. Insel. 1998. Voles and vasopressin: A review of molecular, cellular, and behavioral studies of pair bonding and paternal behaviors. *Progress in Brain Research* 119:483–499.

Wang, Z. X., L. J. Young, Y. Liu, and T. R. Insel. 1997. Species differences in vasopressin receptor binding are evident early in development: Comparative anatomic studies in prairie and montane voles. *Journal of Comparative Neurology* 378:535–546.

Ward, S., H. Lampe, and P. J. B. Slater. 2004. Singing is not energetically demanding for pied flycatchers, *Ficedula hypoleuca*. *Behavioral Ecology* 15:477–484.

Ward, S., J. R. Speakman, and P. J. B. Slater. 2003. The energy cost of song in the canary, *Serinus canaria*. *Animal Behaviour* 66:893–902.

Warkentin, K. M. 1995. Adaptive plasticity in hatching age: A response to predation risk trade-offs. *Proceedings of the National Academy of Sciences of the United States of America* 92:3507–3510.

———. 1999a. The development of behavioral defenses: A mechanistic analysis of vulnerability in red-eyed tree frog hatchlings. *Behavioral Ecology* 10:251–262.

———. 1999b. Effects of hatching age on development and hatchling morphology in the red-eyed treefrog, *Agalychnis callidryas*. *Biological Journal of the Linnean Society* 68:443–470.

———. 2000. Wasp predation and wasp-induced hatching of red-eyed treefrog eggs. *Animal Behaviour* 60:503–510.

———. 2002. Hatching timing, oxygen availability, and external gill regression in the treefrog, *Agalychnis callidryas*. *Physiological and Biochemical Zoology* 75:155–164.

———. 2005. How do embryos assess risk? Vibrational cues in predator-induced hatching of red-eyed treefrogs. *Animal Behaviour* 70:59–71.

———. 2007. Oxygen, gills, and embryo behavior: Mechanisms of adaptive plasticity in hatching. *Comparative Biochemistry and Physiology A* 148:720–731.

Warkentin, K. M., C. R. Buckley, and K. A. Metcalf. 2006a. Development of red-eyed treefrog eggs affects efficiency and choices of egg-foraging wasps. *Animal Behaviour* 71:417–425.

Warkentin, K. M., M. S. Caldwell, and J. G. McDaniel. 2006b. Temporal pattern cues in vibrational risk assessment by red-eyed treefrog embryos, *Agalychnis callidryas*. *Journal of Experimental Biology* 209:1376–1384.

Warkentin, K. M., M. S. Caldwell, T. D. Siok, A. T. D'Amato, and J. G. McDaniel. 2007. Flexible information sampling in vibrational assessment of predation risk by red-eyed treefrog embryos. *Journal of Experimental Biology* 210:614–619.

Warkentin, K. M., C. C. Currie, and S. A. Rehner. 2001. Egg-killing fungus induces early hatching of red-eyed treefrog eggs. *Ecology* 82:2860–2869.

Warkentin, K. M., I. Gomez-Mestre, and J. G. McDaniel. 2005. Development, surface exposure, and embryo behavior affect oxygen levels in eggs of the red-eyed treefrog, *Agalychnis callidryas*. *Physiological and Biochemical Zoology* 78:956–966.

Waters, D. A., and G. Jones. 1996. The peripheral auditory characteristics of noctuid moths: Responses to the search-phase echolocation calls of bats. *Journal of Experimental Biology* 199:847–856.

Weatherhead, P. J., and K. A. Boak. 1986. Site infidelity in song sparrows. *Animal Behaviour* 34:1299–1310.

Weatherhead, P. J., and P. T. Boag. 1995. Pair and extra-pair mating success relative to male quality in red-winged blackbirds. *Behavioral Ecology and Sociobiology* 37:81–91.

Weathers, W. W., and K. A. Sullivan. 1989. Juvenile foraging profficiency, parental effort, and avian reproductive success. *Ecological Monographs* 59:223–246.

Webb, G. J. W., D. Choquenot, and P. J. Whitehead. 1986. Nests, eggs, and embryonic development of *Carettochelys insculpta* (Chelonia: Carettochelidae) from northern Australia. *Journal of Zoology* 1:521–550.

Webster, M. M., and P. J. B. Hart. 2006. Subhabitat selection by foraging threespine stickleback (*Gasterosteus aculeatus*): Previous experience and social conformity. *Behavioral Ecology and Sociobiology* 60:77–86.

Wechsler, B. 1988a. Dominance relationships in jackdaws (*Corvus monedula*). *Behaviour* 106:252–264.

———. 1988b. The spread of food producing techniques in a captive flock of jackdaws. *Behaviour* 107:267–277.

Wedekind, C. 2002. Induced hatching to avoid infectious egg disease in whitefish. *Current Biology* 12:69–71.

Wedekind, C., and R. Müller. 2005. Risk-induced early hatching in salmonids. *Ecology* 86:2525–2529.

Weiss, B. M. K., and K. Kotrschal. 2004. Effects of passive social support in juvenile greylag geese (*Anser anser*): A study from fledging to adulthood. *Ethology* 110:429–444.

Weller, S. J., N. L. Jacobson, and W. E. Conner. 1999. The evolution of chemical defenses and mating systems in tiger moths (Lepidoptera: Arctiidae). *Biological Journal of the Linnean Society* 68:557–578.

Wen, J. Y. M., N. Kumar, G. Morrison, G. Rambaldini, S. Runciman, J. Rousseau, and D. van der Kooy. 1997. Mutations that prevent associative learning in *C. elegans*. *Behavioral Neuroscience* 111:354–368.

Werdenich, D. 2006. Technical and social intelligence in keas, *Nestor notabilis*, exemplified in problem-solving and cooperation tasks. Thesis, University of Vienna.

West, M. J., and A. P. King. 1988. Female visual displays affect the development of male song in the cowbird. *Nature* 334:244–246.

West-Eberhard, M. J. 2003. *Developmental Plasticity and Evolution*. Oxford: Oxford University Press.

Westneat, D. F., and P. W. Sherman. 1997. Density and extra-pair fertilizations in birds: A comparative analysis. *Behavioral Ecology and Sociobiology* 41:205–215.

Westneat, M. W., J. H. Long, W. Hoese, and S. Nowicki. 1993. Kinematics of birdsong: Functional correlation of cranial movements and acoustic features in sparrows. *Journal of Experimental Biology* 182:147–171.

White, D. J. 2004. Influences of social learning on mate-choice decisions. *Learning and Behavior* 32:105–113.

White, D. J., and B. G. Galef Jr. 1999. Mate choice copying and conspecific cueing in Japanese quail, *Coturnix Coturnix japonica*. *Animal Behaviour* 57:465–473.

——. 2000. "Culture" in quails: Social influences on mate choices of female *Coturnix japonica*. *Animal Behaviour* 59:975–979.

White, S. A., and R. D. Fernald. 1997. Changing through doing: Behavioral influences on the brain. *Recent Progress in Hormone Research* 52:455–473.

Whiten, A., and R. Ham. 1992. On the nature and evolution of imitation in the animal kingdom: Reappraisal of a century of research. *Advances in the Study of Behavior* 21:239–283.

Whiten, A., V. Horner, C. Litchfield, A., and S. Marshall-Pescini. 2004. How do apes ape? *Learning and Behavior* 32:36–52.

Whiten, A., and C. P. van Schaik. 2007. The evolution of animal "cultures" and social intelligence. *Philosophical Transactions of the Royal Society of London Series B* 362:603–620.

Whitfield, C. W., M. R. Band, M. F. Bonaldo, C. G. Kumar, L. Liu, J. R. Pardinas, H. M. Robertson, M. B. Soares, and G. E. Robinson. 2002. Annotated expressed sequence tags and cDNA microarrays for studies of brain and behavior in the honey bee. *Genome Research* 12:555–566.

Whitfield, C. W., A. M. Cziko, and G. E. Robinson. 2003. Gene expression profiles in the brain predict behavior in individual honey bees. *Science* 302:296–299.

Wilcox, R. S., R. R. Jackson, and R. Gentile. 1996. Spiderweb smokescreens: Spider trickster uses background noise to mask stalking movements. *Animal Behaviour* 51:313–326.

Wiley, R. H. 1991. Associations of song properties with habitats for territorial oscine birds of eastern North America. *American Naturalist* 138:973–993.

Wiley, R. H., and D. G. Richards. 1978. Physical constraints on acoustic communication in the atmosphere: Implications for the evolution of animal vocalizations. *Behavioral Ecology and Sociobiology* 3:69–94.

Wilkins, A. S. 2002. *The Evolution of Developmental Pathways*. Sunderland, MA: Sinauer.

Wilkinson, G. S., and J. W. Boughman. 1999. Social influences on foraging in bats. In H. O. Box and K. R. Gibson, eds., *Mammalian Social Learning: Comparative and Ecological Perspectives*, 188–204. Cambridge: Cambridge University Press.

Williams, P. D., T. Day, Q. Fletcher, and L. Rowe. 2006. The shaping of senescence in the wild. *Trends in Ecology and Evolution* 21:458–463.

Williams-Guillén, K., I. Perfecto, and J. Vandermeer. 2008. Bats limit insects in a Neotropical agroforestry system. *Science* 320:70.

Williamson, I., and C. M. Bull. 1994. Population ecology of the Australian frog *Crinia signifera:* Egg laying patterns and egg mortality. *Wildlife Research* 21:621–632.

Wilson, B., and L. M. Dill. 2002. Pacific herring respond to simulated odontocete echolocation sounds. *Canadian Journal of Fisheries and Aquatic Sciences* 59:542–553.

Wilson, P. L., M. C. Towner, and S. L. Vehrencamp. 2000. Survival and song-type sharing in a sedentary subspecies of the song sparrow. *Condor* 102:355–363.

Wilson, P. L., and S. L. Vehrencamp. 2001. A test of the deceptive mimicry hypothesis in song-sharing song sparrows. *Animal Behaviour* 62:1197–1205.

Windmill, J. F. C., J. C. Jackson, E. J. Tuck, and D. Robert. 2006. Keeping up with bats: Dynamic auditory tuning in a moth. *Current Biology* 16:2418–2423.

Wingfield, J. C., D. L. Maney, C. W. Breuner, J. D. Jacobs, S. Lynn, M. Ramenofsky, and R. D. Richardson. 1998. Ecological bases of hormone-behavior interactions: The "emergency life history stage." *American Zoologist* 38:191–206.

Winkler, H., B. Leisler, and G. Bernroider. 2004. Ecological constraints on the evolution of avian brains. *Journal of Ornithology* 145:238–244.

Winkworth, A. L., and P. J. Davis. 1997. Speech breathing and the Lombard effect. *Journal of Speech, Language, and Hearing Research* 40:159–169.

Winnington, A. P., R. M. Napper, and A. R. Mercer. 1996. Structural plasticity of identifed glomeruli in the antennal lobes of the adult worker honey bee. *Journal of Comparative Neurology* 365:479–490.

Winslow, J. T., N. Hastings, C. S. Carter, C. R. Harbaugh, and T. R. Insel. 1993. A role for central vasopressin in pair bonding in monogamous prairie voles. *Nature* 365:545–548.

Winston, M. L. 1987. *The Biology of the Honey Bee*. Cambridge: Harvard University Press.

Wisenden, B. D. 2000. Olfactory assessment of predation risk in the aquatic environment. *Philosophical Transactions of the Royal Society of London Series B* 355:1205–1208.

Witelson, S. F., H. Beresh, and D. L. Kigar 2005. Intelligence and brain size in 100 postmortem brains: Sex, lateralization and age factors. *Brain* 129:386–398.

Withers, G. S., N. F. Day, E. F. Talbot, H. E. Dobson, and C. S. Wallace. 2008. Experience-dependent plasticity in the mushroom bodies of the solitary bee *Osmia lignaria* (Megachilidae). *Developmental Neurobiology* 68:73–82.

Withers, G. S., S. E. Fahrbach, and G. E. Robinson. 1993. Selective neuroanatomical plasticity and division of labour in the honeybee. *Nature* 364:238–240.

———. 1995. Effects of experience and juvenile hormone on the organization of the mushroom bodies of honey bees. *Journal of Neurobiology* 26:130–144.

Witte, K., H. E. Farris, M. J. Ryan, and W. Wilczynski. 2005. How cricket frog females deal with a noisy world: Habitat-related differences in auditory tuning. *Behavioral Ecology* 16:571–579.

Witte, K., and B. Noltemeier. 2002. The role of information in mate-choice copying in female sailfin mollies (*Poecilia latipinna*). *Behavioral Ecology and Sociobiology* 52:194–202.

Witte, K., and M. J. Ryan. 1998. Male body length influences mate-choice copying in the sailfin molly *Poecilia latipinna*. *Behavioral Ecology* 9:534–539.

———. 2002. Mate choice copying in the sailfin molly, *Poecilia latipinna*, in the wild. *Animal Behaviour* 63:943–949.

Witthöft, W. 1967. Absolute Anzahl und Verteilung der Zellen im Hirn der Honigbiene. *Zeitschrift für Morphologie und Oekologie der Tiere* 61:160–164.

Wolf, R., and M. Heisenberg. 1991. Basic organization of operant-behavior as revealed in *Drosophila* flight orientation. *Journal of Comparative Physiology A* 169:699–705.

Wolff, J. O., M. H. Freeberg, and R. D. Dueser. 1983. Interspecific territoriality in two sympatric species of *Peromyscus* (Rodentia, Cricetidae). *Behavioral Ecology and Sociobiology* 12:237–242.

Wolff, J. O., and D. W. Macdonald. 2004. Promiscuous females protect their offspring. *Trends in Ecology and Evolution* 19:127–134.

Wollerman, L., and R. H. Wiley. 2002. Background noise from a natural chorus alters female discrimination of male calls in a Neotropical frog. *Animal Behaviour* 63:15–22.

Wooler, R. D., J. S. Bradley, I. J. Skira, and D. L. Serventy. 1990. Reproductive success of short-tailed shearwater *Puffinus tenuirostris* in relation to their age and breeding experience. *Journal of Animal Ecology* 59:161–170.

Wright, T. F., K. A. Cortopassi, J. W. Bradbury, and R. J. Dooling. 2003. Hearing and vocalizations in the orange-fronted conure (*Aratinga canicularis*). *Journal of Comparative Psychology* 117:87–95.

Wright, W. G., D. Kirschman, D. Rozen, and B. Maynard. 1996. Phylogenetic analysis of learning-related neuromodulation in molluscan mechanosensory neurons. *Evolution* 50:2248–2263.

Wunderle, J. M. 1991. Age-specific foraging proficiency in birds. *Current Ornithology* 8:273–324.

Wyles, J. S., J. G. Kunkel, and A. C. Wilson. 1983. Birds, behavior and anatomical evolution. *Proceedings of the National Academy of Sciences of the United States of America* 80:4394–4397.

Wyttenbach, R. A., M. L. May, and R. R. Hoy. 1996. Categorical perception of sound frequency by crickets. *Science* 273:1542–1544.

Yack, J. E. 1988. Seasonal partitioning of atympanate moths in relation to bat activity. *Canadian Journal of Zoology—Revue Canadienne de Zoologie* 66:753–755.

Yack, J. E., and J. H. Fullard. 2000. Ultrasonic hearing in nocturnal butterflies. *Nature* 403:265–266.

Yack, J. E., G. G. E. Scudder, and J. H. Fullard. 1999. Evolution of the metathoracic tympanal organ and its mesothoracic homologue in Macrolepidoptera. *Zoomorphology* 119:93–103.

Yeh, P. J., and T. D. Price. 2004. Adaptive phenotypic plasticity and the successful colonization of a novel environment. *American Naturalist* 164:531–542.

Yoder, J. M., J. L. Dooley, J. F. Zawacki, and M. A. Bowers. 1996. Female aggression in *Microtus pennsylvanicus*: Arena trials in the field. *American Midland Naturalist* 135:1–8.

Young, L. J., and E. A. D. Hammock. 2007. On switches and knobs, microsatellites and monogamy. *Trends in Genetics* 23:209–212.

Young, L. J., R. Nilsen, K. G. Waymire, G. R. MacGregor, and T. R. Insel. 1999. Increased affiliative response to vasopressin in mice expressing the V-1a receptor from a monogamous vole. *Nature* 400:766–768.

Young, L. J., and Z. X. Wang. 2004. The neurobiology of pair bonding. *Nature Neuroscience* 7:1048–1054.

Yurkovic, A., O. Wang, A. C. Basu, and E. A. Kravitz. 2006. Learning and memory associated with aggression in *Drosophila melanogaster*. *Proceedings of the National Academy of Sciences of the United States of America* 103:17519–17524.

Zahavi, A. 1975. Mate selection: A selection for a handicap. *Journal of Theoretical Biology* 53:205–214.

Zajonc, R. B. 1965. Social facilitation. *Science* 149:269–274.

Zelick, R., D. A. Mann, and A. N. Popper. 1999. Acoustic communication in fishes and frogs. In R. Fay and A. N. Popper, eds., *Comparative Hearing: Fish and Amphibians*, 363–411. New York: Springer.

Zelick, R., and P. M. Narins. 1985. Characterization of the advertisement call oscillator in the frog, *Elutherodactylus coqui*. *Journal of Comparative Physiology A* 156:223–229.

Zentall, T. R. 2004. Action imitation in birds. *Learning and Behavior* 32:15–23.

Zentall, T. R., and B. G. Galef Jr. 1988. *Social Learning: Psychological and Biological Perspectives*. Hillsdale, NJ: Erlbaum.

Zhang, S., A. Mizutani, and M. Srinivasan. 2000. Maze navigation by honeybees: Learning path regularity. *Learning and Memory* 7:363–374.

Zhang, Y., H. Lu, and C. I. Bargmann. 2005. Pathogenic bacteria induce aversive olfactory learning in *Caenorhabditis elegans*. *Nature* 438:179–184.

Ziegler, T. A., and R. B. Forward. 2007. Control of larval release in the Caribbean spiny lobster, *Panulirus argus:* The role of chemical cues. *Marine Biology* 152:589–597.

Zuberbühler, K. 2000. Referential labelling in Diana monkeys. *Animal Behaviour* 59:917–927.

———. 2003. Referential signaling in non-human primates: Cognitive precursors and limitations for the evolution of language. *Advances in the Study of Behavior* 33:265–307.

Zuberbühler, K., D. L. Cheney, and R. M. Seyfarth. 1999. Conceptual semantics in a nonhuman primate. *Journal of Comparative Psychology* 113:33–42.

Zuberbühler, K. R. Noe, and R. M. Seyfarth. 1997. Diana monkey long-distance calls: Messages for conspecifics and predators. *Animal Behaviour* 53:589–604.

CONTRIBUTORS

KARIN L. AKRE
Section of Integrative Biology
University of Texas
Austin, TX 78712, USA

MICHAEL D. BEECHER
Departments of Psychology and Zoology
University of Washington
Seattle, WA 98195, USA

JOHN M. BURT
Department of Psychology
University of Washington
Seattle, WA 98195, USA

MICHAEL S. CALDWELL
Department of Biology
Boston University
Boston, MA 02215, USA

NICOLA S. CLAYTON
Department of Experimental Psychology
University of Cambridge
Cambridge CB22 3EB, UK

ISABELLE COOLEN
Laboratoire Evolution et Diversité Biologique
Université Paul Sabatier Toulouse III
UMR CNRS 5174, FR-31062
Toulouse Cedex 4, France

SCOTT DOBRIN
Neuroscience Program
Wake Forest University School of Medicine
Winston-Salem, NC 27157, USA

REUVEN DUKAS
Animal Behaviour Group
Department of Psychology, Neuroscience and Behaviour
McMaster University
Hamilton, ON L8S 4K1, Canada

NATHAN J. EMERY
School of Biological and Chemical Sciences
Queen Mary
University of London
London E1 3NS, UK

SUSAN E. FAHRBACH
Department of Biology
Wake Forest University
Winston-Salem, NC 27109, USA

IRA G. FEDERSPIEL
Sub-department of Animal Behaviour
University of Cambridge
Madingley CB23 8AA, UK

RACHEL L. KENDAL
Department of Anthropology
University of Durham
Durham DH1 3LE, UK

MARK KIRKPATRICK
Section of Integrative Biology
University of Texas
Austin, TX 78712, USA

KEVIN N. LALAND
School of Biology
University of St. Andrews
St. Andrews, Fife KY16 9TS, UK

MARTA B. MANSER
Institute of Zoology
University of Zurich
CH-8057 Zurich, Switzerland

STEPHEN NOWICKI
Department of Biology
Duke University
Durham, NC 27708, USA

ALEXANDER G. OPHIR
Department of Zoology
Oklahoma State University
Stillwater, OK 74078, USA

STEVEN M. PHELPS
Department of Zoology
University of Florida
Gainesville, FL 32611, USA

VLADIMIR V. PRAVOSUDOV
Department of Biology
University of Nevada
Reno, NV 89557, USA

JOHN M. RATCLIFFE
Center for Sound Communication
Institute of Biology
University of Southern Denmark
5230 Odense M, Denmark

MICHAEL J. RYAN
Section of Integrative Biology
University of Texas
Austin, TX 78712, USA

DANIEL SOL
CREAF-CSIC (Center for Ecological Research and Applied Forestries—Spanish National Research Council)
Autonomous University of Barcelona
E-08193 Bellaterra, Catalonia, Spain

KAREN M. WARKENTIN
Department of Biology
Boston University
Boston, MA 02215, USA

INDEX

Page references followed by t *or* f *refer to tables and figures, respectively.*